Advanced Geoscience Remote Sensing

Editor

Hubert O'Kelly

scitus
academics

Advanced Geoscience Remote Sensing
Edited by **Hubert O'Kelly**

ISBN: 978-1-68117-230-9
Library of Congress Control Number: 2016934762

www.scitusacademics.com
Box No. 4766, 616 Corporate Way,
Suite 2, Valley Cottage,
NY 10989

Printed in United States of America on Acid Free Paper

Preface

We perceive the surrounding world through our five senses. Some senses require contact of our sensing organs with the objects. However, we acquire much information about our surrounding through the senses of sight and hearing which do not require close contact between the sensing organs and the external objects. In another word, we are performing remote sensing all the time. Remote sensors can be either passive or active. Passive sensors respond to external stimuli. They record natural energy that is reflected or emitted from the Earth's surface. The most common source of radiation detected by passive sensors is reflected sunlight. Nowadays, advanced remote sensing technology plays tremendous roles to build a quantitative and comprehensive understanding of how the Earth system operates. In a more restricted sense, remote sensing usually refers to the technology of acquiring information about the earth's surface (land and ocean) and atmosphere using sensors onboard airborne (aircraft, balloons) or spaceborne (satellites, space shuttles) platforms. In satellite remote sensing of the earth, the sensors are looking through a layer of atmosphere separating the sensors from the Earth's surface being observed. Hence, it is essential to understand the effects of atmosphere on the electromagnetic radiation travelling from the Earth to the sensor through the atmosphere. The advanced remote sensing technology is also used widely to monitor and survey the natural disasters and man-made pollution. The aim of this book, Advanced Geoscience Remote Sensing, is to provide a platform for researchers and academicians all over the world various new issues and developments in different areas of geoscience and remote sensing.

Table of Contents

1
CHAPTER

REMOTE SENSING APPLICATIONS: BEYOND LAND-USE AND LAND-COVER CHANGE

Chris W. Baynard

Department of Economics and Geography, University of North Florida, Jacksonville, USA

ABSTRACT

Remotely sensed (RS) imagery is increasingly being adopted in investigations and applications outside of traditional land-use land-cover change (LUCC) studies. This is due to the increased awareness by governments, NGOs and Industry that earth observation data provide important and useful spatial and temporal information that can be used to make better decisions, design policies and address problems that range in scale from local to global. Additionally, citizens are increasingly adopting spatial analysis into their work as they utilize a suite of readily available geospatial tools. This paper examines some of the ways remotely sensed images and derived maps are being extended beyond LUCC to areas such as fire modeling, coastal and marine applications, infrastructure and urbanization, archeology, and to ecological, or infrastructure footprint analysis. Given the interdisciplinary approach of such work, this paper organizes selected studies into broad categories identified above. Findings demonstrate that RS data and technologies are being widely used in many fields, ranging from fishing to war fighting. As technology improves, costs go down, quality increases and data become increasingly available, greater numbers of organizations and local citizens will be using RS in important everyday applications.

Keywords: Remote Sensing Applications; Mapping; Monitoring; Detection; Earth Observation

1. INTRODUCTION

A review of the literature regarding the application of remote sensing (RS) data and techniques to solve problems, address policy implications and improve decision making indicates a strong preference for studies focused on land use and land cover change (LULCC). This is not surprising given that earth observation (EO) data provide important and useful spatial and temporal information for studying changes in the natural environment; particularly those caused or affected by human actions, as well as for understanding what these changes are doing to us. As this awareness grows, governments, the public and industry are becoming responsive to the importance of ecosystem goods and services and the monetary value that must necessarily be placed on them [1,2].

For example, new programs are being developed, known as payments for ecosystem services (PES) to conserve natural areas, mostly in developing countries [3]. One such program is known as REDD[1]—reducing greenhouse gas emissions from deforestation and degradation. It draws on financial resources from developed countries to halt deforestation in forest-rich developing countries [4,5].

Toward this aim, advances in RS are improving understanding of social and ecological systems functioning [5], which can range from intact native ecosystems to highly modified ones [6]. A central approach for comprehending dynamic landscapes is to recognize that vegetation is a driving force in terrestrial ecosystems and is often used as a proxy for classification [6]. This is noteworthy, since the type of vegetation growing in a given region provides an indication of climate (namely temperature and rainfall) and thus the type of socioeconomic activities that are likely to be found there. Furthermore, LULCs influence climate change.

Categorizing and examining landscapes, therefore, is a common topic in much of the LUCC literature. Here, remote sensing data are regularly utilized to assess spatial and temporal dynamics usually within a geographic information system (GIS) that is then used to analyze and map these patterns [7]. While much work focuses on using discrete classifications, whereby given land uses and land covers are placed into distinct categories, others prefer to use continuous data. The latter acknowledges that real landscapes do not abruptly end or sharply transit from one type to another. Using continuous data, such as vegetation indexes (e.g. Normalized Difference Vegetation Index, or NDVI), provides a more accurate representation of real world transitions between different landscape categories [8-10].

Consequently, a central focus on LUCC has been on deforestation, namely via agriculture, and logging, since these are direct drivers of landscape change. However, secondary, distal, or underlying drivers leading to such changes have socioeconomic, cultural, political and environmental explanations that vary by scale [11-14] and are harder to discern, since a RS image can show what is happening but not necessarily why.

The use of remote sensing images and derived maps is to better understand distal (secondary) drivers of change, as well their use beyond LUCC leads to multidisciplinary work that includes:

- Monitoring urban expansion, urban sprawl, slums and heat island effects through the use of daytime and nighttime imagery, as well as ground temperature and urban vegetation [15-25].
- Global fisheries management [26], including night time squid fishing efforts in the Pacific [27], increaseing fishing efficiency while reducing costs [28], improving aquaculture, such as shrimp farms [29]; tracking lobster movements [30]; detecting chlorophyll concentrations in the ocean and estimating phytoplankton mass and prey availability [28,31,32]; improving algorithms for classifying and monitoring algal blooms [33,34] and detecting and mapping broad scale coral reef changes [35].
- Modeling wildland fires and gauging potential damage from coal seam fires and detecting burn scars [36-38].
- Producing high-quality bathymetric maps [39].
- Monitoring drinking water quality, turbidity and water quality in coastal and estuarine waters [40-41]; improving efficiency of hydroelectric reservoirs and assessing cumulative environmental alterations resulting from new dams [42,43].
- Detecting thermal plume discharges from nuclear power stations [44].
- Monitoring and mapping salinity distributions in coastal environments [45]; identifying coastline changes and associated impacts [46] and improving edge detection in ocean color images [47]. Also, detecting sea ice for arctic monitoring [48] and forecasting glacier melting [18,49].
- Improving accuracy in archaeological work, detecting buried objects and identifying potential historic sites. Monitoring the evolution of mining induced subsidence [50-56].
- Studying historic debris flows and identifying potential ones at stratovolcanoes [57].
- Identifying villages and urban areas at high-risk of malaria transmission and tracking mosquito habitats [58-61]; modeling seasonal and inter-annual patterns of climatic suitability for mosquitoes [62] and guiding aid efforts to areas with severed

communications such as those following a major earthquake or cyclone [63-66].

- Modeling the impacts of a liquefied natural gas tanker explosion [67].
- Detecting infrastructure alterations related to oil exploration and production activities [68-72]; deforestation and soil contamination [73,74].
- Military operations such as pre-mission planning and post-mission analysis, providing situational awareness for warfighters, developing threat analysis and identifying terrorist hideouts [75,76]; and netcentric warfare, which allows "warfighters to plan, execute, report and visualize a common operating picture" [77].

Given the broad nature of such research, data derived from remote sensing technologies are necessarily complemented with additional or ancillary data. This includes in situ measures; information about ecosystem goods and services; elevation, air and water temperature; rainfall data; hydropower potential; salinity; risk for erosion; the amount of carbon storage; displacement of local communities; habitat fragmentation and loss of wildlife; downstream and drinking water quality and quantity; wind speed; chlorophyll-a concentrations; agricultural production and the impact on food security; and health effects among others.

Having addressed the commonalities that exist in the LUCC literature, this review paper provides an insight into how RS-derived data are being used beyond traditional LUCC maps and images to better understand human-environment interactions, solve problems and address policy issues. Nevertheless, it is impossible to decouple the natural and cultural (human made) landscape from these studies since humans live and operate in particular locations whose specific qualities and histories provide a suite of possibilities to which inhabitants adapt in a variety of ways. This underscores the importance of understanding and implementing geographic thought and spatial analysis.

Thus, while traditional LUCC studies might focus on how humans interact with and transform the landscape, understanding why, and what are some of the socioeconomic effects, which takes us to the secondary drivers. Not surprisingly, given the multidisciplinary nature of this type of the research, a review of this literature indicates that the work being done does not necessarily fall into clean categories such as economics or health or risk assessment. Many of these issues are intertwined. For example, floods in Bangladesh have environmental components (monsoon rains) that are exacerbated by lack of planning, population growth, lack of funding for proper infrastructure, non-enforcement of construction/settlement policies in high-risk areas, bureaucracy in terms of several and often overlapping agencies, land speculation, and loss of arable land among others [7].

This paper, therefore, groups work that is similar into the following categories: Fire Models and Methods; Coastal and Marine Applications; Infrastructure and Urbanization; Archaeology and Remote Sensing; The Ecological Footprint, or the Landscape Infrastructure Footprint, and Ecosystem Goods and Services. These sections also address the types of ancillary data used in these research papers as well as the type of RS data involved and relevant characteristics (when provided). The Conclusion and Future Research sections follow. Given the breadth of this topic, this review provides a general introduction to the following topics and is not exhaustive. Readers are directed to the numerous references at the end of the paper, as well as the future research section, that identifies 13 well-known RS journals.

2. REVIEW

2.1. Fire Models and Methods

Wildland fires pose a major environmental problem for many of the planet's ecosystems and can become an important cause of land degradation and environmental transformation [36-38]. Furthermore they pose a substantial economic risk in terms of lost forested land, loss of soil nutrients which can affect agricultural production [78], affected homes and property, as well as a health risk to nearby populations and a danger to fire fighters. One way to develop management scenarios for wildland fires involves using RS to detect the spatial distribution of fuel types in order to create more robust fire models [37].

Aerial photo interpretation is widely used by forest managers and government agencies, because it provides a compromise between price and precision, and is therefore "one of the most commonly used techniques for mapping vegetation and fuel types" [37]. These authors note that satellites are also being used to study forest fuels, but a main drawback is their inability to penetrate forest canopies. Nevertheless, using QuickBird imagery, some researchers have reported fuel map accuracies of 75% to 81% [37].

Investigators, such as Chuvieco et al. [36] also consider moisture status of vegetation, ignition sources and stakeholder values in their work. Using mainly Landsat TM images, the authors were able to map and classify fire scars over time in the Cape Canaveral National Seashore in Florida using the categories burned and unburned. The objective was to establish sound management practices that balanced managed with natural fire regimes.

Duncan et al. [79] note that remote sensing techniques can prove suitable for fire monitoring as well. For example, for recent fire history at coarse scales, they suggest using MODIS Fire (Moderate Resolution Imaging Spectroradiometer) (see [80,81]), TRMM VIRS (Tropical Rainfall Measuring Mission, Visible and Infrared Scanner—4.4 km^2 pixels) and ATSR-2 (Along-Track Scanning Radiometer—1 km spatial resolution) [79,82], though in their work they relied on LANDSAT TM (30 and 120 m) [83] imagery to help detect burn scars, among other features. But "For longer fire histories, especially when fine detail pattern information is necessary, mapping fire scars from a time series of high resolution imagery is preferred," [79]. Smith et al. [80] concur on the need to use higher spatial resolution satellite imagery than that provided by Landsat ETM+ (15, 30 and 60 m) [83,84]. In particular they refer to the importance of identifying the white ash that remains after fires, which provides an indication of fire severity.

Coal seam fires create another important fire risk, particularly in China, where 70% of the country's energy is derived from coal and where an estimated 20 Mt of uncontrolled fires burn there every year [85]. In such cases Voigt et al. [85] propose an integrated satellite remote sensing approach to detect and monitor "near surface coal seam fires by observing land surface changes induced by fires" that includes digital elevation models (DEMs), radar and MODIS (250, 500 and 1000 m), QuickBird (0.50 cm to 2.4 m), ASTER (15, 30 and 90m) (see [86]) and Landsat 7 ETM+ data [84,85,87]. The monetary costs associated with these fires stem from not only the loss of the burnt coal itself, but also from the loss of accessibility to nearby mining operations [85]. Furthermore, these uncontrolled coal fires lead to environmental stresses through the release of large amounts of toxic and greenhouse gases such as CO, CO_2, CH_4, SO_2, and NO [85].

Prakash et al. [88] used Landsat imagery to identify coals seam fires burning in Alaska based on summertime thermal infrared temperatures. Noting that many of these fires go unnoticed and unreported in inaccessible areas, the authors found that coal seam fire zones exhibited temperatures 5°C to 14°C higher than surrounding areas [88]. Using Landsat ETM+ imagery, the authors focused on band 6 (low-gain band 61), the thermal infrared band, which operates in the 10.4 - 12.5 μm spectral region and has a spatial resolution of 60 m. Thus, high-resolution imagery was not necessary for this approach.

2.2. Coastal and Marine Applications

The SeaWIFS (Sea-viewing Wide Field-of-view Sensors) Project (see [31]) provides quantitative data on bio-optical properties (color) of the world's oceans, based on chlorophyll and other plant pigments, whose higher concentration make the water greener [28,31]. For example, high-resolution airborne data have been used to estimate chlorophyll-a concentrations in Alaskan coastal waters in reference to phytoplankton biomass and prey availability for the Steller sea lion [32]. For these types of studies the authors noted that satellite footprints had two main limitations: chlorophyll-a variations require scale detection less than 1 km, and stray light from land and bottom radiance of shallow waters tend to contaminate near-shore pixels (Montes-Hugo et al., 2005).

Meanwhile, Zagaglia et al. [26] examined the relationship between tropical Atlantic yellowfin tuna and environmental variables obtained from remote sensors in support of global fisheries management and the capture of pelagic species. They examined sea surface temperatures using Advanced Very High Resolution Radiometer (AVHRR) data onboard NOAA satellites (1.09 km) [89,90]; concentration of chlorophyll-a using SeaWIFS data from the SeaStar satellite; sea surface height anomaly information gathered from TOPEX and Poseidon altimeters; and surface wind data collected by scatterometers aboard the European Remote Sensing Satellites ERS-1 and -2. The authors found that tuna catch was strongly associated with the position of the Intertropical Convergence Zone (ITCZ) [26].

In another study, Rajitha et al. [29] discussed how satellite RS technology and GIS were being used to sustainably manage shrimp aquaculture in India. Here, management is important due to mangroves, coastal and marine resources, and agricultural rice lands being used and converted to create shrimp farms. At the same time, shrimp farms create rural employment and economic development in coastal villages, contributing "a major portion of national income through high export earning" [29]. Studies examined by the above authors note the effective use of Indian Remote Sensing Satellites (IRS), with 5.8 m to 1 km m resolution [91], Landsat TM and ETM+ for observing biological productivity and for monitoring coastal water temperature and quality [29].

In an additional application of RS to fishing, Santos [28] noted how "satellite delivered fishery-aid charts can reduce 25% - 50% of US commercial fisheries search time," greatly decreasing fuel consumption and the associated CO_2 emissions, as well as improving fishing efficiency and increasing economic returns. Important data for the rational management of fishing resources includes water temperature and phytoplankton biomass, which the Coastal Zone Color Scanner (CZCS) provided until 1986, followed by the SeaWiFS sensors [28]. Active systems such as Light Detection and Ranging (LIDAR) and Side-Looking Airborne Radar (SLAR) to determine the location and size of fish schools, as well as microwave sensors aboard satellites such as SEASAT to map kelp resources along the California coast have also been implemented [28].

Waluda et al. [27] used satellite-derived imagery for another fishing study, albeit through a different perspective. These researchers utilized DMSP-OLS [39] to quantify squid fishing in the eastern Pacific based on lights. Since commercial fishery of Jumbo flying squid is conducted at night using powerful lights to attract these creatures, Waluda et al. [27] were able to quantify fishing effort based on light signatures of these vessels. "The distribution of lights can be used to observe the distribution and abundance of squid jiggers, and by inference, the distribution of exploited squid stocks, and the location of favorite fishing grounds" [27].

Producing high quality bathymetric maps requires expert personnel and expensive field surveys and high tech equipment, note Ceyhun and Yalçin [39], who found that Aster and QuickBird imagery could be successfully used to reduce the cost and labor needed to produce bathymetric measurements. In situations where detailed mapping was not necessary, Aster images worked fine, noted the authors, but for detecting local depth changes, QuickBird imagery was superior. Yuan and Zhang [92] also used QuickBird images to map and monitor the distribution and growth of submerged aquatic vegetation on a large scale.

Regarding rock lobster fisheries in Tasmania, Lucieer and Pederson [30] processed bathymetric data to derive a digital terrain model (DTM) to help track lobster movement. By then applying landscape classifications (developed by terrestrial ecologists) they quantified the degree and spatial distribution of habitat complexity. They accomplished this by classifying each grid in a bathymetric digital elevation model (DEM) into one of six predetermined morphometric landform classes. This provided more complex seabed characterizations that in turn helped quantitatively predict consequences of different management strategies [30].

RS has proven effective as a tool for detecting and mapping broad scale coral reef changes too [35]. By using pan-sharpening methods on FORMOSAT-2 imagery, the authors transformed 8 m spatial resolution images into higher resolution ones, noting a cost savings over purchasing high spatial resolution images alone. The methods involve using the panchromatic (black and white) band with its higher spatial resolution of 2 m, and resampling the other (lower resolution) bands to produce a composite [35,93]. The estimated brightness values of the lower resolution bands are replaced with the higher panchromatic band and this is possible because the latter covers the same spectral range as the visible (red, green, blue) and near infrared bands [35].

Monitoring, predicting and understanding the availability of water and associated changes in quality entails work by Chang et al. [40], who examine satellite imagery to assess water availability and quantify the hydrological cycle. The authors used GOES (Geostationary Operational Environmental Satellites) [93], Landsat and MODIS data, ground level radar-precipitation data (NEXRAD), as well as past point measurements (changes in water quality at specific locations, river discharge and precipitation), to create a metropolitan water availability index (MWAI) for Tampa, Florida.

Also regarding water availability, Alcântara et al. [42] utilized MODIS data to conduct time and cost effective water quality measures in a Brazilian hydroelectric reservoir (Itumbiara). Using 786 daytime and 473 nighttime images, the researchers computed descriptive statistics (mean, maximum and minimum) to build a time series of day and night monthly mean temperatures used to better understand spatial and temporal variations in the tropical reservoir.

Focusing on the Pearl River Estuary in southern China, Chen et al. [41] used RS data as a tool for ecosystem restoration by focusing on turbidity, defined as an "optical effect that is related to the concentration of total suspended solids (TSS) and the shape and size of other impure elements in water" [41]. The researchers recommend using EO-1 ALI satellite imagery (Earth Observing, Advanced Land Imager—10 and 30 m) [94] for water quality measurements in coastal and estuarine waters.

Li and Damen [46] note that the Pearl River delta "has one of the highest economic development rates of China" leading to loss of agricultural land, sea water intrusion, land subsidence, river siltation and coastal erosion. Combining data from Landsat (MSS, TM and ETM+) (see [95]) and SPOT imagery (see [96]) with topographical and nautical data helped Li and Damen [46] identify coastline changes and related impacts such as the narrowing of river channels, severe flooding, and increased sedimentation, which hinder harbor construction.

AVHRR data was used to detect the thermal plume created by the discharge of warm water from the Daya Bay Nuclear Power Station's cooling system into Daya Bay, China [44]. With a 1.1 × 1.1 km resolution, the authors found that the thermal plumes displayed a seasonal pattern (smaller in winter and larger in summer) and that the temperature difference ranged from 1.0° to 1.5° from non-plume areas.

RS data is also being used to detect sea ice in an effort to monitor arctic conditions. In some cases low-resolution data, 25 × 25 km, is being used to examine sea ice extent and develop regression equations for spring and summer seasons [48]. In this study, Drobot [48] refers to data gathered from the Scanning Multichannel Microwave Radiometer (SMMR), the Special Sensor Microwave/Imager (SSM/I) and AVHRR Polar Pathfinder.

Finally, Wang and Xu [45] point out how RS techniques can be used to monitor salinity in coastal environments by using Landsat TM images to map salinity distribution in Lake Pontchartrain, Louisiana. They found that Landsat TM bands 1, 2, and 4 were positively correlated to salinity levels and that band 2 helped explain up to 20% of the variance. Meanwhile, bands 3 and 5 were negatively correlated to salinity levels and explained about 30% of the variance. The authors also found that hurricanes, such as Katrina, altered spatial patterns of salinity and significantly increased the average salinity levels.

2.3. Infrastructure and Urbanization

Spot 5 Supermode imagery (processed for 2.5 m resolution) has proven useful for urban planning in areas where sprawl is changing quickly due to rapid house construction and residential development, and where up-to-date information such as timely air photos are lacking. This is often the case in developing countries, and with a 60 × 60 km swath, Duriex et al. [21] used object based image analysis and image segmentation to extract buildings, monitor and estimate urban sprawl in the entire Reunion Island, located in the Indian Ocean.

Another way to measure urbanization is via night lights, whereby Sutton [25] used DMSP-OLS[2] and population data as a proxy measure for urban extent. More recently, Small et al. [17] used DMSP-OLS derived data to conclude that the brightness and coverage of stable night lights were correlated with human population density, built area density and economic activity at both the country and global scales [17]. "Night lights provide a means to quantify the size, number and spatial extent of human settlements worldwide" [17].

Landsat MSS, TM and ETM+ imagery was used to study urbanization rates in the Greater Dhaka area of Bangladesh in order to promote sustainable development [7]. The researchers found that as urbanization increased, water bodies shrank; cultivated land, wetlands and vegetation were all reduced. The government did not adequately respond to population growth, permitting settlers to establish themselves in wetlands and low-lying regions prone to flooding [7]. They also noted that land speculation attracted settlers and the subsequent conversion of arable land and natural areas. The authors pointed out that Landsat MSS imagery had a course spatial resolution (79 m) that hampered classification accuracy, thus limiting its use [7,95].

Other data used in this study included: municipal boundaries, road networks, elevation, geomorphic units, topography, demographic, slope and GDP. Important factors included economic development and industrialization, which contributed to rapid urbanization (rise of the ready-made garments industry), while topography affected its direction [7].

Ji et al. [23] formulated urban sprawl metrics that linked construction-based land consumption to remotely sensed land change data in metropolitan Kansas City. They found this method superior to using population data "because usually construction activities, as compared to population change, reflect directly economic opportunities as the major driving force of land alteration," [23]. Their dataset was composed of Landsat imagery (MSS, TM and ETM+), historical photos, and USGS topographic maps. Landscape metrics, such as patch density, largest patch density and the aggregation index of forested and nonforested vegetation were calculated using the FRAGSTATS[3] program. The authors were able to identify slow and fast growing areas and concluded that larger spatial units such as metropolitan areas better reveal landscape effects of urbanization [23].

Bhatta et al. [19] note that defining urban sprawl and measuring it is a complicated process resulting in many variations. They define it as: "characterized by unplanned and uneven pattern of growth, driven by multitude of processes and leading to inefficient resource utilization" [19]. They also suggest using the entropy method to best integrate urban sprawl metrics with remote sensing and GIS. According to the authors "Shannon's entropy (Hn) can be used to measure the degree of spatial concentration or dispersion of a geographical variable (xi) among n zones" [19].

They also note that the imagery used should ideally have a fine enough spatial resolution to represent individual units such as parcels or houses. However, they point out that high-resolution imagery can result in the interpreter identifying high object diversity, which complicates classification algorithms [19]. Nevertheless, their article reviews several approaches to quantifying urban sprawl without applying the entropy method.

Meanwhile, Martinuzzi et al. [22] used Landsat ETM+ imagery, along with NOAA air photos and US Census data and urban-rural classifications to identify high and low urban density patterns in Puerto Rico. To assess accuracy, the authors used a random sample of ground control points for urban and non-urban categories evaluated against the air photos. Using this methodology they found that 11% of the island was covered by urban or built-up surfaces and that "Nearly half of the total development is occurring outside of the solid urban centers, covering one-quarter of the best lands for agriculture, impacting watersheds and reducing urban spaces" [22]. When divided into three regions, the researchers found that 16% of the island was Urban, 36% was Sparsely Populated Rural and 48% was Densely Populated Rural. They reiterated the need for the island to have an effective land use plan in a context where the population density rivals that of New Jersey [22].

In many places, such as in developing countries where municipal records, building permits, road construction and utility infrastructure locations are often not readily available or updated, remote sensing can provide "fundamental observations of urban growth and environmental conditions" [24]. These authors note the primacy for monitoring environmental conditions, more so than for urban planning, particularly where informal settlements are continually expanding. In their study they discuss the use of vegetation indices and temperature data from Landsat 7 imagery.

Lastly, Rajasekar and Weng [20] utilized MODIS and ASTER imagery to monitor the urban heat island in and around Indianapolis, Indiana. They used day and night land surface temperature (LST) MODIS images to create a continuous surface and concluded that they offer great potential to monitor the urban heat island phenomenon (whereby air temperatures in densely urbanized areas are higher than those in the countryside). However, heat islands were easier to distinguish in the summer months of June to August. When using ASTER data, the authors found that "areas with maximum heat signatures were found to have a strong correlation with impervious surfaces" [20].

Rhinane et al. [15] looked at urban ground temperature, building density and "cooler" areas. The latter, were vegetated zones, pointing to the importance of green spaces in controlling the heat island phenomenon.

2.4. Archaeology and Remote Sensing

Compared to what Siart et al., [97] call conventional archaeological GIS applications, the authors in this category propose a multi-method approach to geoarcheaological landscape reconstructions which involve RS, DEM analysis, GPS data, surveying, least-cost analysis, soils, predictive modeling and candidate site selection. Part of their approach involved identifying "aspects of landscape character which might have affected past activity or occupation choices," rather than directly locating archaeological sites or features [97]. One such tactic was to analyze the spatial distribution of Bronze Age transit roads, which would have influenced the location of settlements in the study area of Crete. Part of their methods involved pansharpening Quickbird MS imagery to 0.6 m resolution and then creating iron oxide ratios (band 3 by 1), as well as infrared ratios (band 4 by 3) [97].

The authors concluded that for the Bronze Age inhabitants slope and topography were important determinants for spatial mobility and still useful for predicting potential road networks and new sites [97]. Predicting archaeological site distribution with RS data go back about two decades. In one example Custer et al. [56] used Landsat MSS imagery to classify parts of central Delaware into land cover classes to help predict the location of prehistoric sites.

In another application for archaeology, Daniels et al., [55] tested the use of radar to detect buried subsurface reflectors to a depth of 20 cm. The authors used a scaterometer (operating in the P-band at 441 MHz, 68 cm) mounted above a truck as well as on an airplane to conclude that it's possible to detect buried objects in sandy desert areas up to 4.4 m deep. The authors also highlight numerous potential applications for microwave/VHF radar. They include: tectonic and engineering studies, fluvial geomorphology, glacier covers, subsurface photogrammetry and cartography [55]. In fact NASA radar has been used to explore deep canyons located on the moon [50].

Other projects involving NASA remote sensing and archaeology date back to the 1960s, when black and white and infrared photography from Apollo 11's SO65 multiband experiment was used to identify human-made prehistoric linear features in Arizona [51]. Since the 1970s, infrared images were found to be useful to archaeology because "buried or obscure cultural features may absorb and radiate solar energy in amounts that differ from that of the surrounding soil matrix, thereby revealing the features on the imagery" [51]. Thermal data is also useful, since "heat transfer through the soil will be affected by the presence of buried objects" [53].

Another application of RS for archaeology includes ground-penetrating radar (GPR), which allows non-invasive site exploration to avoid disturbance during excavation of features in sites of interest [50]. Quarto et al. [54] used this technique to find karst caves that contained prehistoric remains in southern Italy, while Hoerle et al. [98] used it to assess the conservation conditions of prehistoric rock art in South Africa. There's even a journal dedicated to this topic aptly named Remote Sensing in Archaeology.

2.5. The Ecological Footprint, or the Landscape Infrastructure Footprint

Few researchers are studying the alterations created by oil and gas companies' infrastructure features on the landscape. These include Janks et al. [73] and Janks and Prelat [74] who studied vegetation health in and around oil fields, deforestation rates as related to oil and agricultural roads and tracked remediation attempts on abandoned well sites using Landsat MSS and TM imagery. Musinsky et al. [72] used Landsat TM, air photos and videography to examine the relationship between oil roads and deforestation in Guatemala.

For other researchers [68-71,99-102] a central goal is to determine "the exact size and extent of the ecological footprint of energy development" [103]. Some of this works applies landscape ecology metrics to quantify the landscape disturbances in oil and gas concessions. These metrics include road or infrastructure density; habitat fragmentation; edge-effect zones; core areas and number of rivers crossed [38,68-69,71,100,101,104-109].

Regarding the landscape infrastructure footprint (LIF) Baynard [70] examined oil concessions in Venezuela using Landsat TM and ETM+ imagery. Change detection was calculated using the NDVI, a commonly used method for identifying biomass, crop estimates and areas prone to drought [8,110-112], as well as climate change, biodiversity and wildlife ecology [113]. It "is considered among the best known indices and widely used to study and map the plants" [15].

For the oil landscape study in Venezuela, the resulting vegetation change maps showing gains, losses or no change helped determine the contribution of LIF to vegetation change. By also including the size of core areas, agricultural land, as well as the number and location of infrastructure intersecting rivers, the concessions were ranked on their environmental performance. This type of performance can be linked to sustainability.

In another study of oil development in remote regions, Baynard et al. [68] examined the spatial relationship between infrastructure pattern related to oil exploration and production (E&P) and parallel activities, surface disturbance and the type of access available in specific oil concessions in eastern Ecuador. This approach combined large-scale Landsat-derived LULC maps, smaller-scale government LULC maps, soils data, protected areas and colonization zones. The authors found that controlledaccess and no-access to oil concessions greatly reduced deforestation rates by keeping settlers from establishing households in these remote regions. Meanwhile, areas where public-access roads, fertile soils and colonization zones overlapped were most prone to deforestation [68].

2.6. Ecosystem Goods and Services

By assessing ecosystem goods and services, economic activity, natural assets and ecological functions can be linked [114]. Goods and services which we obtain from ecosystems include: "provisioning services, such as food and water; regulating services, such as regulation of floods, drought and disease; supporting services, such as soil formation and nutrient cycling; and cultural services, such as recreational, spiritual, and other nonmaterial benefits" [115].

The first step is to acknowledge that ecosystems have value, or natural capital. Then they have to be valued. This can be accomplished by "assessing the contribution of ecosystem services to sustainable scale, fair distribution, and efficient allocation" [115]. This method focuses on economic or utilitarian value; whereby "ecosystems are deemed valuable because they provide environmental goods and services to humans," note Abson and Termansen [116]. It does not focus on intrinsic value, whereby an object has a value for its own sake [116].

Yet once a valuation method has been reached, a main way to fund ecosystem management is to provide payments. Known as payments for ecosystem services (PES) [3], these economic incentives provide a way to manage ecosystems by paying local stakeholders not to cut down trees or convert natural landscapes to other uses such as agriculture or urbanization.

Reducing greenhouse gas emissions from deforestation and degradation, known as REDD, is one such approach, which draws on financial resources from developed countries to halt deforestation in forest-rich developing countries [4,5]. This approach is rather new and being implemented by organizations such as the German Development Bank, the Nature Conservancy and the World Bank, with the duel objectives of conserving and managing [117]. Because of its newness, REDD programs are characterized by uncertainty and incomplete information [4,5]; however, as mentioned earlier, advances in remote sensing are improving understanding of social and ecological systems functioning [5] which can range from intact native ecosystems to highly modified ones [6].

Given that monitoring REDD programs involves detecting LUCC over space and time, RS data and techniques are therefore prime tools to accomplish these objectives. Newer research points to the potential for LIDAR to be used for identifying the forest non-forest boundary, making it "an ideal tool for exact deforestation monitoring", a key requirement for REDD [118].

3. CONCLUSIONS

This paper has described some of the researches that utilize remote sensing to address problems relating to economics (improving efficiency in fishing, aquaculture, water quality); social and health conditions (malaria, earthquakes and typhoons); improved planning of development infrastructure (urban sprawl, the landscape infrastructure footprint); modeling change into the past and future (archaeology, REDD); and risk assessment (reservoirs, population growth). Clearly the type of work reviewed here overlaps more than one category (i.e., urban sprawl problems that can be economic, environmental and social). The topics to which these studies were assigned were based on commonalities within the research.

The paradigm shift, if we may call it that, is that organizations and corporations possessing environmental assets need to understand their roles as land managers and engage in better oversight of the land and people where their assets lie and their economic activities take place. This means paying attention to the triple bottom line of sustainable development: economic, environmental and social issues across all company operations. The way to approach this is to value ecosystem goods and services that are offered by the landscapes under their management—that is, valuing natural capital as one of society's most important assets [115].

A proposed method of implementation is via precision land management (PLM). This set of practices builds on methods used in precision agriculture to integrate layers of data used to promote variable management practices within a given agricultural field [119,120]. In this case, instead of better predicting crop yields, applying fertilizers and pesticides only in the right locations to increase production and efficiency, extractive industries and others operating in natural areas would efficiently manage the activities and infrastructure affecting habitats and people in and around their natural assets.

As Mathieu [43] observes: "Earth Observation (EO) satellites can play a key role in this endeavor, as they are uniquely placed to monitor the state of our environment, in a global and consistent manner, ensuring sufficient resolution to capture the footprint of man-made activities". Through this approach major actors would be in a better position to predict outcomes of the social, environmental and economic interaction that result in a successful operation. By incorporating socioeconomic data and stakeholder input with RS data in a GIS environment, these land managers could "see" what portions of a given real estate are being/might be altered or converted; the relationship to natural areas (for both conservation and remediation purposes); the location and activities of local villagers, indigenous groups and land managers; as well as actions from competing economic actors such as loggers, hunters, gold miners and local resource users.

4. FUTURE RESEARCH

This paper has focused mostly on the application of passive remote sensing satellite imagery. Nevertheless, many applications have not been covered in efforts to provide a general overview—such as CBERS data (see [121]). Recent developments in the use of Radar and LIDAR should be further explored. Regarding radar, for example, Interferometric SAR (InSAR or If SAR) has been used to develop terrain elevation data and subsequent multiple data sets have been compared to monitor land deformation with centimeter precision [122].

If SAR applications can detect whether a vehicle has moved along a road, data fusion methods that combine Geo SAR data with AIS (automatic identification system) ship transponder signals, optical image data and in situ information can detect whether a ship is hiding something [123]. These secret activities could indicate illegal fishing or illegal oil discharging [123].

Other noteworthy developments are real-time in-flight processing, "which is especially important for rapid response, emergency management, and intelligence, surveillance and reconnaissance (ISR) applications" and the introduction of video into commercial airborne photogrammetry [124].

Nevertheless, throughout the papers examined here, Landsat imagery comprises key data sets due to their continuous global coverage and free availability. And now, the recent release of Landsat 8 in February 2013 [125] is a boon to EO researchers, since this satellite overcomes the challenges created by Landsat 7's malfunction in 2003 [126].

Key sources for developments and applications in remote sensing include: Imaging Notes (http://www.imagingnotes.com); Advances in Remote Sensing (http://www.scirp.org/journal/ars/); Remote Sensing of Environment (http://www.journals.elsevier.com);

Remote Sensing (http://www.mdpi.com/journal/remotesensing); International Journal of Remote Sensing (http://www.tandfonline.com/toc/tres20/current#.UboKs5Vpsb0); Applied Remote Sensing Journal (http://www.asciencejournal.net/asj/index.php/ARS); Journal of Applied Remote Sensing (http://spie.org/x3636.xml); Remote Sensing Letters

(http://www.tandfonline.com/action/showMostCitedArticles?journalCode=trsl20#.UcRc1ZVrXE4);

GI Science & Remote Sensing

(http://www.tandfonline.com/action/showMostCitedArticles?journalCode=tgrs20#.UcRdL-5VrXE4); International Journal of Digital Earth (http://www.tandfonline.com/action/showMostCitedArticles?journalCode=tjde20#.UcRdYpVrXE4); ISPRS Journal of Photogrammetry and Remote Sensing (http://www.journals.elsevier.com/isprs-journal-of-photogrammetry-and-remote-sensing/); International Journal of Applied Earth Observation and Geoinformation

(http://www.journals.elsevier.com/international-journalof-applied-earth-observation-and-geoinformation/); and Advances in Space Research

(http://www.sciencedirect.com/science/jounal/02731177).

5. ACKNOWLEDGEMENTS

This work was based on a white paper written for Chevron Corporation's Next Generation Remote Sensing Team. Any errors, omissions or oversights are solely those of the author.

REFERENCES

1. Economist, "Seeing the World for the Trees: An International Deal on Deforestation Makes It Ever More Important to Measure Earth's Woodlands," 2010. http://www.economist.com/node/17730208/print
2. P. J. O'Farrell, J. S. Donaldson and M. T. Hoffman, "The Influence of Ecosystem Goods and Services on Livestock Management Practices on the Bokkeveld Plateau, South Africa," Agriculture Ecosystems and Environment, Vol. 122, No. 3, 2007, pp. 312-324.
3. J. Farley and R. Costanza, "Payments for Ecosystem Services: From local to Global," Ecological Economics, Vol. 69, No. 11, 2010, pp. 2060-2068.
4. P. J. Kanowski, C. L. McDermott and B. W. Cashmore, "Implementing REDD+: Lessons from Analysis of Forest Governance," Environmental Science and Policy, Vol. 14, No. 2, 2011, pp. 111-117.

5. F. Hajek, M. J. Ventresca, J. Scriven and A. Castro, "Regime Building for REDD+: Evidence Form a Cluster of Local Initiatives in South-Eastern Peru," Environmental Science and Policy, Vol. 14, No. 2, 2011, pp. 201-215.
6. G. Yapp, J. Walker and R. Thackway, "Linking Vegetation Type and Condition to Ecosystem Goods and Services, Ecological Complexity, Vol. 7, No. 3, 2010, pp. 292-301.
7. M. Dewan and Y. Yamaguchi, "Land Use and Land Cover Change in Greater Dhaka, Bangladesh: Using Remote Sensing to Promote Sustainable Urbanization," Applied Geography, Vol. 29, No. 3, 2009, pp. 390-401.
8. J. Southworth, D. Munroe and H. Nagendra, "Land Cover Change and Landscape Fragmentation—Comparing the Utility of Continuous Discrete Analyses for a Western Honduras Region," Agriculture Ecosystems and Environment, Vol. 101, No. 2-3, 2004, pp. 185-205.
9. P. Mondal, "Quantifying Surface Gradients with a 2-Band Enhanced Vegetation Index (EVI2)," Ecological Indicators, Vol. 11, No. 3, 2011, pp. 918-924.
10. J, Krishnaswamy, K. S. Bawa, K. N. Ganeshaiah and M. C. Kiran, "Quantifying and Mapping Biodiversity and Ecosystem Services: Utility of a Multi-Season NDVI Based Mahalanobis Distance Surrogate," Remote Sensing of Environment, Vol. 113, No. 4, 2009, pp. 857-867.
11. R. R. Rindfuss, B. Entwisle, S. J. Walsh, C. F. Mena, C. M. Erlien and C. L. Gray, "Frontier Land Use Change: Synthesis, Challenges, and Next Steps," Annals of the Association of American Geographers, Vol. 97, No. 4, 2007, pp. 739-754.
12. B. L. Tuner, E. F. Lambin and A. Reenberg, "The Emergence of Land Change Science for Global Environmental Change and Sustainability," PNAS, Vol. 104, No. 52, 2007, pp. 20666-20671.
13. E. Keys and W. J. McConnell, "Global Change and the Intensification of Agriculture in the Tropics," Global Environmental Change, Vol. 15, No. 4, 2005, pp. 320-337.
14. Veldkamp and E. F. Lambin. "Predicting land-Use Change," Agriculture, Ecosystems and Environment, Vol. 85, No. 1-3, 2001, pp. 1-6.
15. H. Rhinane, A. Hilali, H. Bahi and A. Berrada, "Contribution of Landsat TM Data for the Detection of Urban Heat Islands Areas Case of Casablanca," Journal of Geographic Information System, Vol. 4, 2012, pp. 20-26.

16. H. Rhinane, A. Hilali, A. Berrada and M. Hakdaouli, "Detecting Slums from SPOT Data in Casablanca Morocco Using an Object Based Approach," Journal of Geographic Information System, Vol. 3, No. 3, 2011, pp. 217-224.
17. C. Small, C. D. Elvidge, D. Balk and M. Montgomery, "Spatial Scaling of Stable Night Lights," Remote Sensing of Environment, Vol. 115, No. 2, 2011, pp. 269-280.
18. K. Corbley, "Caribbean Island Maps On-and-Off-Shore Habitats: Balancing Growth with Preservation," Imaging Notes, Vol. 25, No. 3, 2010, pp. 20-21. http://imagingnotes.com/go/article_freeJ.php?mp_id =228
19. B. Bhatta, S. Saraswati and D. Bandyopadhyay, "Urban Sprawl Measurement from Remote Sensing Data," Applied Geography, Vol. 30, 2010, pp. 731-740.
20. U. Rajasekar and Q. Weng, "Urban Heat Island Monitoring and Analysis Using a Non-Parametric Model: A Case Study of Indianapolis," ISPRS Journal of Photogrammetry and Remote Sensing, Vol. 64, No. 1, 2009, pp. 86- 96.
21. L. Duriex, E. Lagabrielle and A. Nelson, "A Method for Monitoring Building Construction in Urban Sprawl Areas Using Object-Based Analysis of Spot 5 Images and Existing GIS Data," Photogrammetry and Remote Sensing, Vol. 63, No. 4, 2008, pp. 399-408.
22. S. Martinuzzi, W. A. Gould and O. M. R. González, "Land Development, Land Use, and Urban Sprawl in Puerto Rico Integrating Remote Sensing and Population Census Data," Landscape and Urban Planning, Vol. 79, No. 3-4, 2007, pp. 288-297.
23. W. Ji, J. Ma, R. W. Twibell and K. Underhill, "Characterizing Urban Sprawl Using Multi-Stage Remote Sensing Images and Landscape Metrics," Computers, Environments and Urban Systems, Vol. 30, No. 6, 2006, pp. 861-879. doi.org/10.1016/j.compenvurbsys.2005.09.002
24. R. B. Miller and C. Small, "Cities from Space: Potential Applications of Remote Sensing in Urban Environmental Research and Policy," Environmental Science and Policy, Vol. 6, No. 2, 2003, pp. 129-137.
25. P. C. Sutton, "A Scale-Adjusted Measure of "Urban Sprawl" Using Nighttime Satellite Imagery," Remote Sensing of Environment, Vol. 86, No. 3, 2003, pp. 353-369.

26. C. R. Zagaglia, J. A. Lorenzzetti and J. L. Stech, "Remote Sensing Data and Longline Catches of Yellowfin Tuna (Thunnus albacores) in the Equatorial Atlantic," Remote Sensing of Environment, Vol. 93, No. 1-2, 2004, pp. 267- 281.
27. C. M. Waluda, C. Yamashiro, C. D. Elvidge, V. R. Hobson and P. G. Rodhouse, "Quantifying Light-Fishing for Dosidicus gigas in the Eastern Pacific Using Satellite Remote Sensing," Remote Sensing of Environment, Vol. 91, No. 2, 2004, pp. 129-133.
28. M. P. Santos, "Fisheries Oceanography Using Satellite and Airborne Remote Sensing Methods: A Review," Fisheries Research, Vol. 49, No. 1, 2000, pp. 1-20.
29. K. Rajitha, C. K. Mukherjee and R. V. Chandran, "Applications of Remote Sensing and GIS for Sustainable Management of Shrimp Culture in India," Aquacultural Engineering, Vo. 36, No. 1, 2007, pp. 1-17.
30. V. Lucier and H. Pederson, "Linking Morphometric Characterization of Rocky Reef with Fine Scale Lobster Movement," Photogrammetry and Remote Sensing, Vol. 63, No. 5, 2008, pp. 496-509.
31. C. Feldman, "Background of SeaWiFS Project," http://oceancolor.gsfc.nasa.gov/SeaWiFS/BACKGROUND/SEAWIFS_BACKGROUND.html
32. M. A. Montes-Hugo, K. Carder, R. J. Foy, J. Cannizzaro, E. Brown and S. Pegau, "Estimating Phytoplankton Biomass in Coastal Waters of Alaska Using Airborne Remote Sensing," Remote Sensing of Environment, Vol. 98, No. 4, 2005, pp. 481-493.
33. Simon and P. Shanmugam, "An Algorithm for Classification of Algal Blooms Using MODIS-Aqua Data in Oceanic Waters around India," Advances in Remote Sensing, Vol. 1, No. 2, 2012, pp. 35-51.
34. S. P. Tiwari and P. Shanmugam, "An Optical Model for the Remote-Sensing of Absorption Coefficients of Phytoplankton in Oceanic/Coastal Waters," Advances in Remote Sensing, Vol. 1, Session 17, 2012, pp. 633-637.
35. H. Hanaizumi, M. Akiba, H. Yamano and T. Matsunaga, "A Pan-Sharpening Method for Satellite Image-Based Coral Reef Monitoring With Higher Accuracy," Proceedings of the 11th International Coral Reef Symposium, Ft. Lauderdale, 7-11 July 2008, pp. 626-630.
36. E. Chuvieco, I. Aguado, M. Yebra, H. Nieto, J. Salas, M. P. Martín, et al., "Development of a Framework for Fire Risk Assessment Using Remote Sensing and Geographic Information System Technologies," Ecological Modelling, Vol. 221, No. 1, 2010, pp. 46-58.

37. L. A. Arroyo, C. Pascual and J. A. Manzanera, "Fire Models and Methods to Map Fuel Types: The Role of Remote Sensing," Forest Ecology and Management, Vol. 256, No. 6, 2008, pp. 1239-1252.
38. FAO, "Fire Management-Global Assessment: A Thematic Study Prepared in the Framework of the Global Forest Resources Assessment 2005," FAO Forestry Paper 151, FAO, Rome, 2006. http://www.fao.org/docrep/009/a0969e/a0969e00.HTM
39. Ö. Ceyhun and A. Yalçin, "Remote Sensing of Water Depths in Shallow Waters via Artificial Neural Networks," Estuarine, Coastal and Shelf Science, Vol. 89, No. 1, 2010, pp. 89-96.
40. N. Chang, Y. J. Yang, J. A. Goodrich and A. Daranpob, "Development of the Metropolitan Water Availability Index (MWAI) and Short-Term Assessment with MultiScale Remote Sensing Techniques," Journal of Environmental Management, Vol. 91, No. 6, 2010, pp. 1397-1413.
41. S. Chen, L. Fang, L. Zhang and W. Huang, "Remote Sensing of Turbidity in Seawater Intrusion Reaches of Pearl River Estuary—A Case Study in Modaomen Water Way, China," Estuarine, Coastal and Shelf Science, Vol. 82, No. 1, 2009, pp. 119-127.
42. E. H. Alcântara, J. L. Stech, J. A. Lorenzzetti, M. P. Bonnet, X. Casamitjana, A. T. Assireu, et al., "Remote Sensing of Water Surface Temperature and Heat Flux over a Tropical Hydroelectric Reservoir," Remote Sensing of Environment, Vol. 114, No. 11, 2010, pp. 2651-2665.
43. P. P. Mathieu, "Sustainable Development Using Earth Observations: Accounting from Above Forestry, Hydropower and Mining," Imaging Notes, Vol. 25, No. 4, 2010, pp. 24-27.
44. D. Tang, D. R. Kester, Z. Wang, J. Lian and H. Kawamura, "AVHRR Satellite Remote Sensing and Shipboard Measurements of the Thermal Plume from the Daya Bay, Nuclear Power Station, China," Remote Sensing of Environment, Vol. 84, No. 4, 2003, pp. 506-515.
45. F. Wang and Y. J. Xu, "Development and Application of a Remote Sensing-Based Salinity Prediction Model for a Large Estuarine Lake in the US Gulf of Mexico Coast," Journal of Hydrology, Vol. 360, No. 1-4, 2008, pp. 184- 194.
46. X. Li and M. C. J. Damen, "Coastline Change Detection with Satellite Remote Sensing for Environmental Management for the Pearl River Estuary, China," Journal of Marine Systems, Vol. 82, 2010, pp. S54-S61.

47. R. Vinuchandran and D. Ramakrishnan, "Detection of Fronts from Ocean Colour Monitor Images Using Entropic Technique: A Case Study of Mesoand MicroScale Chlorophyll Mapping in Bay of Bengal, India," Advances in Remote Sensing, Vol. 2, 2013, pp. 71-76.
48. S. D. Drobot, "Using Remote Sensing Data to Develop Seasonal Outlooks for Arctic Regional Sea-Ice Minimum Extent," Remote Sensing of Environment, Vol. 111, No. 2-3, 2007, pp. 136-147.
49. C. Gordon, "Tracking Glacial Activity in Norway with Photogrammetry Software," Imaging Notes, Vol. 22, No. 1, 2007, pp. 24-29. http://imagingnotes.com/go/article_free.php?mp_id=90
50. C. Coleman and M. Quijada, "NASA Earth Scientists Advance Space Archaeology," NASA, 2010. http://www.nasa.gov/topics/earth/features/space-archaeology.html
51. M. J. Giardino, "A History of NASA Remote Sensing Contributions to Archeology," Journal of Archaeological Science, Vol. 38, No. 9, 2011, pp. 2003-2009.
52. L. Liang-jun and W. Yan-bin, "Application of RemoteSensing-Image Fusion to the Monitoring of Mining Induced Subsidence," Journal of China University of Mining and Technology, Vol. 18, No. 4, 2008, pp. 531- 536. doi:10.1016/S1006-1266(08)60289-8
53. Rowlands and A. Sarris, "Detection of Exposed and Subsurface Archaeological Remains Using Multi-Sensor Remote Sensing," Journal of Archaeological Science, Vol. 34, No. 5, 2007, pp. 795-803.
54. R. Quarto, D. Schiavone and I. Diaferia, "Ground Penetrating Radar Survey of a Prehistoric Site in Southern Italy," Journal of Archaeological Science, Vol. 34, No. 12, 2007, pp. 2071-2080.
55. J. Daniels, D. G. Blumberg, L. D. Vulfson, A. L. Kotlyar, V. Freiliker, G. Ronen and J. Ben-Asher, "Microwave Remote Sensing of Physically Buried Objects in the Negev Desert: Implications for Environmental Research," Remote of Environment, Vol. 86, No. 2, 2003 pp. 243-256.
56. J. F. Custer, T. Eveleigh, V. Klemas and I. Wells, "Application of LANDSAT Data and Synoptic Remote Sensing to Predictive Models for Prehistoric Archaeological Sites: An Example from the Delaware Coastal Plain," American Antiquity, Vol. 51, No. 3, 1986, pp. 572-588.
57. J. K. Crowley, B. E. Hubbard and J. C. Mars, "Analysis of Potential Debris Flow Source Areas on Mount Shasta, California, by Using Airborne and Satellite Remote Sensing Data," Remote Sensing of Environment, Vol. 87, No. 2-3, 2003, pp. 345-358.

58. GMES (Global Monitoring for Environment and Security), "How Satellites Track a Mass Killer," 2012. http://esamultimedia.esa.int/docs/EarthObservation/GMES_Brief_Malaria_Issue23_December2012.pdf
59. N. Hassan, N. E. Nogoumy and H. A. Kassem, "Characterization of Landscape Features Associated with Mosquito Breeding in Urban Cairo Using Remote Sensing," The Egyptian Journal of Remote Sensing and Space Science, 2013, pp. 1-7, in Press.
60. B. G. Jacob, J. E. Funes, E. J. Muturi and R. J. Novak, "Using Imaging Technologies to Control Malaria," Imaging Notes, Vol. 22, No. 1, 2007, pp. 18-23. http://imagingnotes.com/go/article_free.php?mp_id=91
61. P. M. Masouka, D. M. Claborn, R. G. Andre, J. Nigro, S. W. Gordon, T. A. Klein and H.-C. Kim, "Use of IKONOS and Landsat for Malaria Control in the Republic of Korea," Remote Sensing of Environment, Vol. 88, No. 1-2, 2003, pp. 187-194.
62. T.-W. Chuang, G. M. Henebry, J. S. Kimball, D. L. VanRoekel-Patton, M. B. Hildreth and M. C. Wimberly, "Satellite Microwave Remote Sensing for Environmental Modeling of Mosquito Population Dynamics," Remote Sensing of Environment, Vol. 125, 2012, pp. 147-156.
63. J. Bevington, B. Adams and R. Eguchi, "GEO-CAN Debuts to Map Haiti Damage," Imaging Notes, Vol. 25. No. 2, 2010, pp. 26-30. http://imagingnotes.com/go/article.php?mp_id=208
64. R. Ann, "The National System for Geospatial Intelligence: Enhancing the Value and Reach of GEOINT," Pathfinder, Vol. 9, No. 1, 2011, pp. 8-9. http://www.scribd.com/doc/127404947/1-NGA-Pathfinder-Jan-Feb-2011?secret_password=2hpcs5ns0depbofzqk4f
65. E. Batchelor, "Canada Responds to Haiti Earthquake," Pathfinder, Vol. 8, No. 6, 2010, p. 16. http://www.scribd.com/doc/127400218/6-2010-Nov-Dec?secret_password=1zmpyovoizcw0ltcll1x
66. N. Cutsforth, "CRS Impact: Myanmar Cyclone Relief and Coral Reef Monitoring," Imaging Notes, Vol. 25, No. 3, 2010, pp. 15-16. http://imagingnotes.com/go/article_freeJ.php?mp_id=227
67. J. Jones, "NGA Lights Way in Boston Harbor," Pathfinder, Vol. 9, No. 1, 2011, p. 17. http://www.scribd.com/doc/127404947/1-NGA-Pathfinder-Jan-Feb-2011?secret_password=2hpcs5ns0depbofzqk4f

68. C. W. Baynard, J. M. Ellis and H. Davis, "Roads, Petroleum and Accessibility: The Case of Eastern Ecuador," GeoJournal, 2012, pp. 1-16. http://link.springer.com/content/pdf/10.1007%2Fs10708-012-9459-5.pdf
69. C. W. Baynard, "Evaluating Disturbance of E&P Access Roads," ESRI News for Petroleum, 2012, pp. 6-7. http://www.esri.com/library /newsletters/petroleum-perspectives/petrol-spring-2012.pdf
70. C. W. Baynard, "The Landscape Infrastructure Footprint of Oil Development: Venezuela's Heavy Oil Belt," Ecological Indicators, Vol. 11, No. 3, 2011, pp. 789-810. doi:10.1016/j.ecolind.2010.10.005
71. C. W. Baynard, "Using Geospatial Technologies to Quantify Environmental Sustainability Performance," Petroleum GIS Perspectives, 2009-2010, pp. 4-5. http://www.esri.com/library /newsletters/petroleum-perspectives/petrol-winter-2009-2010.pdf
72. J. N. Musinsky, C. C. S. Reining, A. B. Rosenfeld and J. B. Thomsen, "An Analysis of Human Settlement along the Xan Oil Road in Laguna del Tigre National Park, Guatemala," Society of Petroleum Engineers, 1998. doi:10.2118/46824-MS
73. J. S. Janks, A. E. Prelat and J. D. Wieser, "Remote Sensing Technology Applications to Environmental Assessments in the Oriente Region of Ecuador," Second Thematic Conference on Remote Sensing for Marine and Coastal Environments, New Orleans, 31 Jan-2 Feb 1994, pp. 633-644.
74. J. S. Janks and A. E. Prelat, "Environmental Impact Assessment Using Remote Sensing Technology: Methodology and Case Studies," Society of Petroleum Engineers, 1994.
75. S. Pontius and K. Boyer, "In Fields Afar, NGS Product May Be Warfighters' Only Friend," National Geospatial Intelligence Agency, Springfield, 2007.
76. J. D. Krasnow, "Scene Visualization Advancements Great Boon to Partners," Pathfinder, Vol. 10, No. 3, 2012, p. 20. https://www1.nga.mil/MediaRoom/Press%20Kit/Documents/Pathfinder%20Magazines/2012/2012_May-Jun.pdf
77. S. Marchant, "Toolkit Empowers Warfighters for NetCentric Warfare," National Geospatial Intelligence Agency, Springfield, 2007.

78. S. Lavorel, M. D. Flannigan, E. F. Lambin and M. C. Scholes, "Vulnerability of Land Systems to Fire: Interactions among Humans, Climate, the Atmosphere, and Ecosystems," Mitigation and Adaptation Strategies for Global Change, Vol. 12, No. 1, 2007, pp. 33-53.
79. B. W. Duncan, G. Shao and F. W. Adrian, "Delineating a Managed Fire Regime and Exploring Its Relationship to the Natural Fire Regime in East Central Florida, USA: A Remote Sensing and GIS Approach," Forest Ecology and Management, Vol. 258, No. 2, 2009, pp. 132-145.
80. A. M. S. Smith, N. A. Drake, M. J. Wooster, A. T. Hudak, Z. A. Holden and C. J. Gibbons, "Production of Landsat ETM+ Reference Imagery of Burned Areas within Southern African Savannahs: Comparison of Methods and Application to MODIS, "International Journal of Remote Sensing, Vol. 28, No. 2, 2007, pp. 2753-2775.
81. MODIS, "Thermal Anomalies/Fire," 2010. http://modis-land.gsfc.nasa.gov/fire.html
82. ESA, "ATSR," 2011. http://earth.esa.int/ers/eeo4.80/
83. GLCF, "Landsat Imagery," 2013. http://www.landcover.org/data/landsat/
84. NASA, "The Enhanced Thematic Mapper Plus," 2013. http://landsat.gsfc.nasa.gov/about/etm+.html
85. S. Voigt, A. Tetzlaff, J. Zhang, C. Künzer, B. Zhukov, G. Strunz, et al., "Integrating Satellite Remote Sensing Techniques for Detection and Analysis of Uncontrolled Coal Seam Fires in North China," International Journal of Coal Geology, Vol. 59, No. 1-2, 2004, pp. 121-136.
86. Aster, "The Aster Instrument," 2004. http://asterweb.jpl.nasa.gov/instrument.asp
87. Digital Globe, "Basic Imagery," 2013. http://www.digitalglobe.com/node/1544#overview
88. A. Prakash, K. Schaefer, W. K. Witte, K. Collins, R. Gens and M. P. Goyette, "Remote Sensing: GIS Based Investigation of a Boreal Forest Coal Fire," International Journal of Coal Geology, Vol. 86, No. 1, 2011, pp. 79-86.
89. NOAA NSIS, "Advanced Very High Resolution Radiometer-AVHRR," 2013. http://noaasis.noaa.gov/NOAASIS/ml/avhrr.html
90. NOAA OSO, "NOAA's Environmental Satellites: A History," 2013. http://www.oso.noaa.gov/history/index.htm
91. ANTRIX, "Indian Remote Sensing Satellite Programme," 2006. http://www.pecad.fas.usda.gov/pdfs/2006/MUrthi_Indian_Remote_Sensing_SaPro-2006-TRIAL.pdf

92. L. Yuan and L. Zhang, "Mapping Large-Scale Distribution of Submerged Aquatic Vegetation Coverage Using Remote Sensing," Ecological Informatics, Vol. 3, No. 3, 2008, pp. 245-251.
93. A. Garzelli and F. Nencini, "Panchromatic Sharpening of Remote Sensing Images Using a Multiscale Kalman Filter," Pattern Recognition, Vol. 40, No. 12, 2007, pp. 3568-3577.
94. USGS, "Earth Observing 1 (EO-1)," 2011. http://eo1.usgs.gov
95. USGS, "Landsat: A Global Land-Imaging Project," 2010. http://pubs.usgs.gov/fs/2010/3026/pdf/FS2010-3026.pdf
96. SPOT, "Spot Satellite Imagery: Accuracy and Coverage Combined," 2013. http://www2.astrium-geo.com/files/pmedia/public/r233_9_geo_0013_spot_en_low.pdf
97. C. Siart, B. Eitel and D. Panagiotopoulos, "Investigation of Past Archaeological Landscapes Using Remote Sensing and GIS: A Multi-Method Case Study from Mount Ida, Crete," Journal of Archaeological Science, Vol. 35, No. 11, 2008, pp. 2918-2926.
98. S. Hoerle, F. Huneau, A. Salomon and A. Denis, "Using Ground-Penetrating Radar to Assess the Conservation Condition of Rock-Art Sites," Comptes Rendus Geoscience, Vol. 339, No. 8, 2007, pp. 536-544.
99. M. Wilbert, J. Thomson and N. W. Culver, "Analysis of Habitat Fragmentation from Oil and Gas Development and Its Impacts on Wildlife: A Framework for Public Land Management Planning," The Wilderness Society, Washington DC, 2008. http://www.fws.gov/southwest/ES/Documents/Oil-Gas-Fragmentation-Wilbert%20et%20al%202008.pdf
100. Wilderness Society, "Habitat Fragmentation from Roads: Travel Planning Methods to Safeguard Bureau of Land Management Lands," The Wilderness Society, Washington DC, 2006. https://partners.tws.org/wildscience/Publications1/Habitat%20Fragmentation%20from%20Roads.pdf
101. J. L. Thomson, T. S. Schaub, N. W. Culver and P. C. Aengst, "Wildlife at a Crossroads: Energy Development in Western Wyoming-Effects of Roads on Habitat in the Upper Green River Valley," The Wilderness Society, Washington DC, 2005. http://wilderness.org/ sites/default/files/wildlife-at-crossroads-report.pdf
102. P. Morton, C. Weller and J. T. Thomson, "Energy and Western Wildlands: A GIS Analysis of Economically Recoverable Oil and Gas," The Wilderness Society, Washington DC, 2002.

103. P. Morton, C. Weller, J. Thomson, M. Haefele and N. Culver, "Drilling in the Rocky Mountains: How Much and at What Cost?" The Wilderness Society, Washington DC, 2004.

104. NOAA DMSP, "OLS-Operational Linescan System," 2011. http://www. ngdc.noaa.gov/dmsp/sensors/ols.html

105. E. Ivits, M. Cherlet, W. Mehl and S. Sommer, "Estimating the Ecological Status and Change of Riparian Zones in Andalusia Assessed by Multi-Temporal AVHHR Datasets," Ecological Indicators, Vol. 9, No. 3, 2009, pp. 422-431.

106. R. R. Schneider, J. B. Stelfox, S. Boutin and S. Wasel, "Managing the Cumulative Impacts of Land Uses in the Western Canadian Sedimentary Basin: A Modeling Approach," Conservation Ecology, Vol. 7, No. 1, 2003, pp. 1-11. http://www.ecologyandsociety.org/vol7/iss1/art8/

107. R. T. T. Forman, D. Sperling, J. A. Bissonnette, A. P. Clevenger, C. D. Cutshall, V. H. Dale, et al., "Road Ecology: Science and Solutions," Island Press, Washington DC, 2003.

108. J. L. Stein, J. A. Stein and H. A. Nix, "Spatial Analysis of Anthropogenic River Disturbance at Regional and Continental Scales: Identifying the Wild Rivers of Australia," Landscape and Urban Planning, Vol. 60, No. 1, 2002, pp. 1-25.

109. S. J. Dyer, J. P. O'Neill, S. M. Wasel and S. Boutin, "Avoidance of Industrial Development by Woodland Caribou," The Journal of Wildlife Management, Vol. 65, No. 3, 2001, pp. 531-542.

110. R. T. T. Forman and R. D. Deblinger, "The Ecological Road-Effect Zone of a Massachusetts (U.S.A.) Suburban Highway," Conservation Biology, Vol. 14, No. 1, 2000, pp. 36-46.

111. J. R. Jensen, "Introductory Digital Image Processing: A Remote Sensing Perspective," 3rd Edition, Pearson Prentice Hall Inc., Upper Saddle River, 2007.

112. C. C. Funk and M. E. Brown, "Intra-Seasonal NDVI Change Projections in Semi-Arid Africa," Remote Sensing of Environment, Vol. 101, No. 2, 2006, pp. 249-256.

113. S. Aronoff, "Remote Sensing for GIS Managers," ESRI Press, Redlands, 2005.

114. J. N. Hird and G. J. McDermid, "Noise Reduction of NDVI Time Series an Empirical Comparison of Selected Techniques," Remote Sensing of Environment, Vol. 113, No. 1, 2009, pp. 248-258.

115. O. A. Somorin, I. J. Visseren-Hamakers, B. Arts, D. J. Sonwa and A. M. Tiani, "REDD+ Policy Strategy in Cameroon: Actors, Institutions and Governance," Environmental Science and Policy, 2013, In Press.
116. S. Liu, R. Constanza, S. Farber and A. Troy, "Valuing Ecosystem Services: Theory, Practice, and the Need for Transdisciplinary Synthesis," Annals of the New York Academy of Sciences, Vol. 1185, 2010, pp. 54-78.
117. D. J. Abson and M. Termansen, "Valuing Ecosystem Services in Terms of Ecological Risks and Returns," Conservation Biology, Vol. 25, No. 2, 2011, pp. 250-258.
118. B. Blom, T. Sunderland and D. Murdiyarso, "Getting REDD to Work Locally: Lessons Learned from Integrated Conservation and Development Projects," Environmental Science and Policy, Vol. 13, No. 2, 2010, pp. 164-172.
119. B. Koch, "Status and Future of Laser Scanning, Synthetic Aperture Radar and Hyperspectral Remote Sensing Data for Forest Biomass Assessment," ISPRS Journal of Photogrammetry and Remote Sensing, Vol. 65, No. 6, 2010, pp. 581-590.
120. J. Sims, "Rapidly Serving Imagery for Precision AG and County and State Governments," Imaging Notes, Vol. 24, No. 1, 2009, pp. 30-31. http://imagingnotes.com/go/article_free.php?mp_id=157
121. S. K. Seelan, S. Laguette, G. M. Casady and G. A. Seielstad, "Remote Sensing Applications for Precision Agriculture: A Learning Community Approach," Remote Sensing of Environment, Vol. 88, No. 1-2, 2003, pp. 157- 169.
122. J. C. N. Epiphanio, "CBERS: Remote Sensing Cooperation between Brazil and China," Imaging Notes, Vol. 23, No. 2, 2008, pp. 16-19. http://www.imagingnotes.com/go/article_free.php?mp_id=134
123. M. Luccio, "Aerial Cameras: Focus Shifts to Productivity," Imaging Notes, Vol. 25, No. 4, 2010, pp. 49-54. http://www.imagingnotes.com/go/article_freeJ.php?mp_id=243
124. M. Luccio, "Radar Gets it Done: Neither Clouds nor Gloom of Night Blind a Radar Satellite," Imaging Notes, Vol. 25, No. 2, 2010, pp. 33-38. http://www.imagingnotes.com/go/article.php?mp_id=209
125. USGS, "Landsat 8 (LDCM) History," 2013. http://landsat.usgs.gov/about_ldcm.php
126. USGS, "SLC-Off Products: Background," 2013. http://landsat.usgs.gov/products_slcoffbackground.php

2
CHAPTER

MORPHODYNAMIC ENVIRONMENT IN A SEMIARID MOUTH RIVER COMPLEX CHOAPA RIVER, CHILE

Joselyn Arriagada[1, 2, 3], María-Victoria Soto[2] and Pablo Sarricolea[2]

[1] *Proyecto FONDECYT, Chile*

[2] *Departamento de Geografía, Facultad de Arquitectura y Urbanismo, Universidad de Chile, Chile, Santiago, RM, Chile*

[3] *Laboratoire Environnements et Paléoenvironnements Océaniques et Continentaux (EPOC), Université Bordeaux 1, Francia, 351 Cours de la Libération, Talence, France*

1. INTRODUCTION

The knowledge about mouth systems, estuarine environments and associated processes, seen from the viewpoint of geomorphologic science in Chile, focusing on the contributions of Araya-Vergara [1, 2,3] for the central Chile and Paskoff [4], in the case of semi-arid.

The traditional European and American studies on deltas and estuaries can not address the reality of the most complex estuaries, through evolutionary and taxonomic schemes clear. Such is the case of the mouths on the coast of central Chile, which cannot be studied if it is enriched and modified some molds of the coastal geomorphology as a discipline [4].

According to work done by Pritchard and Caspers [1] indicates that the estuary is a term hydrological and not geomorphological, also provides a definition accepted by the estuarine specialists, defined as a coastal water body that has a free connection with the open sea and from which seawater is measurably diluted with freshwater derived from continental drainage. For its part, Caspers [1] emphasizes the importance of the tidal factor for the classification of the estuary.

Because estuarine delta expression, understood as the accumulation of banks that grow from the inside of the estuary to the sea, is applied to a delta developed into an estuary, it is necessary to seek ways of the land on which was developed estuary, and then pursued a concept of geomorphology. Thus, Araya-Vergara [3] concluded that this form in its most typical commonly is an estuary. Consequently, intends to call delta in *ria* to the internal deltas formed in valleys flooded by the sea.

Notably Davies [3] conceptualizes the estuarine delta as part of a continuum of types of mouth, forming the transitional phase between the estuary and the delta as such (understood as a typically delta marine and distal).

One of the conclusions reached by Araya-Vergara [3, 5] in central Chile, refers to the strong influence morphoclimatic that seems to define the domain of these environments mouth. The author acknowledges the following areas depending on the significance morphoclimatic:

Distal zone of deltas. Corresponds to the Morphoclimatic influence of the desert with summer contributions halogen derivatives of influences height.

Prograded *rias* zone. These coincide with the semi-arid conditions of the Norte Chico, from the Copiapo River (27 º 20 'lat. S) to the Maipo river inclusive (33 º 47 lat. S).These paleo – *rias*determined by Paskoff [5]. This reflects the higher solid load is attributed to the rivers of arid areas.

Estuarine delta zone. These seem to correspond to the transition area between the semi-arid and humid, from the mouth of the Rapel River (34 º lat. S) and the Bio-Bio (37 ° lat. S) inclusive, where the river power is important to calibrate wide channel lower course and to provide plenty of substance delta, but also marine energy is sufficient to impede the progress of the progradation into the sea.

Rias Zone. This is connected with the moist and wet conditions in the Región de los Lagos (38° lat. S).

One way to characterize an estuarine system is through the characteristics of its key components. In this context, you can set the forms constituents of estuarine systems, which allows determining a zoning classification of these [3]:

A fluvial or proximal zone, with meandering estuarine.

An interior or media delta zone, with media and distributary banks.

Lagoon or distal zone, with distal lake shore or barrier semiblock by spits or spit outer barrier (Fig. 1).

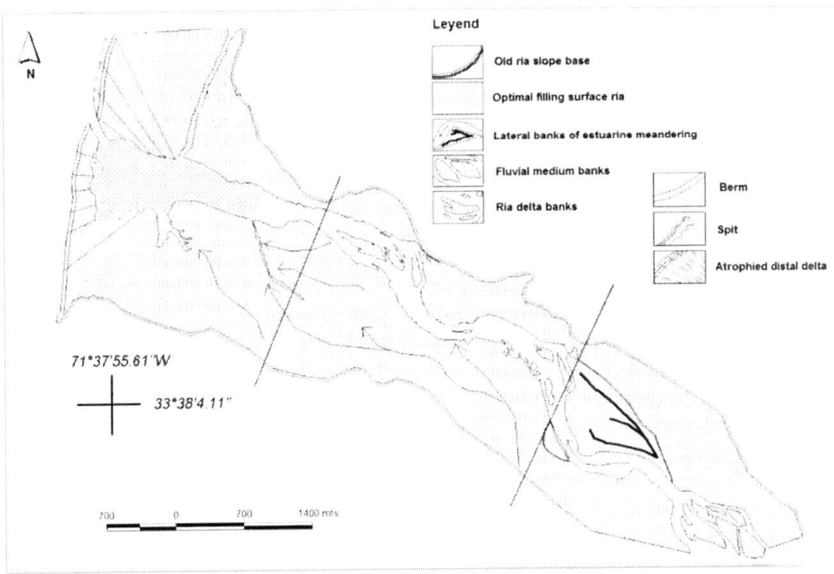

FIGURE 1. Essential Components, Maipo estuarine system. From: Arriagada [6]

Proximal zone: indicating the presence of estuarine meanders, they owe their properties to the action of currents ebb and flow, which are related to the action of river and sea action. Estuarine meanders have an essential feature, which is reflected in the way you present: wide at the media and narrow at the ends, also the banks lateral has a shape of cusp.

Media zone: is determined by a change in the pattern of the channel downstream from the area of estuarine meanders. Clearly denoted start means estuarine banks, which have a form characterized by the pointed shape of its ends.

Distal: identify the existence of an estuarine lagoon, which is located between the distal part of the delta area and a spit in the mouth. The spit or barrier are built with the contributions of the littoral drift (in the case of the Chilean coast, it makes sense from south to north), which carries sediment from the existing extension ridge further south coast, the waves push these sediments towards the coast as an arrow, with its free end facing the waves and giving elongated in the same sense that the beach from which the materials. Its curvature, as a *hook*, is explained by the refraction of waves.

This classification has been applied to the case of estuarine systems of the Aconcagua, Maipo, Itata, among others. The case of the Aconcagua, considered by Cortez [7], Martinez and Cortez [8], has provided interesting results in relation to chemical, physical and mixing processes. In applying the model of Cooper [9] the system of Aconcagua river mouth, it was established its status as an estuary dominated by fluvial processes and the waves.

The Maipo River estuarine complex was analyzed by Arriagada [6], work from which it is concluded that estuarine dynamics, associated with changes in the essential forms has been greatly affected by the construction of the San Antonio port, constituting this element of anthropogenic influence in a major evolutionary factor. In this case, according to the analysis developed by applying the model of Cooper [9] concluded that the estuary has a tendency to be dominated by fluvial action. However, there is evidence to indicate the alternation of dominant effects, depending on the season.

Other estuarine systems have been analyzed Rapel, the Itata, the Maule and the Mataquito [3]. In each of them have been identified and characterized the essential forms of estuarine environments, where the presence of banks and spits cuspidate is ubiquitous in them.

Then, the most studied systems from the perspective geodynamics and where the model has been applied, corresponded to central Chile, with few information on systems related to semi-arid mouth.

1.1. Supply and Mass Transfer

The study of the evolution of coastal systems, must include not only the characterization of the estuaries, but must also incorporate the behavior and evolution of the mass associated dune system. This aspect is relevant because they are indicators of sediment supply and dynamics of coastal environments in general, and estuaries in particular systems [10, 11, 12].

In this aspect, Aqueveque [13] notes the development of three main generations of dunes, ancient and modern means, which are dominated by the degree of evolution and morphology, called *continuumdunaris*. Also, Araya-Vergara [14] adopts the term transmudation for the spatial variation of the dune system, and transmutation, to the change in the morphology of the dune system.

Thus, the mass balance of inner and outer beach is explained by the orientation of the beach, as well as the type of surf zone [14, 15]. This has been found in genetic- evolutionary studies of barjan belts, categories of changes dunes and wave interaction system in large beach coves.

In this situation, one should expect that, in its evolution, the dunes adjacent to estuarine systems present a variation in the mass of dunes, backed by the dynamics and the associated contribution to the estuary.

In addition to the above, we can infer that the dynamic conditions of estuarine systems may be influenced by the processes associated with the basins. Such conditions are of global importance, as they are localized areas where most human settlements, whose impacts are significant morphological [16].

An example of these impacts was what happened at the mouth of the Maipo, where human influence changed the coastline less than 100 years [6]. Other cases have been studied by Federicci and Rodolfi [17, 18], who stated that the rapid retreat of the coastline in Ensenada de Atacama (Ecuador), is due to erosion caused by El Niño events, but also the indirect action of the destruction of wetlands (mangroves).

Araya-Vergara [5] taxonomy points to the classification of estuaries prograded from Copiapo to the Maipo River. These are associated with higher solid load of rivers is attributed to the semi-arid zones.

Consequently, the hypothesis behind this research, notes that in the mouth of Choapa system are recognized morphological traits that account for the evolution of these systems flows into the context of a domain morphoclimatic transitional semi-arid to temperate.

2. MATERIALS AND METHODS

2.1. Study Area

In this research, we analyze the mouth of the river system Choapa (Region of Coquimbo) (Fig. 2). The rationale for the study of these systems is based on the differentiation of morphoclimatic domains that support these environments, which highlights the semiarid condition.

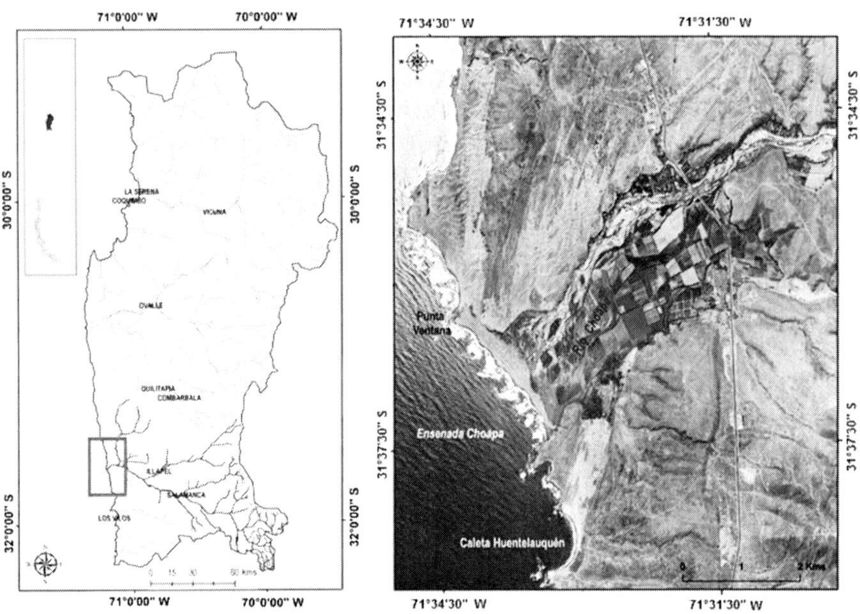

FIGURE 2. Study Area, Choapa river mouth in the national and regional context.

From the point of view of the climatic characteristics, the study area is located in the steppe climate with abundant cloudiness; geologically the area is inserted into the structural domain premesozoico correspond to the basement, where you can distinguish Choapa Metamorphic Complex [19].

There is also the presence of Huentelauquén formation, interpreted as a deposit of open marine platform [19]. It highlights the unconsolidated

Quaternary deposits: aeolian, coastal, alluvial and colluvial. The sand dunes have a transverse component and indicate acontribution of wind SSW-NE direction. These correspond to barjanes and transverse coastal sand dunes.

The geographical context into which you insert the specific problems of the mouth of Choapa, is related to a watershed with land use changes and intensity, but less rapidly than in the case of the Copiapo and Aconcagua river. For Choapa, recently begun to observe the process of conversion of agriculture and traditional farming to fruit farminggoats [20].

2.2. Methodology

The recognition of the essential systems linked to Choapa and Copiapo river mouths and sustained took place on the following concepts morphological from recognized estuarine systems theory for the domain morphoclimatic semiarid in Chile:

a. The conceptual basis of the systematization of Araya-Vergara [3]. Among the essential forms to be identified, indicating the estuarine meanders and banks, cuspidate lateral banks, and spit estuarine lagoon, located in the proximal, media and distal, respectively.

The proximal part has meandering estuaries, characterized by fluvial influence that still operates in this area, in the media zone, it means banks are estuarine or cuspidate, due to the influence of river flows, as well as the marine influence. In the distal zone is possible to identify the presence of the spit, as well as estuarine lagoon.

Was realized the identification of the forms cited by photo interpretation of aerial photographs of a period of 30 years, as well as recognition in the field. The final information was mapped by using the software ArcGIS 9.2.

a. The zoning of each estuarine system was carried out according to the scheme of Araya-Vergara [3], in this way, we use a general conceptual geomorphological model for all estuarine systems, classifying them in the following areas: proximal, media and distal. It included an analysis of the components in each zone.

Analysis of morphodynamics temporal-spatial of estuarine complex through the identification of significant changes that occurred over a period covering the last 30 years, using the topographical maps and photo interpretation of aerial photographs (30 years).

Status of change and evolution of the systems analyzed.

The patterns of change and evolution trend of the morphology surrounding the estuary, was systematized by producing map geomorphology of each system, for the period of time as indicated. Thus, together with analysis of the essential forms of estuarine system, emphasis was placed on the dynamics of the dune systems. The main method of analysis refers to the mapping principle follow the design of dune ridges, which are classified according to the notions of sequence [21] and *continuum dunaris* [14, 22], combining families to determine the dunes (Fig. 3)

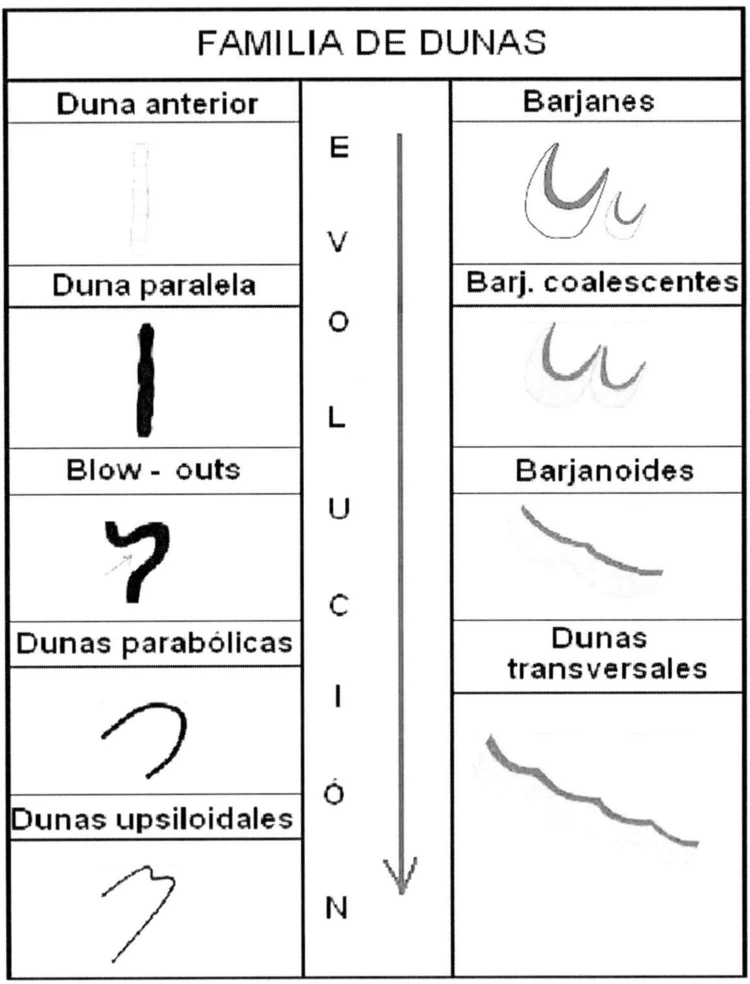

FIGURE 3. Family of dunes and continuum dunaris. Source: Prepared by the author, from notions of Verstappen [21], Araya-Vergara [14, 22].

Patterns of change and development trends of these estuarine systems, related to the domain morphoclimatic in the context of headland bay beach systems in which they reside.

In the characterization of the near shore environment shows the importance of processes derived from swell. This results in temporal-spatial distribution of the energy of breaking waves. An extensive bibliography on the subject is reviewed by Martínez [23], who delivers new observations to Central Chile. Considered essential advances in active sensing and morphodynamic [24, 25, 26], however working with passive sensors.

Analysis of wave-dominated beaches was performed according to the scheme of Wrigtht & Short [27], as amended by Araya-Vergara [15] for use in aerial photographs. This taxonomy classifies the beach as dissipative, intermediate and reflective. The nomenclature used is summarized in Table 1.

TABLE 1. Classification of type of wave-dominated beach [28, 29].

Type of breaker zone	Nomenclature
Reflective	R
Low tide terrace	LTT
Transverse bars and rip	TBR
Rhythmic bars and rip	RBB
Long shore bar and trough	LBT
Dissipative	D

To detect change between 1997 and 2008 used three images Landsat 5 Thematic Mapper (TM) (table 2). Landsat TM operate in circular, sun synchronous, near polar orbits at a nominal altitude of 705 km. The inclination of the orbit is 98.2 degrees. The TM sensors view the entire earth every 16 days with a complete cycle of 233 orbits. Full scene products contain 5728 lines of 6120 pixels, corresponding to a ground area of 172 km by 184 km. The TM is a sensor, which records a continuous strip image of the earth in 7 spectral bands, between 0.45 and 12.5 microns. The satellite images were corrected geometric and atmospheric correction in all the bands required. The spectral water index is a single number derived from an arithmetic operation of two or more spectral bands.

TABLE 2. Used images.

Acquisition Date	Satellite Images	Resolution
6/26/1997	Landsat 5 TM	30 m
6/21/2001	Landsat 5 TM	30 m
9/12/2008	Landsat 5 TM	30 m

An appropriate threshold of the index is then established to separate water bodies from other land-cover features based on the spectral characteristics. Adopting the format of the normalized difference vegetation index (NDVI), McFeeters [30] and Kuleli et al. [31] developed the normalized difference water index (NDWI).

Where ρ *green* and ρ *NIR* are the reflectance of green and NIR bands, respectively. The NDWI value ranges from +1 to -1. McFeeters [30] set zero as the threshold. The NDWI data derived from Landsat TM (Thematic Mapper) of the years 1997, 2001 and 2008. They were processed in the software IDRISI Taiga for correcting some of the distortions mentioned by Chuvieco [32] needed in order to obtain a clear index. After the georeferencing, an atmospheric correction using the COST model was used. This model was chosen because his accuracy in semiarid regions, and specially for his accuracy on the Landsat TM blue band. (Fig.4). Moreover, the figure 5 ilustrates the flow of this method.

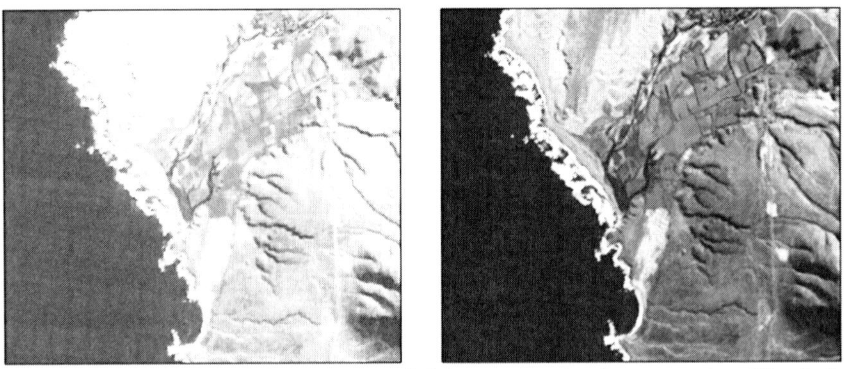

FIGURE 4. Example of the COST model for atmospheric correction. The left image is a false color composition using the raw bands of the TM image for the mouth of the Choapa River for 2001. The right image is the same composition using the same bands after being corrected by the COST model on the software IDRISI Taiga.

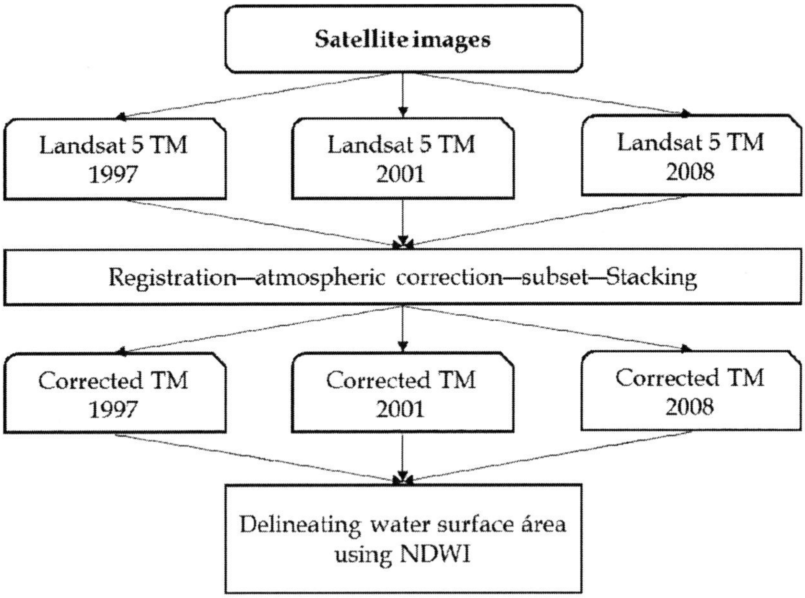

FIGURE 5. Flow diagram showing the methodology steps

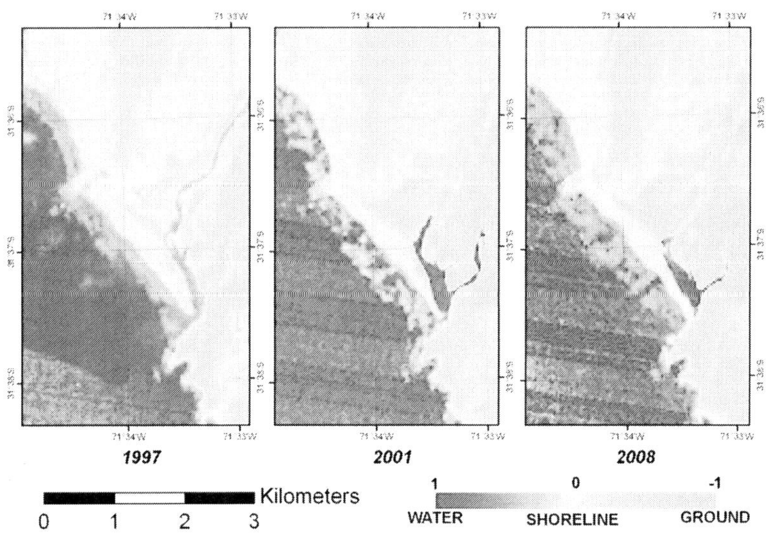

FIGURE 6. NDWI raw values on the corrected TM images for the years 1997, 2001 and 2008.

The result of the NDWI for each pixel, in which positive values represent water, and negative values represent ground (Fig. 6).

However, this index tend to show wet sand pixels as water due the reflectivity in the band 2, but these values, while positive, are close very close to zero, so a calibration of the water-ground break value was performed though photo interpretation, reaching the breaking value of -0.045 for the 1997 image, 0.053 on the 2001 image, and 0.0048 for 2008. The resulting calibrated water-ground break line is shown on the Figure 7.

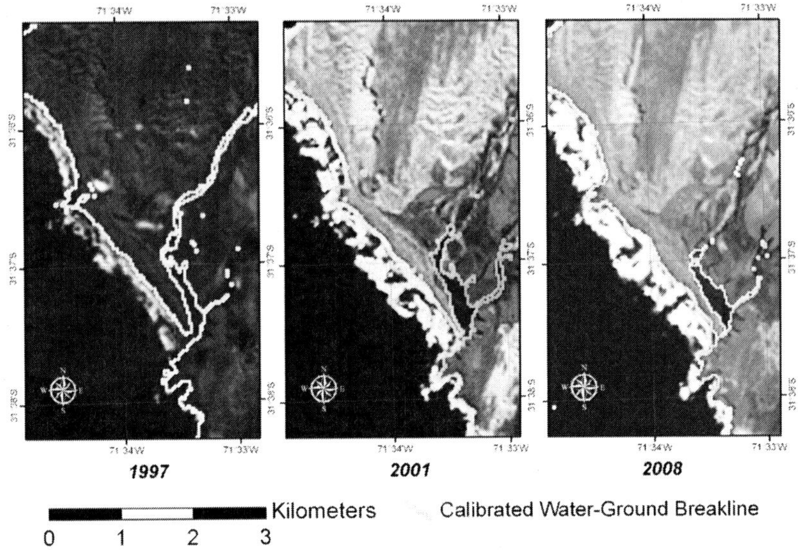

FIGURE 7. Resulting Water-Ground limit after calibrating the NDWI raw break value.

Finally, a reclassification of these values was made on the software ArcGis 9.2 creating polygons for water and ground categories, allowing a direct comparison for the years used on the images.

3. RESULTS

3.1. Geodynamics of The Choapa Mouth System (1978 To 2008)

As shown in Figure 8, for 1978 highlights the presence of a channel in the proximal diffluent, where a glimpse of the river influence that dominates the environment in this sector, where there are large lateral river banks extension. Then in the media part of the system there are the rotation from fluvial media banks and lateral fluvial banks. Note the presence of abandoned channels on the terrace river banks and in the estuarine side. In the distal part, called attention to the extension of the fluvial-marine terrace, which extends aprox. 2 km.

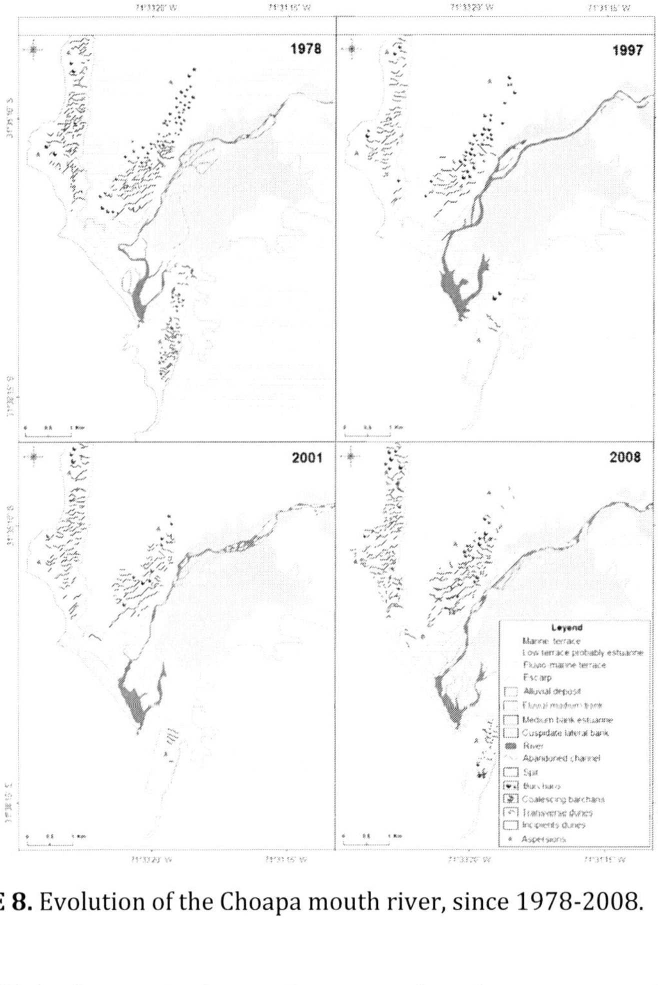

FIGURE 8. Evolution of the Choapa mouth river, since 1978-2008.

For 1997, in the proximal part, the estuary has the same type of channel had meandrante to 1978, this continues to the middle of the estuary, where the channel is meandering and winding. Towards the middle, it is possible to clearly distinguish estuarine character means banks, which are detected cuspidate, just before the river course becomes estuarine lagoon. Estuarine lagoon (distal) presents a greater extent compared to 1978. Relative to the sand dune system located south of the estuary include the presence of barjans and certain transverse sand dunes.

For the year 2001 in the distal zone is appreciated the meandering channel, as in previous years. In the middle part, is where the biggest differences you notice it tends to notice a greater number of banks means, indicating a greater sediment load, as a geodynamic behavior like braided channel. This extends to the edge of the middle, where the river channel again becomes meandering. In the distal part, the lagoon estuary is clearly distinguishable. We highlight the presence of abandoned channels, which could be indicators of lower river load.

The situation of 2008 is not much different to the situation of another year. In the proximal part, it shows the meandering course is appreciate where fluvial banks. On the average, banks still appear fluvial as well as the presence of banks estuarine media. The river course is still meandering, reaching its mouth.

The results of the NDWI comparison (Fig. 9), they are completely consistent as they clearly show the variations and some changes on the cannels and banks. However, the most evident change shown by the comparison is the growth of the estuarine lagoon.

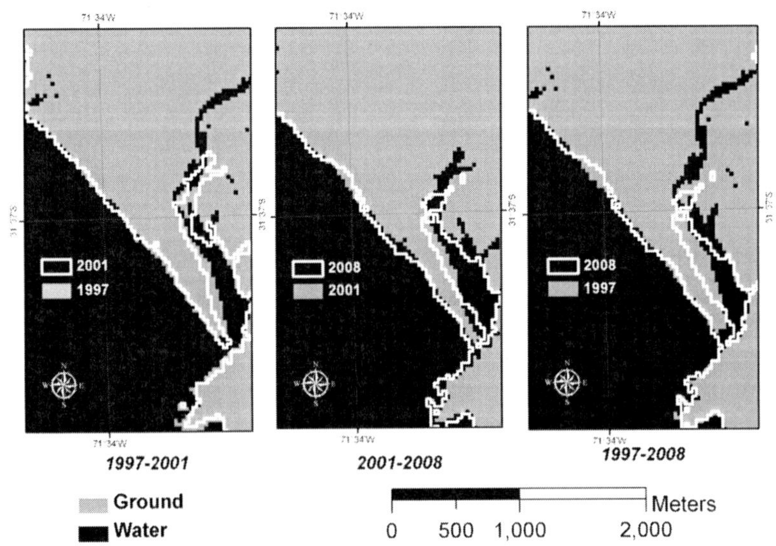

FIGURE 9. NDWI comparison between 1997, 2001 and 2007. On the 1997-2008 comparison, the estuarine lagoon variation and growth is accurately detected by the NDWI.

3.2. Characterization of Choapa Cove (1978 To 2008)

For the year 1978, in the case of Choapa system, the most striking are the dunes. According to thefigure 10, there is a predominance of transverse and barjanes sand dunes. In the coastal sector, there are some cross-sectional and longitudinal patterns less important, these dunes overlie the marine terrace, which is covered. Towards the north-inner estuary, dune field is quite distinguishable, repeating the pattern of tranverse dune, but more remarkable is the barjanic sand fiel, which is considerably large.

The system located to the west, corresponds to the evolution of barjanics sets, so it is assumed to be the oldest, on the contrary, the system located to the east (individual sand dunes barjanic) are recent and are moving in the direction NE.

For 1997, the sand dunes express changes in morphology and in the case of the northern section - coastal estuary, there appears to be an increase in the different dunes, where the cross pattern becomes much more noticeable (Fig 9). Barjan remain, but at lower densities. In the same figure, we see the same patterns and barjanics cross in the north-interior, although it refers to a decrease in the number of barjanic units. In turn, the spray area includes increasing magnitude, so one might infer that the barjan have evolved and in turn, transverse dunes have been dismantled.

In 2001, major changes are in the dunes, as well as for 1978-1996. This year, they become barjanic and transverse barjanoides sand dunes, in the north, we can differentiate the extension of coverage for spraying (Fig. 9). There is a low representation of barjan. In the dune field located south of the estuary, it is noteworthy that the area near the estuary is devoid of mass contribution, and that is clearly seen that there is associated to vegetation, which has no clear traces of individual dunes.

For the year 2008 in the northern part there is a progressive decrease in barjans, reflected in the increased presence of barjanoides, as well as the largest area of spraying, these dune systems (barjanoides, barjans coalescing) cover large areas. Generate distinct forms. There is a continuous supply of material to the marine terrace, which is reflected not only in the designated spray, but also in the genesis and stabilization of dunes.

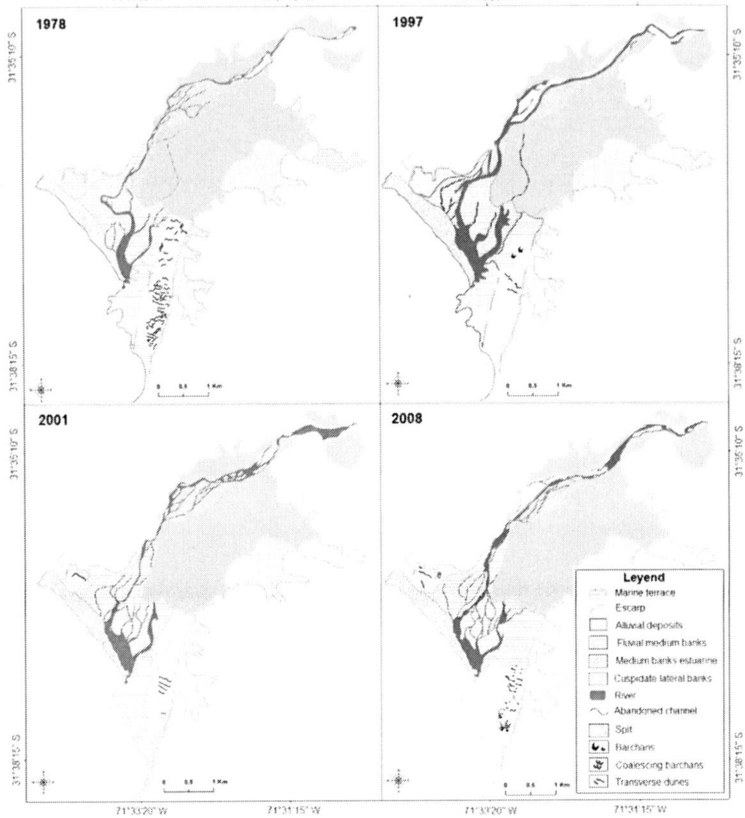

FIGURE 10. Dynamics Choapa Cove, since 1979 to 2008

4. DISCUSSION

4.1. Geodynamic System

In the Choapa mouth river system, the presence of the essential forms differ according to the area to be analyzed, so that in the proximal part, is a clear presence meanderinge course, while in the middle is not as noticeable presence estuarine banks. The depositional landforms in the middle (banks) have a great dynamic in the study period, varying in shape and location, a situation that hindered the categorization of these banks, in its identification as a river or estuarine. In the distal, zoning Araya-Vergara [3] is quite applicable, because the presence of estuarine lagoon and spit presence of essential forms to correspond to the characterization of this area. This condition is similar to the conditions analyzed in central Chile, as is the case of the Maipo [6], Aconcagua [7], Rapel [1, 3] and Itata [33].

From the point of view of the relative influence of the waves, tide and fluvial component [9, 34] in this estuary river influence seems to prevail, as evidenced by the presence of the spit, it demonstrate a continuous supply of sediment mass, which seems to be more evident in the last year analyzed. This condition is comparable to the situation in the Maipo [6] where the spit remains almost intact over a long period of time. However, consider the case of the Maipo and Aconcagua morphoclimatic environments are restricted to the case of Choapa different, yet similar conditions exist particularly in the presence of the spit.

4.2. The Relationship with The Zone Surf Dynamic

The Choapa Cove present the sand dune complexes identified by Araya-Vergara [13]. There is a predominance sand dunes systems from of genetic barjanic, which also presents transmudation conditions and transmutation [14]. However, Choapa Cove there is a high wave energy because the width of the surf zone is enough. This becomes important as the contribution of mass: in the case of Choapa Cove is possible to see the *continuum* of dunes, led to the sand dunes

These processes are quite notorious during the study period (30 years), indicating a strong dynamics related to sediment this is where the component of the wave and surf zone type is important (Fig. 11), main component is transversal.

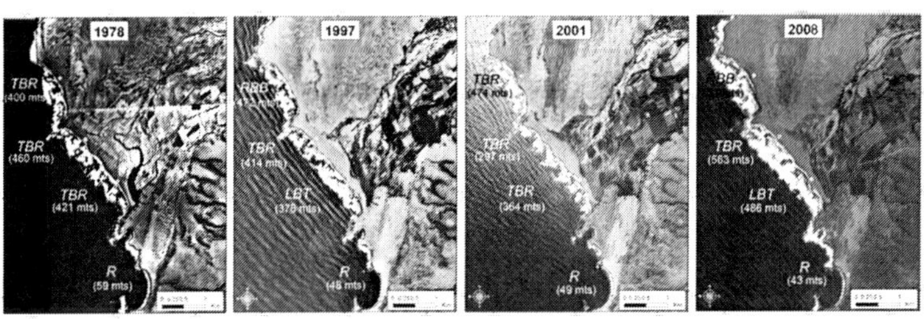

FIGURE 11. Space-time dynamics of wave-dominated beaches

5. CONCLUSIONS

In relation to the Choapa estuarine system, it presents conditions similar to those found in central Chile, mainly in the distal zone. Presents an estuarine lagoon flanked by a spit, which is quite stable in the middle section you can find river banks and estuaries, in alternancy. In the proximal part, the river is clearly a meandering stream.

You could say that the Choapa estuarine system realize greater sediment supply, because the river course change in the distal part in the review period: meandering change to a braided dynamic. So you can say that the system supplies material to the coast on a continuous basis over a period of 30 years.

In the case of Choapa mouth system, evolution has been much more pronounced: by 2007, the sand dunes barjanes change to coalescent barjan and barjanoides, and even become quite transverse dunes. That is, in this system can be seen in 30 years the *continuum dunaris* from origin barjanic coastal sand dunes.

It is also important to note the change associated with the estuary Choapa, in 1978 is presented as meandering in its distal part, but in 2007, there are differential patterns that denotes a greater amount of sediment, resulting in an barided river in its distal zone. In addition, it adds to the change in the direction of the river. In the Choapa river system there are basic shapes, such as media cuspidate banks, lateral cuspidate banks and spit, which presents constant throughout the period under review.

For this reason, it opens the discussion about the limits proposed by Araya-Vergara [3] regarding the classification of estuaries and the possible conclusion that morphologically mouth system can be termed as estuarine delta and not as part of prograted rias system, by displaying dynamics similar characteristics with the mouths of central Chile, as the Aconcagua and Maipo rivers.

The calculation of the NDWI is a valid method for an automatic delimitation of a shoreline or boundary between ground and water corpses (like the estuarine lagoon) as is shown successful using the presented materials and being consistent with the overall data. However, the accuracy of this index is dependent of the pixel resolution, the preciseness of the geometric and atmospheric corrections, but even more dependent of the calibration phase of the break value between ground and water. This break value differs in every image but is always close to 0. A care photo interpretation can achieve a satisfactory boundary line between ground and water and obtain the data needed for an easy comparison between years.

6. ACKNOWLEDGEMENTS

To FODECYT and the Faculty of Architecture and Urbanism of the University of Chile.

REFERENCES

1. Araya-Vergara J. Contribución al estudio de los procesos estuariales en las desembocaduras de los ríos Rapel y Maipo. Revista Informaciones Geográficas 1970; 20:17-38.
2. Araya-Vergara J, Castro C, Andrade B. Análisis de la turbidez fluvial en el mar mediante teledetección y control terrestre. In: Instituto Geográfico Militar de Chile, editors. Primer Symposium Internacional sobre Percepción Remota. 1978, Apr. Santiago de Chile.
3. Araya-Vergara J. El concepto de delta en ría y su significado en la evolución litoral (ejemplo de Chile Central). Informaciones Geográficas Chile. 1981; 28:19-42.
4. Paskoff R. Recherches geomorphologiques dans le Chili Semi-aride. Vol 1. Bordeaux: Bordeaux Biscaye; 1970.
5. Araya-Vergara J. Análisis de la localización y de los procesos y formas predominantes de la línea litoral de Chile. Informaciones Geográficas de Chile. 1982; 29:35-55.
6. Arriagada J. Cambios en el sistema estuarial del Maipo y su relación con obras portuarias, Chile Central. Memoria de título de Geógrafo, Universidad de Chile; 2005.
7. Cortéz C. Observaciones dinámicas y geomorfológicas en el estuario del Aconcagua, Chile Central. Memoria de título de Geógrafo, Universidad de Chile; 2002.
8. Martínez C, Cortéz C. Características hidrográficas y sedimentológicas en el estuario del río Aconcagua, Chile Central. Revista de Geografía Norte Grande. 2007; 37:63-74.
9. Cooper JA. Sedimentary processes in the river-dominated Mvoti stuary, South Africa. Geomorphology. 1994; 9:271-300.
10. Arriagada J. Geomorfología comparada en la zona semiárida de Chile: casos Copiapó y Choapa. Tesis de Magister. Universidad de Chile: 2009.

11. Arriagada J, Soto MV, Castro CP. Dinámica del complejo estuarial del Choapa: istema estuario-playa-duna, región de Coquimbo. In: Resumen XXIX Congreso Nacional y XIV Internacional de Geografía. 2008 Oct. Temuco, Chile.
12. Soto MV, Arriagada J, Castro CP, Märker M, Rodolfi G. Aspectos geodinámicos de n paleo estuario del desierto marginal de Chile. Río Copiapó. Revista de Geografía Norte Grande. 2010; 46:123-135.
13. Aqueveque C. Análisis de suelos desarrollados en dunas litorales antiguas de Chile entral. Memoria de título de Geógrafo, Universidad de Chile; 2008.
14. Araya-Vergara J. The evolution of modern coastal dune systems in central Chile. n: International Geomorphology 1986 Part II. V.Gardiner. John Wiley & Sons Ltd.; 987. p. 1231-1244.
15. Araya-Vergara J. Sistema de interacción oleaje-playa frente a los ergs de Chanco y Arauco, Chile. Gayana Oceanología. 1996; 4(2):159-167.
16. Verhaar PM, Biron PM, Ferguson RI, Hoey TB. A modified morphodynamic model or investigating the response of Rivers to short-term climate change. Geomorphology. 2008; 101(4):674-682.
17. Federici P, Rodolfi G. Rapid shorelines retreat along the Esmeraldas Coast, Ecuador: natural and man-induced processes. Journal of Coastal Conservation. 001; 7:163-170.
18. Federici P, Rodolfi G. Geomorphological features and evolution of the Ensenada de Atacames (Provincia de Esmeraldas, Ecuador). Journal of Coastal Research. 2004; 20(3):700-708.
19. Rivano S, Sepulveda P. Hoja Illapel, Región de Coquimbo, 1: 250.000. In: Servicio Nacional de Geología y Minería,editors, Serie Geología Básica (69). 1986. Santiago de Chile.
20. Castro CP, Meza M, Soto MV. Pérdida de calidad de suelos en el periurbano de ciudades intermedias. Salamanca y Quillota. In: Anales XXIX Congreso Nacional y XIV Internacional de Geografía. 2008 Oct. Temuco, Chile.
21. Verstapppen H. On dune types, families and sequences in areas of unidirectional winds. Gottiner Geogr. Abh. 1972; 60:341-353.
22. Araya-Vergara J. Significance of barchans in Beach-dune Systems interactions in Central Chile. Thalassas. 1986; 4(1):23-27.
23. Martínez C. 2001. El efecto de ensenada en los procesos litorales de las ensenadas de Valparaíso, Algarrobo y Cartagena, Chile Central. Tesis de Magíster, Universidad de Chile; 2001.

24. Marghany M. Modelling shoreline rate of changes using holographic interferometry. nternational Journal of the Physical Sciences. 2011; vol. 6(34):7694–7698.
25. Marghany M. DEM Reconstruction of Coastal Geomorphology from DINSAR. In Murgante B. et al., editors Lecture Notes in Computer Science (ICCSA 2012), Part II, LNCS 7335; 2012. p. 435–446.
26. Marghany M. DInSAR technique for three-dimensional coastal spit simulation from Radarsat-1 fine mode data. Acta Geophysica. 2013 Apr.; vol. 61(2);478-493.
27. Short AD. A note on the control of beach type and change, with S. E. Australian Examples. Journal of Coastal Research. 1987; 3(3): 387-395.
28. Wright LD, Short AD. Morphodynamic variability of surf zones and beaches: A síntesis. Marine Geology. 1984; 56 (1): 93-118.
29. Short AD. Waves-dominated beachs. In Short A, editors. Handbook of beach and shoreface morphodynamics. Chichester; John Wiley & Sons; 1999. p. 173-191.
30. Mcfeeters SK. The use of Normalized Difference Water index (NDWI) in the delineation of open water features. International Journal of Remote Sensing. 1996; vol 17(7):1425–1432.
31. Kuleli T, Guneroglu A, Karsli F, Dihkan M. Automatic detection of shoreline change on coastal Ramsar wetlands of Turkey. Ocean Engineering. 2011; 38:1141-1149.
32. Chuvieco E. Teledetección Ambiental. La observación de la tierra desde el espacio. Barcelona: Ariel; 2002.
33. Castro CP. (1987). Influencia del río Itata en el desarrollo de líneas de costa arenosas en la parte sur de Chile Central. Memoria de Título de Geógrafo. Universidad de Chile; 1987.
34. Dalrymple R, Zaimln B, Boyd R. Estuarine facies models. Conceptual basis and stratigraphic implications. Journal of Sedimentary Research. 1992; 62: 1130-1146.

3
CHAPTER

REMOTE SENSING AND GEOSPATIAL TECHNOLOGICAL APPLICATIONS FOR SITE-SPECIFIC MANAGEMENT OF FRUIT AND NUT CROPS: A REVIEW

Sudhanshu Sekhar Panda [1,*], Gerrit Hoogenboom [2,†] and Joel O. Paz [3]

[1]*Department of Science, Engineering, and Technology, Gainesville State College, 3820 Mundy Mill Road, Oakwood, GA 30566, USA*

[2]*Department of Biological and Agricultural Engineering, The University of Georgia, Griffin, GA 30223, USA*

[3]*Department of Agricultural and Biological Engineering, Mississippi State University, Mississippi State, MS 39762, USA*

ABSTRACT

Site-specific crop management (SSCM) is one facet of precision agriculture which is helping increase production with minimal input. It has enhanced the cost-benefit scenario in crop production. Even though the SSCM is very widely used in row crop agriculture like corn, wheat, rice, soybean, etc. it has very little application in cash crops like fruit and nut. The main goal of this review paper was to conduct a comprehensive review of advanced technologies, including geospatial technologies, used in site-specific management of fruit and nut crops. The review explores various remote sensing data from different platforms like satellite, LIDAR, aerial, and field imaging. The study analyzes the use of satellite sensors, such as Quickbird, Landsat, SPOT, and IRS imagery as well as hyperspectral narrow-band remote sensing data in study of fruit and nut crops in blueberry, citrus, peach, apple, etc. The study also explores other geospatial technologies such as GPS, GIS spatial modeling, advanced image processing techniques, and information technology for suitability study, orchard delineation, and classification accuracy assessment. The study also provides an example of a geospatial model developed in ArcGIS ModelBuilder to automate the blueberry production suitability analysis. The GIS spatial model is developed using various crop characteristics such as chilling hours, soil permeability, drainage, and pH, and land cover to determine the best sites for growing blueberry in Georgia, U.S. The study also provides a list of spectral reflectance curves developed for some fruit and nut crops, blueberry, crowberry, redblush citrus, orange, prickly pear, and peach. The study also explains these curves in detail to help researchers choose the image platform, sensor, and spectrum wavelength for various fruit and nut crops SSCM.

Keywords: site-specific management; orchards; land-use land cover; geospatial technology; geographic information system; remote sensing; chill hour; high value crops

1. INTRODUCTION

Growth observation, impact assessment, and timely strategic response to small variations in crop production is known as precision agriculture. It has been used in a wide range of agricultural activities including field crop production, dairy farming, horticulture, and forest management. Site-specific crop management (SSCM) is one facet of precision agriculture which involves spatial referencing, crop and climate monitoring, attribute mapping, decision support systems, and differential action. SSCM is carried out with a greater degree of precision through the use of geospatial technologies. Geospatial technology is a combination of four essential tools: remote sensing, geographic information systems (GIS), global positioning systems (GPS), and information technology or data management [1,2,3,4,5,6,7]. SSCM has become very common in management of field and row crops during recent years. Yet, its application for non-traditional horticultural crops has not been very widespread [8]. Companies, such as

Lanworth (http://lanworth.com/), which gather intelligence on natural resources, use geospatial technologies to map, analyze, and forecast agriculture and forest output. This is based on crop and forest growth parameters which is a large amount of data, terabytes in size. It helps in the SSCM of agricultural and forest products and increases the yield in a sustainable manner.

The application of aerial or satellite imaging, along with GPS and GIS, is the first step towards the successful application of SSCM for fruit and nut crops [9]. The use of SSCM for horticultural crops such as fruit and nut crops has potential for increasing net returns and optimizing resource use [8]. The delineation of orchards and spatial analysis using geospatial technology can provide additional information for management decision making, such as the determination of fruit yield, the quantification and scheduling of precise and proper fertilizer, irrigation needs, and the application of pesticides for pest and disease management. Ultimately, it will improve profits for producers [8]. The use of remote sensing has become common for the general detection of the growth and health of orchards on a larger scale. However, digital imaging technology is increasingly being used for intensive site-specific management of orchards as well, for instance: estimating the amount of fruits on individual trees, fruit quality, and also leaf area index or crown cover.

The overall goal of this paper is to provide a review of studies that use geospatial technology and especially remote sensing for conducting SSCM in fruit and nut crops, including oranges, peaches, pecans, apples, grapes, blueberries, as well as other crops around the world. Another objective of this paper is to help fruit and nut crop managers and other stakeholders to understand the applications of geospatial technology for proper management of their crops and subsequently increase their productivity.

2. SSCM in Agricultural (Row) Crops

Research on precision agriculture has been conducted since the late 1980s. However, a rapid development with respect to SSCM of agricultural crops, especially row crops, occurred during the nineties and the early years of the 21st century. Initial research mostly included yield forecasting and recommendations to address the low yield levels for specific locations in individual fields. Research on precision management of agricultural crops includes maize [2,3,4,10,11,12], wheat [5,6], soybean [11,13,14], cotton [15,16,17], sugar beet [18], barley [19], sorghum [20,21], oats [22] and others that are common in the United States and other countries across the world [23,24,25,26,27]. Most studies used remotely sensed imageries, including panchromatic, multi-spectral or hyperspectral satellite, aerial, and video photographs. They also used other geospatial techniques such as image processing, GIS, and GPS for the spatial-based yield estimation of different crops. A few examples of SSCM for traditional row crops are discussed below:

Diker et al. [10] suggested the potential application of aerial photos and GIS to monitor spatial and temporal variability within a field and to estimate the potential yield of maize. They used four vegetation indices, the normalized difference vegetation index (NDVI), green NDVI (GNDVI), IR/R, and IR/G, to compare yield monitor data, SPAD (Soil and Plant Analyzer Development) chlorophyll meter reading, and leaf area index (LAI). Gopalapillai et al. [11] used digital multi-spectral images to derive spatial weed density for a soybean field. For image calibration, they used image indices such as NDVI, SRI, normalized NIR, and red and green band information. In their study, surprisingly, the correlation between image indices and weed density changed with different resolutions. High resolution imageries pinpointed the weed locations with single pixels. Thus, it reduced the number of mixed pixels (mixels) of crop and weed together. At a 5.3 m pixel resolution, the correlation between the indices and the weed density was highest. They also found a correlation between planting date, soil types, and herbicide applications with respect to weed density.

Seidl et al. [13] found that by incorporating remote sensing information through normalized vegetation index (NVI) with modeling they could improve the prediction of soybean yield with the CROPGRO-Soybean model using spatial variability information. Goel et al. [14] reported a strong correlation between the digital information, e.g., spectral data of the aerial image, and soybean crop physiological parameters such as chlorophyll fluorescence, leaf greenness, leaf area index (LAI), photosynthesis rate, and plant height. In another application, Goel et al. [14] used multi-spectral (24 wave band) airborne optical remote sensing for detecting weed infestation in site-specific managed field crops. They used the multi-spectral band, with a range of 475.12 nm to 910.01 nm, and their results indicated that the wave band centered on 675.98 and 685.17 nm in the red region and 743.93 to 830.43 nm in the NIR region, had good potential for weed classification in a maize field. The Schiffes multiple range tests provided a p-value that was less than 0.05 to support their findings.

Plant et al. [15] used the aerial images of a cotton field to determine final yield potential based on NDVI. They found that the spatio-temporal pattern of NDVI reflected the nitrogen stress factor of the cotton plant. However, the NDVI also tended to indicate that the stress factor did not result in a significant yield reduction. Teixeira and Ustin [16] determined that NDVI values developed from aerial images can be good indicators for estimating cotton yield, while the differences in NDVI could provide guidance in locating specific areas where yield was reduced.

Yang and Anderson [20] used color-infrared (CIR) images obtained in 1995 and 1996 with a three-camera digital video imaging system from two sorghum fields in southern Texas to estimate crop yield. They found significant correlation coefficients (r) between grain yield and the NIR-band, R-band, G-band, and NDVI information of the digital video image based on stepwise regression. Yang and Everett [21] used color-infrared digital images to predict the yield of commercial sorghum fields. They found a correlation between sorghum yield and the green ratio (NIR/ Green) value of the digital image data.

Humburg and Stange [18] obtained five multi-spectral aerial images on different dates from an experiment consisting of ten different fertility treatments and spectrometer readings from the same plots to determine the correlation between the spectral indices and sugarbeet quality variables, such as recoverable sugar. They found that the green NDVI had the best correlation of more than 86% compared to the normal gray values of individual bands (red, green, or blue). Lamb et al. [22] used airborne digital imaging for evaluating the accuracy of mapping weeds in seedling crops (wild oats). They found a correlation of 71% between the NDVI and SAVI images and the weed data obtained from the field.

With respect to the application of SSCM in other regions of the world, Ray et al. [17] used NDVI derived from spectral information that was obtained from the LISS I sensor located on an Indian Remote Sensing (IRS) satellite to predict cotton yield for a very large area, i.e., the Sundarnagar district of Gujarat in India. Their model to estimate yield, provided very promising results with only a 1.2% relative deviation from the actual yield. Sun [27] used advanced "3S-technology" (RS, GIS, and GPS) to monitor the growth of crops and to provide recommendations for crop management in China. He analyzed a range of information sources: aerial photographs, meteorological satellite data, Landsat TM data, GIS, ground survey data for local growth and development monitoring, crop planting area, crop yield, and other relevant information pertaining to China. He concluded that remote sensing has potential for dynamic monitoring and yield estimation of many types of crops for large regions. However, his study did not provide any information with respect to the accuracy of his yield predictions.

3. PRECISION AGRICULTURE FOR MANAGEMENT OF TRADITIONAL AGRONOMIC CROPS

Many studies [1,2,3,4,5,6,7,8,9,10,11,12,13,14,15,16,17,18,19,20,21,22,23, 24,25,26,27] have been conducted worldwide to determine the variation in crop yield using remotely-sensed measurements. However, more recent SSCM studies have focused on the determination of the spatial variability of soil nutrients, water stress, location-based crop diseases, and other parameters associated with in-field variability [28,29,30,31,32,33,34, 35,36,37,38]. Some research has also been performed to determine the potential impact of climate change on the spatial variability of crop production, especially on a larger scale [28,29]. Although these studies are not pure SSCM studies, they are intended to understand and improve crop production. Researchers are now using advanced image processing techniques such as optical sensing and well-developed machines fitted with sensors suitable for intense analysis of the factors that control the variability in crop yield [34].

Pests and diseases can have a potential impact on crop yield as well as product quality. The application of SSCM for detecting the spatial variability of pest and disease infestations for different crops is now very common. For instance, Kulkarni et al. [30] used SSCM to detect soybean cyst nematode (SCN; Heterodera glycine) in a soybean field because it is difficult to determine the onset of SCN as its symptoms are not dramatic. Airborne visible-infrared imagery of the field was acquired and linear regression analysis of the green normalized difference vegetation index (GNDVI) derived from aerial imagery, was used to obtain an estimation of the infestation that was over 70% accurate. There are other studies as well that used SSCM to determine the onset or localization of pest and disease infestations [31,32,33].

For potatoes, it is important to determine the in-season potato crop nitrogen status (CNS) to be able to obtain a high yield. A review study conducted by Goffart et al. [34] suggested that the most important and available methods for CNS assessment be evaluated for their accuracy, precision, sensitivity, and feasibility. These assessments include the determination of the nitrate concentration of petiole sap, the measurement of leaf chlorophyll concentration using a hand-held chlorophyll meter, the measurement of crop light reflectance through a hand-held radiometer using passive sensors, and other methodologies. According to Goffart et al. [32], more recent methods that are still being developed are based on near, ground-based, air-borne or space-borne remote sensing. This information is then incorporated into decision support systems to provide recommendations for N fertilization management of potatoes [32].

Reyniers et al. [34] stated that optical measurement platforms such as satellites, airplanes, and hand-held sensors do not meet the requirements for good site-specific nitrogen management recommendations. They therefore developed a tractor-mounted automated sensor system to automatically detect crop nitrogen stress status "on the go". They found a deviation that was less than 10% from the actual nitrogen level in their study with wheat and recommended that the spectrograph red edge measurement was best suitable for estimating crop nitrogen uptake during growing season.

Panda et al. [37] developed a yield-based water use model to predict spatial water requirements for corn. The water use was based on: crop yield information that was estimated from vegetation aerial images taken between late July and early August of 1997–2001; bare soil image data of same years; land elevation data; weather data such as temperature and accumulated growing degree days (AGDD), solar radiation, rainfall; and the dates and amounts of irrigation applications. These crop growth-influencing parameters were integrated from germination until the date the aerial image acquisition was taken. The model was developed using a radial basis functional network (RBFN). A comprehensive review paper by Kitchen [38] describes the use of emerging technologies for real-time and integrated agricultural decision making for irrigation scheduling.

There are ample SSCM studies for traditional row crop agriculture and advancements are being made over time [1,2,3,5,6,7,13,14,15,16,17,18,19,20,21,22]. Along with the advancements in technology, the practical applications of modeling, mechanical development, and technological usage are also increasing. However, these technologies are not as greatly developed for SSCM of fruit crops.

4. SSCM IN FRUIT AND NUT CROPS

Many studies have recently been conducted on the micromanagement of horticultural crops for large orchards and small farms. Horticultural crops, including fruit and nut crops, are high value crops for which SSCM might have potential for increasing net returns and optimizing resource use [8]. Oranges, peaches, pecans, apples, grapes and blueberries are major components of the overall crop production system in the U.S. These fruit and nut crops need advanced geospatial technology for micromanagement.

SSCM for fruit and nut crops would utilize soil sampling, yield monitoring, and geospatial technology, such as remote sensing, GIS, GPS, and information technology [8,9]. In order for GIS to work efficiently in SSCM model development, comprehensive plant related databases need to be established. The key agricultural factors that contribute towards fruit and nut crop growth and production are different soil characteristics, climatic conditions, and land-use type. The soil characteristics involved in fruit and nut crop production are the soil pH, nutrient levels, drainage efficiency, texture, permeability and water holding capacity. Climatic factors that contribute toward fruit and nut crop growth are temperature, solar radiation, rainfall, chilling hours, growing degree days, etc. All these physical factors must be part of a geospatial database and correlate to the corresponding spatial-based yield for fruit and nut crops for the development of a SSCM model.

5. WEATHER STUDY FOR SSCM IN FRUIT AND NUT CROPS

According to Rao [39], weather conditions can either hasten or delay phenological phases of horticultural crops. It is especially a major factor for crops grown in northern latitudes where there is a shorter period of warm days. For example, higher temperatures in the tropics have resulted in increased plant growth due to longer growing seasons while growth is reduced in the colder climatic conditions that are prevalent at northern latitudes like Europe and North America. The key annual phenological phases of vegetation dynamics are green-up, maturity, senescence and dormancy [40] which is also applicable to horticultural crops. These phenological phases of horticultural crops can be observed remotely as continuous changes in reflectance of vegetation. Advance image processing approaches can be applied to infer information from the changing spectral reflectance values to the changes in phenological development.

6. REMOTE SENSING IN FRUIT AND NUT CROP SSCM

Remotely sensed images are a quick and inexpensive means to differentiate fruit trees or orchards from other land-uses [8]. The imageries are the best possible resource to demarcate the spatial variation of crops, including horticultural crops. Several studies have been conducted over the years to delineate or classify forests or shrubs, e.g., horticultural plants, based on satellite and aerial images for precision agriculture decision making [8,42,43,44,45,46,47,48,49,50].

Many fruit and nut orchards are close to, within, or adjacent to, forested land cover [8]. Differentiating the plants from forest trees is a difficult task, but it can be achieved with advanced image processing. The total reflectance of a forest canopy is the combination of illuminated and shaded components of the tree crown as well as the background because of the bidirectional behavior of forest canopy [41]. With advanced image processing techniques and use of high resolution multispectral imagery, most fruit and nut plants can be distinguished from the mixed vegetation of tall forest trees and dwarf grass or other small shrubs if their unique spectral characteristics are known, especially with respect to the land covers such as forests or grasses [8].

7. SPECTRAL CHARACTERISTICS OF FRUIT AND NUT CROPS

Given that fruits and nuts are bushes or trees with green leaves, their spectral signatures are somewhat similar to other healthy vegetation. Mura [51] suggested that near infrared (NIR) is very useful for vegetation survey and mapping because a very steep gradient of percent reflectance (of incident light) is produced at the 0.7–0.9 μm band. This can only be found for vegetation, not for any other land cover. It is well-known that for vegetation land cover, a strong absorption is dominated by carotenoid pigments (centered at 0.48 μm (blue)), chlorophylls (centered at 0.68 μm (red)), and a green peak centered at 0.56 μm where the cell pigments are least absorptive [52]. A sharp rise in the curve at 0.7 μm is due to cellular reflectance rather than chlorophyll absorption that happens in the visible range of the spectrum. The high reflectance between 0.7 and 1.3 μm (NIR) band of the electromagnetic spectrum is dominated by the cell-wall/airspace interface and by refractive index discontinuities of cellular constituents of the leaves [52]. The molecular vibrations in the leaves provide stronger absorption in the longer wavelength after 1.3 μm.

It is, therefore, essential to use images of the visible and NIR bands of the electromagnetic spectrum. As a result, any multispectral image (with NIR bands) can serve the purpose of distinguishing fruit and nut crops from other land covers. The spectral behavior of different fruit and nut crops are discussed below to obtain specific information for analysis.

Spectral signatures for American blueberry (rabbiteye (vaccinium virgatum), southern highbush (vaccinium darrowii), and northern highbush (vaccinium corymbosum)) plants are yet to be developed. A study was conducted in the Fäjemyren peatland in southern Sweden in the temperate climatic zone to develop spectral signatures of Lingonberry or Lingon Blueberry (vaccinium myrtillus) [39]. Spectral signatures of the Lingon blueberry and bearberry plants suggest that their shrubs have a very low reflectance as compared to grasses and forest trees because of their medium stature and the fact that they mix easily with surrounding shadows [39]. Blueberry is distinguishable at the 680–700 nm (red) band width and the 800–900 nm (NIR) band width range. A spectral reflectance curve for blueberry is shown in Figure 1a along with feathermoss and crowberry (Empetrum) [39]. For this study, in the southern Swedish peatland, at select locations, 286 vegetation spectra (including Ligum Blueberry, feathermoss, and crowberry) were collected in the month of September and 116 and 63 spectra were recorded of the same species in the months of October and November, respectively [39]. The locations for sampling were strategized using stratified random sampling [39]. The three months spectra helped in studying the temporal variations in the sampled vegetations. The researchers calibrated the handheld field spectroradiometer using white spectralon panel and then followed by recording of the vegetation spectrum with the bare head of the instrument with a spectral sampling interval of 1.58 nm [39]. They recorded the spectra as an average over ten sample spectra. According to the researchers, the accuracy of the spectral curve were affected by a wide range of factors such as water level, soil reflectance, time of day and year, cloud cover, humidity and temperature, etc. However, for the time being, the blueberry spectral signature curve is the only available one to refer to.

NASA Langley Research Center, Hampton, Virginia had developed or collected spectral reflectances for 156 natural targets including a few fruit and nut crops, such as redblush citrus, oranges, prickly pear, and peach. The data or curves were developed with a uniform 0.01 or 0.02 μm [52]. According to NASA Reference Publication 1139, the spectral curves of different targets were selected for field measurements made with high-resolution scanning spectrophotometer and laboratory spectral measurements in the case of limited field measurement data [52]. It is not clearly specified which measurement methods were used for the fruit and nut crops' spectral curve development. The sampling size and process, the temporal measurement procedure, and the locations of the samples are unknown for these fruit and nut crops' spectral curve development. But these curves should be used for reference because they were compiled by the reputed organization, NASA.

Redblush citrus is distinguishable at the 600–650 nm (Green) band width with a high absorption, at the 950–1,000 nm (NIR) band width range with a steep absorption rate, and at the 1,000–1,150 nm (NIR) band width range with the highest reflection rate. A spectral reflectance curve for redblush citrus is shown in Figure 1(b) as obtained from NASA Reference Publication 1139 [52].

The leaves of orange trees can be distinguishable at the 650–700 nm band width where they have the highest absorption and at the 800–1,100 nm (NIR) band width range with the highest reflection rate. Figure 1c shows the spectral reflectance curve to help researchers in SSCM decision making [52]. Prickly pear has the highest absorption at the 700 nm band and the highest reflectance at the 750–900 nm (NIR) band width (Figure 1(d)) [52]. The leaves of peaches have almost the same reflectance curve as the leaves of oranges with the highest absorption at the 650 nm band and the highest reflectance at 750–1,200 nm (NIR) band width (Figure 1(e)) [52]. Therefore, other digital image characteristics such as the textural feature should be used to distinguish peach trees from orange trees.

These spectral reflectance curves are useful for researchers engaged in developing SSCM for individual fruit and nut crops through the use of remote sensing. The curves would guide researchers to decide which band images they need to procure to obtain distinguishable digital information. Again, these curves also point towards obtaining the digital number range (0–255) for a crop at a particular band using the reflectance percentage (Y-axis info of the curves) with respect to the wavelengths (X-axis info of the curves).

Spectral reflectance curves at different wavelengths for other fruit or nut trees such as pecan, apple, grape, and other crops (not shown in the study) are not available yet. Research is needed in order to develop these curves to support SSCM for these crops.

8. IMAGE PLATFORM TO USE FOR FRUIT AND NUT PLANT SSCM

Knowledge for the selection of an image platform, basically the image resolution, is essential for any kind of SSCM study [8]. The scope of the study is a primary indicator of the image resolution requirement. If the study is for a large area of coverage, e.g., determining the blueberry orchards for the entire state of Georgia, U.S., Landsat imagery at a 30-meter resolution is possibly sufficient. With a Landsat image, a fruit or nut orchard (large size) can be easily delineated. But small orchards (< 2.5 ha) surrounded by forests or tall grasses cannot be differentiated easily [8]. Therefore, high resolution imageries such as SPOT panchromatic (10-m resolution) and MSS (20-m resolution) imagery are crucial. If individual plants are to be differentiated for SSCM, very high resolution imagery such as QuickBird (2.15-m resolution) images or National Agricultural Imagery Program (NAIP; 1-m resolution) orthophotographs are best suited.

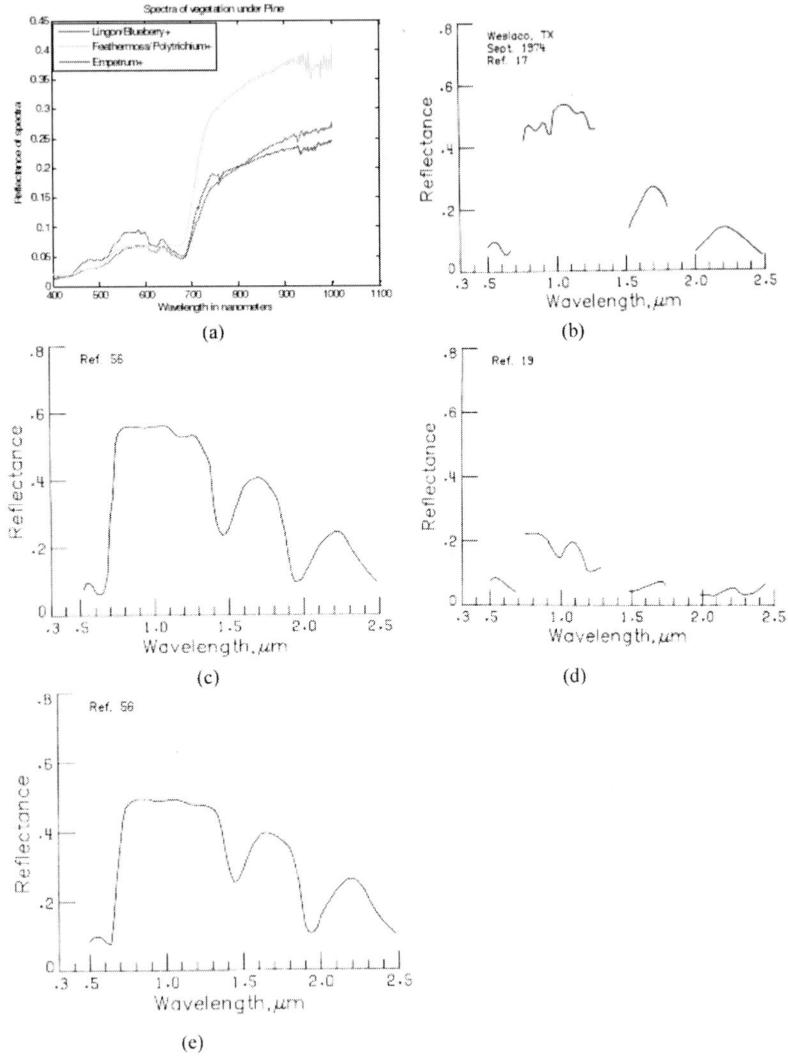

FIGURE 1. Spectral reflectance curves of different fruit and nut crops. Spectral reflectance curve for: **(a)** blueberry, feathermoss crowberry [39]; **(b)** redblush citrus [52]; **(c)** orange leaves [52]; **(d)** prickly pear [52]; **(e)** peach leaves [52].

Shrivastava and Gebelein [48] performed a study in Florida to delineate citrus groves for economic assessment, using Landsat Enhanced Thematic Mapper Plus (30 m) imagery. Their results showed a significant correlation between citrus production/income with remotely sensed imagery-derived citrus area coverage. Sharma and Panigrahy [49] used Indian Remote Sensing Satellite (IRS) LISS III and IRS AWiFS (23 m and 55 m, respectively) imagery data to develop an apple orchard database for the entire state of Himachal Pradesh in India and were successful in the delineation using a relatively low resolution imagery. The same low resolution IRS imageries were also used for identifying and prioritizing the suitable sites for passion fruit plantations in the hilly districts of the north-eastern states of India.

O'Connel and Goodwin [46] used remotely-sensed high resolution aerial photographs to identify the tree canopy of a peach orchard for yield forecasting and the estimation of crop water requirements. In their case, they needed to identify individual trees to perform SSCM and hence they used high resolution images. Blueberry orchards are generally established in an area that has just been cleared from shrubs and forest trees, and, therefore, it is difficult to distinguish blueberry bushes from other trees and shrubs with low resolution imagery such as 30-m Landsat TM or ETM. Panda et al. [54] used very high resolution 1-meter multi-spectral NAIP imagery to identify blueberry orchards in three southeastern counties (Bacon, Brantley, and Camden) in Georgia, U.S. (Figures 2(a–c)). Panda et al. [8] also used 2.15-m resolution QuickBird imagery for distinguishing blueberry plants from pine forest and dwarf and tall grasses (Figures 3(a,b)). Torres et al. [57] conducted a study to distinguish olive tree orchards using remote sensing images by clustering assessment techniques. They used 0.25- to 1.5-m resolution imageries to perform the advanced SSCM process for an olive orchard that was only 2 hectares in size.

As discussed earlier, multispectral imageries with visible and NIR bands would be sufficient for studying fruit and nut plants because the differentiable spectral characters are available mostly in the visible and NIR bands (Figure 2, Figure 3, Figure 4, Figure 5 and Figure 6), while some fruit crops have a very long range for the highest spectral reflectance (800–1,100 nm band). Therefore, AVIRIS imagery that records reflected light in 224 visible and near-infrared wavelengths from a National Aeronautics and Space Administration (NASA) high-altitude aircraft could be used because this high-spectral resolution allows detection of subtle spectral features that are diagnostic of specific minerals and plant characteristics. There is a great potential for airborne hyperspectral imagery use in estimating fruit yield and detecting plant stress as supported by several studies [62,63].

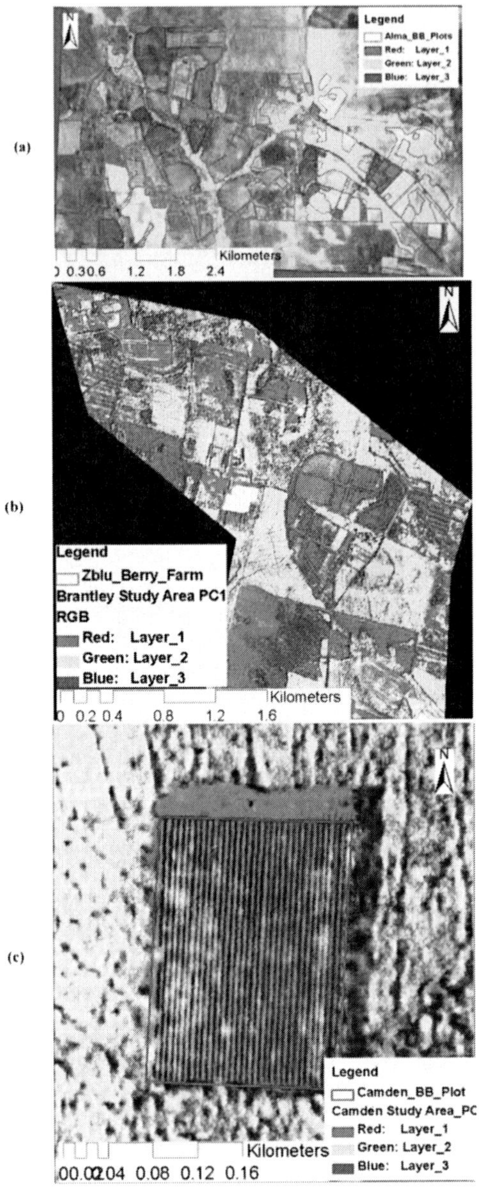

FIGURE 2. Principal component image of the three study areas in Georgia, U.S: Bacon **(a)**, Brantley **(b)**, and Camden County **(c)**, along with the GPS recorded blueberry farms at different stages.

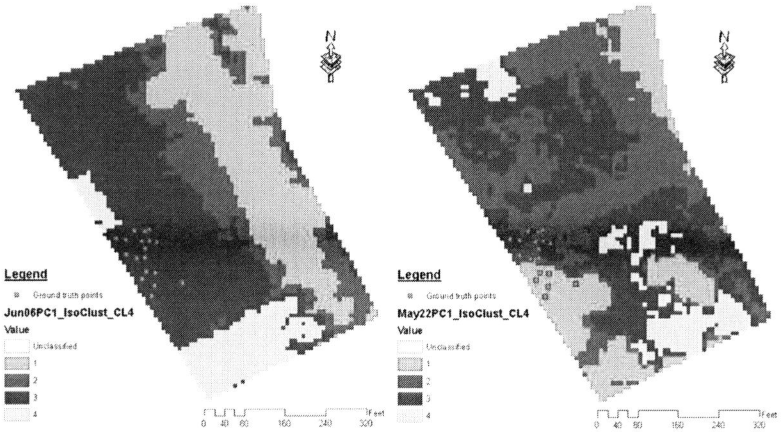

(a) (b)

FIGURE 3. Ground truth points of blueberry plants in unsupervised classified image of study area (blueberry plants) shown on May 22, 2006 **(a)** and June 06, 2006 **(b)** with ISODATA clustered classified QuickBird image.

Researchers have used various bands of the electromagnetic spectrum to study different aspects of diverse horticultural crops. Sutherland et al. [60] have used the TIR band imagery to detect frost on horticultural crops for subsequent decision-making. Polo et al. [61] have used a low-cost tractor-mounted scanning Light Detection and Ranging (LIDAR) system for non-destructive recordings of tree-row structure in orchards and vineyards to determine the tree-row volume and total surface area (crown cover). The use of LIDAR imagery data for horticultural SSCM has an immense scope because it can determine plant heights along with spectral characteristics. Walklate et al. [62] followed a similar procedure with LIDAR to determine the canopy cover of semi-dwarf Cox apple trees. Use of airborne LIDAR in studying pesticide spray applications on horticultural plants has become an important topic recently due to the emergence of organic farming [63].

9. IMAGE PROCESSING APPROACH FOR FRUIT AND NUT CROPS SSCM

To extract digital information from remote sensing images and to correlate this with fruit and nut crop SSCM, one must use image-processing techniques and build models [64]. Image processing techniques generally involve classification and algebraic manipulation to determine the spatial variability in the image [65]. High resolution images contain an enormous amount of information and can, therefore, be computationally intensive when analyzed. Thus, it is desirable to reduce the dimensionalities of the image data while at the same time extracting relevant and significant information [8]. As fruit and nut crops have similar spectral characteristics to forested vegetation, supervised classification is not the best procedure to distinguish these crops in imagery. Therefore, unsupervised clustering is one of the most desirable techniques for image segmentation, especially for analyses where unexpected land cover types can be obtained [8]. Three visible bands, i.e., red, green, and blue, in multi-spectral imagery contain digital information about land uses which differ significantly because each band has a different reflectance pattern. Therefore, it is essential to use all three bands to extract pertinent information about land use/land cover types from remotely sensed images [66].

Since 1978, scientists working in remote sensing have argued that the vegetation dynamics can be detected with the red and infrared bands of the electromagnetic spectrum [67]. However, vegetation such as shrubs can be easily distinguished with the green band along with the red band [8]. In many cases, classified individual (R-, G-, and B-) band images show poor classification accuracy because the data from all of the spectral bands involve a certain degree of redundancy [68]. The principal component analysis (PCA) approach is one of the best techniques to analyze correlated multidimensional data so that all the multispectral band information regarding the vegetation reflectance would be intact [8].

Along with these image processing techniques, most of the researchers who work in this area use other advanced image processing and geospatial technologies to study horticultural plants for decision support. Advanced image classification techniques such as ISODATA, WARD-minimum variance technique, k-means clustering, Bayesian classification algorithm, fuzzy multivariate cluster analysis, and artificial neural classification techniques like back-propagation neural networks and self-organizing map techniques are being used to delineate horticultural crops in satellite images [8]. Panda et al. (2010) has studied the application of different advanced image clustering algorithms to distinguish blueberry plants from bare soil/prepared land, plant shadow, prepared soil, grass and small shrubs [69]. Four different algorithms such as ISODATA, K-means, SOM, and Fuzzy ARTMAP were used in the classification process in two different locations (Nahunta and Woodbine) in Georgia, U.S. [69]. Figure 4(a–h) provide a comparative result of the study. Table 1 and Table 2 show the comparative results obtained from the study on these two locations. These studies support the importance of remote sensing for the management of fruit and nut crops. The use of remote sensing in horticultural SSCM has increased in recent years but there is a vast capacity for improvement.

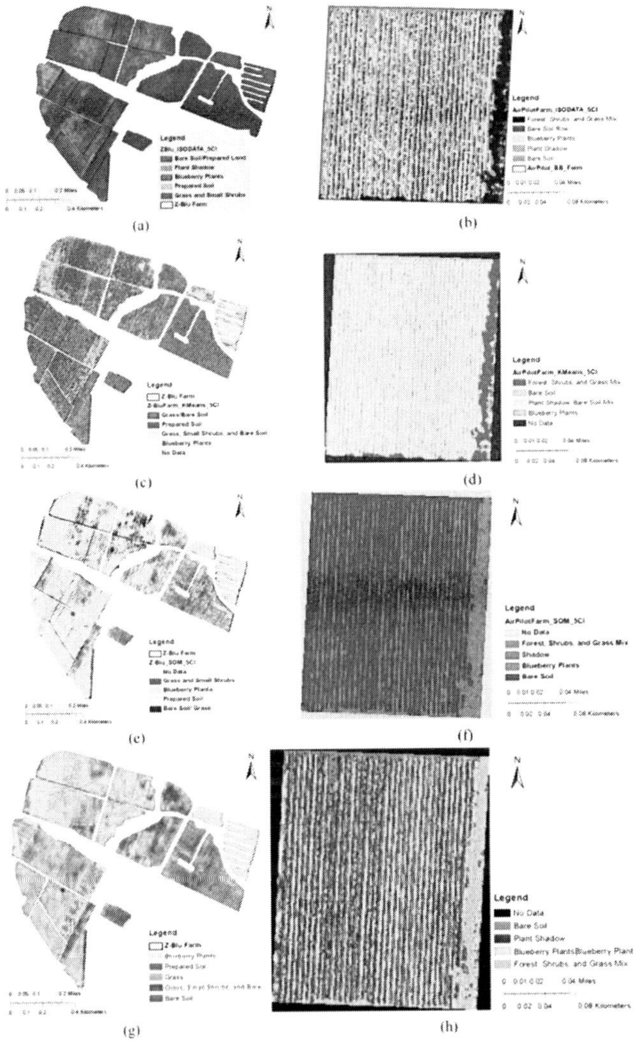

FIGURE 4. Comparison of different classification algorithms in blueberry plants differentiation from other land-use in blueberry orchards of Z-Blu Farm in Nahunta, GA and Air-Pilot Farm in Woodbine, GA, U.S. ISODATA clustering segmentation of the NAIP image of the Z-Blu Farm **(a)** and of the Air Pilot Farm **(b)**; K-means clustering segmentation of the NAIP image of the Z-Blu Farm **(c)** and Air Pilot Farm **(d)**; SOM clustering segmentation of the NAIP image of the Z-Blu Farm **(e)** and Air Pilot Farm **(f)**; and ISODATA clustering segmentation of the NAIP image of the Z-Blu Farm **(g)** and Air Pilot Farm **(h).**

TABLE 1. The classification accuracy report of four image segmentation algorithm applications on Z-Blu Farm 2009 NAIP imageries.

Image segmentation algorithms	Producer's accuracy %	User's accuracy %	Overall accuracy %	Kappa Statistics
ISODATA	65	85	80	0.38
K-means	70	88	82	0.42
SOM	82	94	88	0.53
Fuzzy ARTMAP	80	91	85	0.50

TABLE 2. The classification accuracy report of four image segmentation algorithm applications on Air-Pilot Farm 2009 NAIP imageries.

Image segmentation algorithms	Producer's accuracy %	User's accuracy %	Overall accuracy %	Kappa Statistics
ISODATA	80	92	86	0.52
K-means	85	90	88	0.56
SOM	88	95	92	0.72
Fuzzy ARTMAP	88	94	91	0.70

10. GIS USE IN FRUIT AND NUT CROP SSCM

Geographic Information Systems can also provide an important contribution towards fruit and nut crop SSCM. Study area delineation, orchard boundary digitization for ground truth, and preparation of other spatial data in addition to remotely sensed imagery are some of the tasks that could be accomplished with GIS. Spatial data processing is most essential for SSCM related model development such as yield prediction, water-demand forecasting, or the determination of soil nutrient availability. Figure 5 provides an example of GIS and remote sensing usage in delineating blueberry orchard boundaries and describing the extent of different blueberry crop growth types in a study area [54]. This study was conducted in the proximity of Nahunta City of Brantley County in Georgia, U.S. An advanced blueberry orchard, Z-Blu Farm is delineated inside the study area using GIS tools. Individual plot areas were calculated (not shown in the figure) using the Calculate Geometry Tool of ArcGIS software. The same tool also helped in calculating the total area of the study area, i.e., 5.88 km^2 [54]. IDRISI Andes (Clarks Lab, Clarke University, Worcester, MA, U.S.) software was used to classify the study area NAIP (1-m resolution) imagery using Self-organizing Map (SOM) image classification technique. The study image was classified with 15 classes and a classification dendrogram was developed using WARD minimum variance technique to merge clusters with close cluster mean values. Four distinct classes (Mature Blueberry/Forest, Intermediate Blueberry/Shrubs, Young Blueberries/Grassland, and Prepared Field/Bare Soil) were observed in the study area, which helped in further advanced SSCM of the crop. GIS also facilitated to find the total area under each class. For example, in this study, Mature Blueberry/Forest class covered 2.87 km^2 or 48.78% of the total study area. The other three classes, Intermediate Blueberry/Shrubs, Young Blueberries/Grassland, and Prepared Field/Bare Soil covered 0.52 km^2, 0.58 km^2, and 1.91 km^2, respectively, or 8.76%, 9.86%, 32.60% of the total study area, respectively. Ground control points (GCP) are shown in Figure 5, which were used for the classification accuracy assessment.

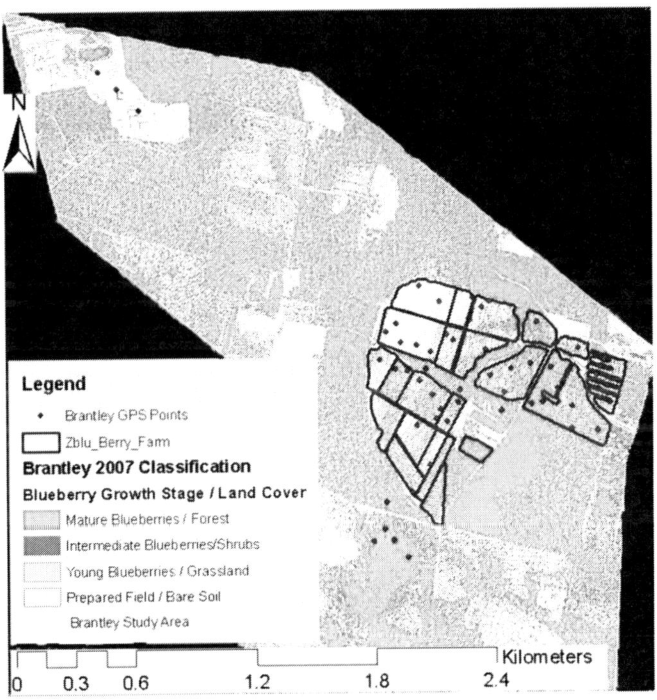

FIGURE 5. SOM classified image of Brantley County study area along with ground truth points (50 points) and GPS-recorded blueberry orchards at different stages.

The advanced application of GIS can help in designing geospatial models to conduct suitability analysis for a particular fruit or nut crop. Figure 6 represents a geospatial model that was developed to find suitable locations in the state of Georgia in the United States for blueberry production [53].

The model used five pertinent factors: chilling hours, soil permeability, soil drainage ability, soil pH, and the land cover types in Georgia to obtain suitable locations for blueberry plantations. The study also demarcated the area where blueberry is being produced and which area has potential to plant blueberry in Georgia [53]. Figure 7 shows the suitable locations based on individual contributing factors used in the analysis. Figure 8 represents the suitable locations for blueberry production in Georgia [53]. Similar models can be developed for any fruit or nut crops suitable location analysis in either the United States or other parts of the world using similar or different contributing factors.

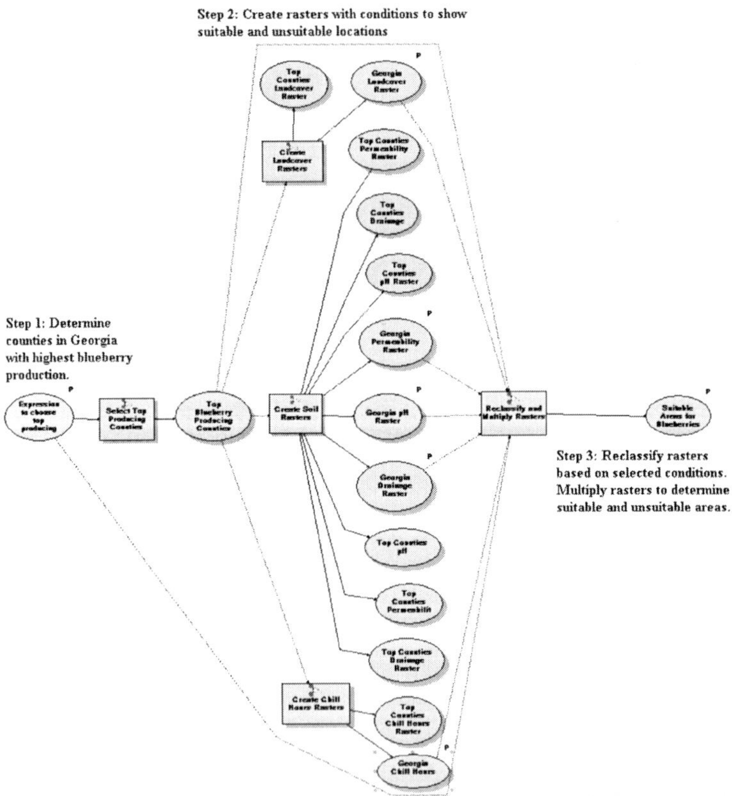

FIGURE 6. Geospatial model developed in ArcGIS Model Builder for the blueberry crop suitability analysis in the State of Georgia, U.S.

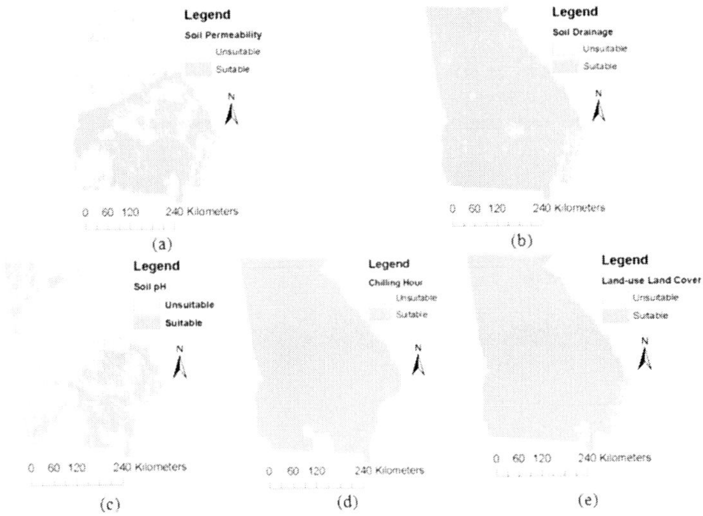

FIGURE 7. Maps showing the suitable areas in Georgia for various blueberry production parameters: soil permeability **(a)**, soil drainage **(b)**, soil pH **(c)**, chilling hours **(d)**, and land-use **(e)**.

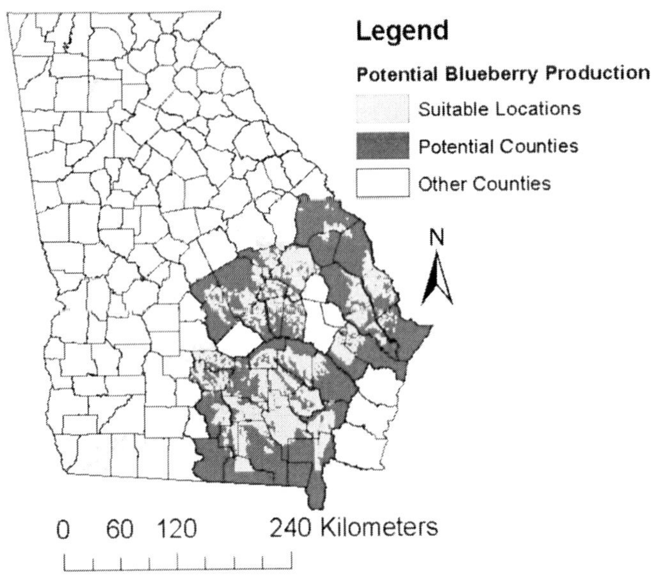

FIGURE 8. Blueberry production suitability map for Georgia.

11. GPS USE IN FRUIT AND NUT CROP SSCM

Global Positioning Systems is the most common technology used for any SSCM and precision agriculture. With GPS, ground truth points are recorded which are then used for differential correction and exported as GIS shapefiles for accuracy assessment of remotely sensed image classification. A polygon layer for the orchard boundary can also be created with the use of a GPS to support fruit and nut crop image classification and accuracy assessment using ground truth polygons instead of points. Figure 3 provides examples of collected GCPs and blueberry orchard boundary using Trimble GeoXH (Trimble Navigation Limited, Sunnyvale, CA, U.S.) GPS instrument, which were used in classification accuracy assessment.

12. OTHER PHYSICAL CHARACTERISTICS ANALYSIS OF FRUIT AND NUT CROP SSCM

Similar to other crops, successful cultivation of fruit and nut crops has specific requirements with respect to weather suitability such as chilling hours, growing degree days (GDD), and rainfall, soil characteristics such as nutrient availability, pH level, drainage, and permeability, and land-use and land cover (LULC) types [8]. These can all be analyzed to determine the perfect physical conditions for growth of a crop and to delineate potential areas for fruit and nut crop production.

In the case of blueberries, the rabbiteye varieties normally require a smaller number of chilling hours, i.e., 150–600 hours, compared to 200–700 hours for southern highbush varieties and more than 1,000 hours for northern highbush varieties [70]. Blueberries prefer an acidic, well-drained soil with an average pH of 4.0–5.5 and an optimum pH of 4.8 [71]. A well-drained soil profile with a minimum depth of 45 cm is suitable for blueberry production [72]. High soil permeability is a necessity for blueberry production because water-saturated soil conditions for a prolonged period of time can harm the bushes and can cause a severe infestation of Phytophthora root rot [72]. For blueberry production, soils with an organic matter of at least 3% are preferred, although rabbiteye varieties can grow in soils that have as little as 1% organic matter [4]. Blueberries grow best on land that has recently been cleared of forests or has never been planted with other crops. It is also preferred to grow blueberries close to forest cover for frost protection [73]. Other fruit and nut crops have similar physical requirements for optimum growth and development. However, further information is needed in order to be able to develop an extensive database for suitability analysis and fruit and nut crop SSCM.

13. ADVANCED TECHNIQUES FOR SSCM OF FRUIT AND NUT CROP

We are now in the irrefutable era of global warming. Therefore, basic SSCM application for crop management need to be changed to reflect the erratic weather pattern that is happening now due to El Nino and La Nina effect [74]. The erratic weather pattern includes rainfall pattern, cold and heat wave movement, and sudden freeze condition in non-freezing locations, etc. Therefore, advanced determination of rainfall pattern and freezing conditions are essential for SSCM application in crop production and especially fruit and nut crop production.

In the next few decades water scarcity will be acute and, therefore, advanced water management techniques should be developed to meet the need of water usage in agriculture [75]. As mentioned in Section 3, Panda et al. [37] used the soil characteristics, climate factors, and aerial images of the crop field to develop crop yield based water use model to predict the total water requirement for corn. Similar models could be developed to predict the water requirements for the fruit and nut crops. Thus, judicious site specific water use will help in conserving water. It will also help in the SSCM of the fruit and nut crop in case of erratic weather pattern. Wired magazine in its 16.07 issue [76] mentioned how the Lanworth Company, through its advanced data management technology, reduced the corn crop production estimation by a whopping 200 million bushels in the United States which was conducted by United States Department of Agriculture (USDA). Therefore, in the future, the use of data mining [2], image mining [77], and advanced database management technologies will help in the advancement of the SSCM in fruit and nut crops.

14. CONCLUSIONS

This paper provides a comprehensive review on the development and application of SSCM for fruit and nut crops. The review also confirmed that not many studies (compared to row crop agriculture) have been conducted so far on fruit and nut crops' SSCM. This review paper provides one-stop-information on available technology for the fruit and nut crop SSCM. The study provides a detailed application of geospatial and information technology in the blueberry crop management with a case study in Georgia, U.S., and in a few study locations in southeast Georgia. They can be replicated as described or modified based on local requirements and needs. This study also provides a detailed analysis of several characteristics responsible for blueberry production and describes, through references, how to develop geospatial models using those characteristics to plan for its management. Similar models can be developed using related crop characteristics to manage the crop.

Schupp et al. [78] provides an overview of current progress and the potential for future innovations for specialty crops which includes fruit and nut crops. They stated that "Video, laser, and satellite imaging technologies will become pertinent to future improvements in orchard production and associated equipment. First, it is important to develop a tree system that is compatible with imaging technologies—a high density planting, trained to a continuous planar fruiting wall. Engineers have indicated that this type of system will lend itself more easily to upcoming commercial technologies, both mechanical and imagery."

According to Schupp et al. [78], imaging technology has the potential to cross multiple areas of tree fruit production, such as crop load assessment, including blossom or green fruit counts, yield estimation, determination of insect presence or disease infection and associated eradication, soil moisture content determination for enhanced irrigation system design, estimation of fertilizer, pesticide, and herbicide application rates, and the development of assisted or automated pruning and harvesting strategies. Schupp et al. [78] state that different technological approaches such as satellite imagery, real-time laser images, and video imagery application will prove successful for fruit and nut crop SSCM.

Finally, this review paper would encourage researchers to advance the SSCM in fruit and nut crops as is being done in other row crop cases.

ACKNOWLEDGEMENT

This work was supported in part by the Georgia Space Grant Program, managed by the Georgia Institute of Technology on behalf of the National Aeronautics and Space Administration, and by State and Federal Funds allocated to Georgia Agricultural Experiment Stations Hatch project GEO1682.

REFERENCES

1. Casanova, D.; Epema, G.F.; Goudriaan, J. Monitoring rice reflectance at field level for estimating biomass and LAI. Field Crop Res. **1998**, 55, 83–92.

2. Panda, S.S. Data Mining Application in Production Management of Crop (Paper 1). Ph.D. Dissertation, North Dakota State University, Fargo, ND, USA, 2003.
3. Magri, A. Soil test, aerial image and yield data as inputs for site-specific fertility and hybrid management under maize. Precis. Agric. **2005**, 6, 87–110.
4. Baez-Gonzalez, A.D.; Kiniry, J.R.; Maas, S.J.; Tiscareno, M.L.; Macias, J.C.; Mendoza, J.L.; Richardson, C.W.; Salinas, J.G.; Manjarrez, J.R. Large-area maize yield forecasting using leaf area index based yield model. Agron. J. **2005**, 97, 418–425.
5. Lobell, D.B.; Ortiz-Monasterio, J.I.; Asner, G.P.; Naylor, R.L.; Falcon, W.P. Combining field surveys, remote sensing, and regression trees to understand yield variations in an irrigated wheat landscape. Agron. J. **2005**, 97, 241–249.
6. Otter-Nacke, S.; Godwin, D.C.; Ritchie, J.T. Testing and Validating the CERES-wheat Model in Diverse Environments; AGGRISTARS YM-15-00407. Johnson Space Center: Houston, TX, USA, 1986.
7. Li, T.; Feng, Y.; Li, X. Predicting crop growth under different cropping and fertilizer management practices. Agr. Forest Meteorol. **2009**, 149, 985–998.
8. Panda, S.S.; Hoogenboom, G.; Paz, J. Distinguishing blueberry bushes from mixed vegetation land-use using high resolution satellite imagery and geospatial techniques. Comput. Electron. Agr. **2009**, 67, 51–59.
9. Sevier, B.J.; Lee, W.S. Precision Farming Adoption by Florida Citrus Producers: Probit Model Analysis; Circular # 1461; Institute of Food and Agricultural Sciences, University of Florida: Gainesville, FL, USA, 2005.
10. Diker, K.; Bausch, W.C.; Heermann, D.F. Monitoring Temporal Changes of Irrigated Corn by Aerial Images; ASAE Paper No. 01-1144; American Society of Agricultural and Biological Engineers: St. Joseph, MI, USA, 2001.
11. Gopalapillai, S.; Tian, L.F.; Tang, L.; Goering, C.E. Identification and Mapping of Weed Density in a Soybean Field Using DMS Image; ASAE Paper No. 003058; American Society of Agricultural and Biological Engineers: St. Joseph, MI, USA, 2000.
12. Gopalapillai, S.; Tian, L. In-field variability detection and yield prediction in corn using digital aerial imaging. Trans. ASAE **1999**, 42, 1911–1920.

13. Seidl, M.S.; Batchelor, W.D.; Paz, J.O. Integrating remotely sensed images with a soybean model to improve spatial yield simulation. Trans. ASAE **2004**, 47, 2081–2090.
14. Goel, P.K.; Prasher, S.O.; Patel, R.M.; Smith, D.L.; Dittomasho, A. Use of Remote Sensing for Monitoring Growth of Corn and Soybean Crops in the Presence of Weeds; ASAE Paper No. 001094; American Society of Agricultural and Biological Engineers: St. Joseph, MI, USA, 2000.
15. Plant, R.E.; Munk, D.S.; Roberts, B.R.; Vargas, R.L.; Rains, D.W.; Travis, R.L.; Hutmacher, R.B. Relationship between remotely sensed reflectance data and cotton growth and yield. Trans. ASAE **2000**, 43, 535–546.
16. Teixeira, A.S.; Ustin, S.L. Use of Biweekly High Resolution Aerial Images in Precision Farming; ASAE Paper No. 003051; American Society of Agricultural and Biological Engineers: St. Joseph, MI, USA, 2000.
17. Ray, S.S.; Pokharna, S.S.; Ajai, S. Cotton yield estimation using agrometeorological model and satellite-derived spectral profile. Int. J. Remote Sens. **1999**, 20, 2693–2702.
18. Humburg, D.; Stange, K. Modeling of Sugarbeet Quality Variables from Canopy Spectral Indices in a Fertility Trial; ASAE Paper No. RRV00-402; ASAE: St. Joseph, MI, USA, 2000.
19. Taylor, J.C.; Wood, G.A.; Thomas, G. Mapping yield potential with remote sensing. Precis. Agric. **1997**, 1, 713–720.
20. Yang, C.; Anderson, G.L. Mapping grain sorghum yield variability using airborne digital videography. Precis. Agric. **2000**, 2, 7–23.
21. Yang, C.; Everett, J.H. Relationship between yield monitor data and airborne multi-spectral digital imagery. In Proceedings of the Fifth International Conference on Precision Agriculture, Madison, WI, USA; 2000.
22. Lamb, D.W.; Weedon, M.M.; Rew, L.J. Evaluating the accuracy of mapping weeds in seedling crops using airborne digital imaging: Avena spp. in seedling triticale. Weed Res. **1999**, 39, 481–492.
23. Doraiswamy, P.C.; Moulin, S.; Cook, P.W.; Stern, A. Crop yield assessment from remote sensing. Photogramm. Eng. Remote Sensing **2003**, 69, 665–674.
24. Tan, G.; Shibasaki, R. Global estimation of crop productivity and the impacts of global warming by GIS and EPIC integration. Ecol. Model. **2003**, 168, 357–370.

25. Baez-Gonzalez, A.D.; Chen, P.; Tiscareno-Lopez, M.; Srinivasan, R. Using satellite and field data with crop growth modeling to monitor and estimate corn yield in Mexico. Crop Sci. **2002**, 42, 1943–1949.
26. Yang, P.; Tan, G.X.; Zha, Y.; Shibasaki, R. Integrating remotely sensed data with an ecosystem model to estimate crop yield in north China. In Proceedings of XXth ISPRS Congress, Commission VII, Istanbul, Turkey; 2004; pp. 150–156.
27. Sun, J. Dynamic monitoring and yield estimation of crops by mainly using the remote sensing technique in China. Photogramm. Eng. Remote Sensing **2000**, 66, 645–650.
28. Simelton, E.; Fraser, E.; Challinor, A.; Termansen, M. Socioeconomic factors that influence farmers' adaptive capacity to drought—A global and multi-region perspective. In Proceedings of 8th International Conference of the European Society for Ecological Economics, University of Ljubljana, Slovenia, June 29–July 2, 2009.
29. Hazell, P.; Wood, S. Drivers of change in global agriculture. Phil. Trans. Roy. Soc. B-Biol. Sci. **2008**, 363, 495–515.
30. Kulkarni, S.; Bajwa, S.G.; Robbins, R.T.; Costello, T.A.; Kirkpatrick, T.L. Aerial remote sensing to detect soybean cyst nematodes (Heterodera glycines) in a resistance rotation study. Trans. ASABE **2008**, 51, 1511–1517.
31. Kulkarni, S.S.; Bajwa, S.G.; Rupe, J.; Kirkpatrick, T. Spatial correlation of crop response to soybean cyst nematode (Heterodera glycines). Trans. ASABE **2008**, 51, 1451–1459.
32. Cui, D.; Zhang, Q.; Li, M.; Zhao, Y.; Hartman, G.L. Detection of soybean rust using multispectral image sensor. Sens. Instrum. Food Qual. Safety **2009**, 3, 49–56.
33. Willers, J.L.; Jenkins, J.N.; McKinion, J.M.; Gerard, P.; Hood, K.B.; Bassie, J.R.; Cauthen, M.D. Methods of analysis for georeferenced sample counts of tarnished plant bugs in cotton. Precis. Agric. **2009**, 10, 189–212.
34. Goffart, J.P.; Olivier, M.; Frankinet, M. Potato crop nitrogen status assessment to improve N fertilization management and efficiency: Past–present–future. Potato Res. **2008**, 51, 355–383.
35. Kalra, N.; Chander, S.; Pathak, H.; Aggarwal, P.K.; Gupta, N.C.; Sehgal, M.; Chakraborty, D. Impacts of climate change on agriculture. Outlook Agr. **2007**, 36, 109–118.

36. Reyniers, M.; Vrindts, E.; Baerdemaeker, J.D. Fine-scaled optical detection of nitrogen stress in grain crops. Opt. Eng. **2004**, 43, 3119–3129.
37. Panda, S.S.; Panigrahi, S.; Steele, D.; Ames, D.P. Precision water management in corn using automated crop yield modeling. Agr. Water Manage. **2008**. In Review. [Google Scholar]
38. Kitchen, N.R. Emerging technologies for real-time and integrated agricultural decisions. Comput. Electron. Agr. **2008**, 61, 1–3.
39. Rao, P. Remote Sensing for Assessing Vegetation Dynamics and Productivity of a Peatland in Southern Sweden. M.S. Thesis, International Institute for Geo-Information Science and Earth Observation, Enschede, The Netherlands, 2007.
40. Zhang, X.; Friedl, M.A.; Schaaf, C.B.; Strahler, A.H.; Hodges, J.C.F.; Gao, F.; Reed, B.C.; Huete, A. Monitoring vegetation phenology using MODIS. Remote Sens. Environ. **2003**, 84, 471–475.
41. Peltoniemi, J.I.; Kaasalainen, S.; Näränen, J.; Rautiainen, M.; Stenberg, P.; Smolander, H.; Smolander, S.; Voipio, P. BRDF measurement of understory vegetation in pine forests: Dwarf shrubs, lichen, and moss. Remote Sens. Environ. **2005**, 94, 343–354.
42. Holmgren, P.; Thuresson, T. Satellite remote sensing for forestry planning: A review. Scand. J. Forest Res. **1998**, 13, 90–110.
43. Franco-Lopez, H.; Ek, A.R.; Bauer, M.E. Estimation and mapping of forest stand density, volume, and cover type using the k-nearest neighbor method. Remote Sens. Environ. **2001**, 77, 251–274.
44. Li, D.; Di, K.; Li, D. Land use classification of remote sensing image with GIS data based on spatial data mining techniques. In Proceedings of XIXth ISPRS Congress, Commission III, Amsterdam, The Netherlands, 2000; In International Archives of Photogrammetry and Remote Sensing. ISPRS: Vienna, Austria, 2000; Volume 33, pp. 238–245.
45. Haapanen, R.; Ek, A.R.; Bauer, M.E.; Finely, A.O. Delineation of forest/nonforest land use classes using nearest neighbor method. Remote Sens. Environ. **2003**, 89, 265–271.
46. O'Connell, M.G.; Goodwin, I. Spatial variation of tree cover in peach orchard. Acta Horticul. (ISHS) **2005**, 694, 203–205.
47. Izzawati, E.D.; Wallington, E.D.; Woodhouse, I. Forest height retrieval from commercial X-band SAR products. IEEE Trans. Geosci. Remote Sens. **2006**, 44, 863–870.

48. Shrivastava, R.J.; Gebelein, J.L. Land cover classification and economic assessment of citrus groves using remote sensing. ISPRS J. Photogramm. **2006**, 61, 341–353.
49. Sharma, A.; Panigrahy, S. Apple orchard characterization using remote sensing and GIS in Shimla district of Himachal Pradesh. In Proceedings of Remote Sensing and Photogrammetry Annual Conference 2007, Newcastle, UK, September 11–14, 2007.
50. van Aardt, J.A.N.; Randolph, H.W.; Scrivani, J.A. Lidar-based mapping of forest volume and biomass by taxonomic group using structurally homogenous segments. Photogramm. Eng. Remote Sensing **2008**, 74, 1033–1044.
51. Murai, S. Remote Sensing Note; Japan Association of Remote Sensing: Tokyo, Japan, 1995.
52. Bowker, D.E.; Davis, R.E.; Myrick, D.L.; Jones, W.T. Spectral Reflectance of Natural Targets for Use in Remote Sensing Studies; NASA Reference Publication 1139-N85-30450. NASA: Hampton, VA, USA, 1986; p. 188.
53. Reed, J.; Panda, S.S.; Hoogenboom, G.; Paz, J. A suitability analysis model for determining potential locations for blueberry production in Georgia using geospatial technology. Photogramm. Eng. Remote Sensing **2010**. Submitted.
54. Panda, S.S.; Hoogenboom, G.; Paz, J. Blueberry orchard delineation with high-resolution imagery and self-organizing map neural image classification. Precis. Agric. **2009**. In Review.
55. SAC. Identification and Prioritization of Suitable Sites for Passion Fruit Using Remote Sensing and GIS in Aizwal District, Mizoram; Project Report RSAM/SAC/RESA/NEHORT/PR/01/03. Space Application Centre: Gujarat, India, 2003.
56. SAC. Village Level Passion Fruit Site Suitability Maps of Champai and Klasib Circles, Aizwal District, Mizoram, Atlas; Project Report RSAM/SAC/RESA/NEHORT/PR/02/03. Space Application Centre: Gujarat, India, 2003.
57. Torres, L.G.; Pena-Barragan, J.M.; Lopez-Granados, F.; Jurado-Exposito, M.; Fernandez-Escobar, R. Automatic assessment of agro-environmental indicators from remotely sensed images of tree orchards and its evaluation using olive plantations. Comput. Electron. Agr. **2008**, 61, 179–191.
58. Tajeda, P.J.Z.; Berjon, A.; Miller, J.R. Stress detection in crops with hyperspectral remote sensing and physical simulation models. In

Proceedings of Airborne Imaging Spectroscopy Workshop, Bruges, Belgium, October 8, 2004.
59. Ye, X.; Sakai, K.; Garciano, L.O.; Asada, S.I.; Sasao, A. Estimation of citrus yield from airborne hyperspectral images using a neural network. Ecol. Model. **2006**, 198, 426–432.
60. Sutherland, R.A.; Hannah, H.E.; Cook, A.F.; Martsolf, J.D. Remote sensing of thermal radiation from an aircraft—An analysis and evaluation of crop freeze protection methods. J. Appl. Meteorol. **1981**, 20, 813–820.
61. Polo, J.R.R.; Sanz, R.; Llorens, J.; Arnó, J.; Escolà, A.; Ribes-Dasi, M.; Masip, J.; Camp, F.; Gràcia, F.; Solanelles, F.; Pallej, T.; Val, L.; Planas, S.; Gil, E.; Palacín, J. A tractor mounted scanning LIDAR for the non-destructive measurement of vegetative volume and surface area of tree-row plantations: A comparison with conventional destructive measurements. Precis. Agric. **2008**, 102, 128–134.
62. Walklate, P.J.; Richardson, G.M.; Baker, D.E.; Richards, P.A.; Cross, J.V. Short-range LIDAR measurement of top fruit tree canopies for pesticide application research in United Kingdom. Proc. SPIE **1997**, 3059, 143–151.
63. Cross, J.V.; Walklate, P.J. The UK PACE scheme for adjusting the dose to suit apple crops. Agr. Eng. Int.: CIGR EJournal **2008**, 6. Manuscript ALNARP 08 003.
64. Panda, S.S. Data Mining for Precision Management of Crops (Paper 2). Ph.D. dissertation, North Dakota State University, Fargo, ND, USA, 2003.
65. Senay, G.B.; Ward, A.D.; Lyon, J.G.; Fausey, N.R.; Nokes, S.E. Manipulation of high spatial resolution aircraft remote sensing data for use in site specific farming. Trans. ASAE **1998**, 41, 489–495.
66. Jain, A.K.; Murty, M.N.; Flynn, P.J. Data clustering: A review. ACM Comput. Survey. **1999**, 31, 264–323.
67. Deering, D.W. Rangeland Reflectance Characteristics Measured by Aircraft and Spacecraft Sensors. Ph.D. Dissertation, Texas A&M University, College Station, TX, USA, 1978; p. 338.
68. Fung, T.; LeDrew, E. Application of principal component analysis to change detection. Photogramm. Eng. Remote Sensing **1987**, 53, 1649–1658.
69. Panda, S.S.; Hoogenboom, G.; Paz, J. Image classification algorithm efficiency comparison: A contextual study of blueberry orchard delineation. Comput. Electron. Agr. **2010**. In Review. [Google Scholar]

70. Krewer, G.; NeSmith, D.S. Blueberry cultivars for Georgia. 2006. Available online: http://www.smallfruits.org/Blueberries/production/06bbcvproc_Nov 0206.pdf (Accessed on 26 February 2010).
71. Krewer, G.; Meyers, S. Commercial Blueberry Culture; Circular 713; University of Georgia, Cooperative Extension Service, The University of Georgia: Athens, GA, USA, 1989.
72. Williamson, J.G. Dooryard blueberry production in Florida. Proc. Fla. State Hort. Soc. **1994**, 107, 399–403.
73. Williamson, J.; Lyrene, P. Blueberry Gardener's Guide; Circular 1192; Horticultural Sciences Department, Florida Cooperative Extension Service, Institute of Food and Agricultural Sciences, University of Florida: Gainesville, FL, USA, 2004.
74. Panda, S.S. Rainfall Patterns. In Encyclopedia of Global Warming and Climate Change; Philander, S.G., Ed.; Sage Publications: Los Angeles, CA, USA, 2008; pp. 846–847.
75. Swain, G. Water Scarcity: Preventing Water Shortages. Trends in America February **2008**.
76. Paynter, B. Feeding the Masses: Data in, Crop Predictions Out. Wired Magazine **2008**.
77. Panda, S.S. Spatial image mining. In Encyclopedia of Geographical Information Science; Sekhar, S., Xiong, H., Eds.; Springer: New York, NY, USA, 2008; pp. 475–479.
78. Baugher, T.; Schupp, J.; Travis, J.; Hull, L.; Ngugi, H.; Krawczyk, G.; Harsh, M.; Reichard, K.; Ellis, K.; Remcheck, J.; Crassweller, R.; Marini, R.; Harper, J.; Kime, L.; Heinemann, P.; Liu, J.; Lewis, K.; Hoheisel, G.; Jones, V.; Glenn, M.; Miller, S.; Tabb, A.; Park, J.; Slaughter, D.; Johnson, S.; Landers, A.; Reichard, G.; Singh, S.; Bergerman, M.; Kantor, G.; Messner, W. Speciality Crop Innovations: Progress and Future Directions; Specialty Crop Innovations Progress Report; College of Agricultural Sciences, Penn State University: University Park, PA, USA, 2009; pp. 1–16.

4

CHAPTER

USE OF REMOTE SENSING DATA AND GIS TOOLS FOR SEISMIC HAZARD ASSESSMENT FOR SHALLOW OILFIELDS AND ITS IMPACT ON THE SETTLEMENTS AT MASJED-I-SOLEIMAN AREA, ZAGROS MOUNTAINS, IRAN

Hojjat Ollah Safari [1], Saied Pirasteh [2,4], Biswajeet Pradhan [2,3,*] and Ladan Khedri Gharibvand [4]

[1] *Department of Geology, College of sciences, Golestan University, Gorgan, Iran*

[2] *Institute of Advanced Technology, University Putra Malaysia, Malaysia*

[3] *Institute for Cartography, Faculty of Forestry, Hydro & Geosciences, Dresden University of Technology, 01062, Dresden, Germany*

[4] *Department of Civil Engineering, Dezful Branch of Azad University, Dezful, Iran*

ABSTRACT

Masjed-i-Soleiman (MIS) is situated in the northern part of the Dezful embayment, which is in the Zagros fold–thrust belt with high seismic activities. MIS faces a shallow buried anticline, formed by the shallowest oilfield with a thick gas cap. The cap rocks of this oilfield are highly fractured, which has resulted in leakages from the gas cap. In this paper, we have used remote sensing techniques and image interpretation for the identification of the Niayesh, Lahbari, Andika and MIS fault zones in the studied area. Further, the study exploited seismic potential mapping using the remote sensing techniques. The relationships between the structural controls and localized gas leakage are assessed within the GIS environment. Additionally, field observation data corroborated that the leakages (and seepages) are smashed within the intersection of Niayesh and MIS fault zone, which belongs to the high fractured hinge zone of the MIS anticline. As a result, the reactivation of these active faults may cause large earthquakes with a maximum magnitude of between $6.23 < Ms < 7.05$ (Richter scale) and maximum horizontal acceleration $0.26 < a < 0.55$ g. Finally, the authors concluded that this anticipated earthquake may cause large scale fracturing of cap rocks, releasing a large volume of H_2S gas from the uppermost layer of the reservoir.

Keywords: seismic hazard assessment; remote sensing; GIS; Zagros Mountain

1. INTRODUCTION

The Zagros Orogenic Belt of Iran is one of the most prolific petroliferous areas, approximating 12% of the proven global oil reserves. It is well documented that a major portion of the oil reserve in the Zagrous Orgenic Belt is contained within the reservoirs of the Asmari (Oligo-Miocene) and Bangestan Group (Upper Cretaceous) anticlinal structures [1]. In the folded belt of the Zagros Mountains, a sequence of Precambrian to Pliocene shelf sediments is deposited with a thickness of 12 km. This has undergone folding from the Miocene to more recent times [2]. Most of these sections (6,000–7,000 m), which form the Cambrian to Miocene rocks, represent a single structural lithological unit popularly known as the Competent group [2]. It is bounded by the detachment zones (above) and evaporite deposits (below).

The main types of salt structures in the Persian Gulf Basin are salt domes, salt pillows, salt walls, salt piercements, rim anticlines, turtleback structures, disharmonic folds, orogenic fold fillings and dissolution drapes [3]. Diapiric oil fields, which accounts for 60% of the 600 billion barrels of total oil reserves of the Persian Gulf Basin, have grown continuously since Late Jurassic-Permian.

During the last decade, remote sensing and GIS techniques have been increasingly used for various applications in geosciences. More recently, the satellite data and GIS techniques have been used to detect and map the structural features [4,5,6,7,8,9] and also sub-surface deep seepages [6].

This study emphasizes seismic potential mapping for the shallow oilfield in the Masjed-i-Soleiman area. Earthquakes are one of the natural hazards that cause damage to the structures, property and livestock. The most frequent occurring natural phenomena related to earthquakes are faulting, landslides, liquefaction and tsunamis. One of the most important damages caused by earthquakes occurs predominantly in shallow oilfields, and is mainly due to the faulting in ductile shallow cap rocks. It subsequently causes leakage of natural gas (through the uppermost hydrocarbon layer at the top of reservoir) from the oilfields. The sudden release of large-scale natural gas causes environmental pollution and firing of the adjacent area, leading to potential loss of life and property, or both.

The Masjed-i-Soleiman city is situated over the shallow buried MIS oilfield (the first explored oilfield in the Middle East) [10]. From field observations, it was evident that the cap rock of this oilfield was fractured causing gas and oil leakages mostly in the residential areas. The population density surrounding these leakage areas is quite high. The continuous oil production from this oilfield over the last decades (i.e., from 1908 to date) has caused the formation of a thick gas cap, predominantly H_2S, in the uppermost part of this hydrocarbon reservoir. Consequently, many oil seepages have changed to gas leakages [11]. Fault activities have caused the large scale fracturing of the cap rock in the oil reservoir. Therefore, a sudden large-scale release of natural gas is predictable in the MIS.

In this research, with the aid of remote sensing techniques and image interpretation tools, the main fault zones were recognized. Further, the seismic potential of these fault zones were determined. Finally, the relationships of these structures, with respect to the localizing of gas leakage, have been studied in the GIS environment.

2. STUDY AREA AND GEOLOGICAL SETTING

The Zagros fold-thrust belt was formed during several episodes of shortening, due to closure of the Tethys Ocean, and subsequent collision of the Arabian and Eurasian plates in the Late Cretaceous to Recent age [12]. This seismically active fold-thrust belt can be divided into Thrust zone

(High Zagros), simply folded belt and Coastal plain [13,14,15,16,17,18,19]. The simply folded belt is divided into Fars Arc, Izeh zone and Dezful embayment by the Kazerun and Izeh fault zones. Dezful embayment appears to be a discrete structural unit, with boundaries defined by the Mountain front thrust fault to the northwest, the Kazerun-Borazjan and Izeh transverse faults to the east and southeast and the Zagros Fore deep (Frontal) fault to the southwest [20,21].

The MIS area lies within the coordinates: 31°42′37″N to 32°03′46″N and 49°02′31″E to 49°25′44″E and is located in the northern part of the Dezful embayment (Figure 1). Topographically, MIS area is located in the piedmont zone of the Zagros Mountains (averaging 360 m). The main part of the study area is situated in and around the Tembi river valley and adjacent regions (Figure 1).

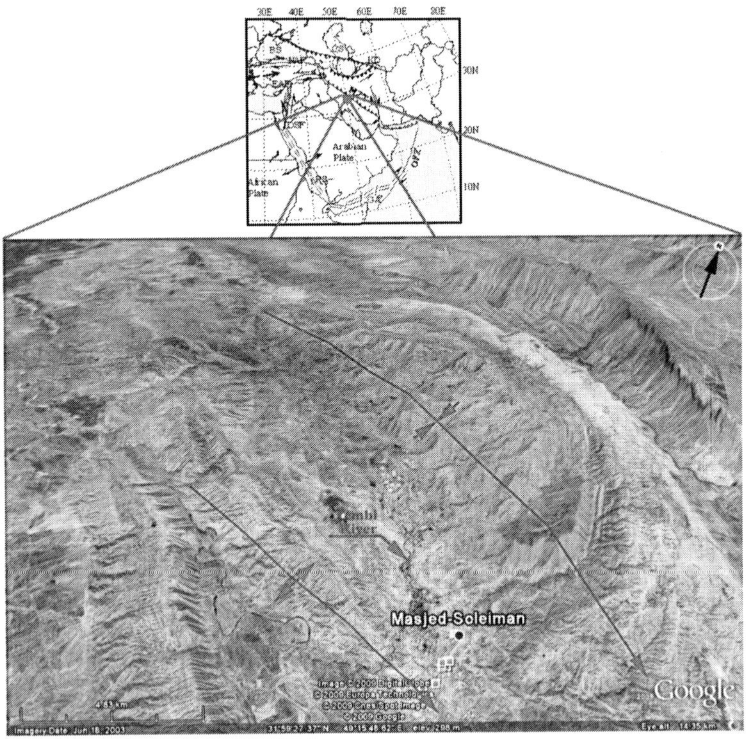

FIGURE 1. Geographical location of Masjed-i-Soleiman area (Google earth 2009 with modifications).

One of the most important features of this embayment is the buried shallow depth Asmari brittle formation (in Oligo-Miocece age, as hydrocarbon reservoir) under the ductile Gachsaran Formation (in Miocene age, as cap rock of oilfield). This hydrocarbon reservoir is situated in shallow depth (lesser than 300 m in depth) (NIOC 1988) and, therefore, a shallow oilfield has been formed in this area. The outcropped stratigraphic units (Figure 2) comprise of Gachsaran, Mishan, Aghajari (with lahbari member) and Bakhtiari Formations. The main stratigraphic characteristics of Gachsaran and non-outcropped Asmari Formations are described in the following section.

2.1. Asmari Formation

This formation is of Oligo-Miocene age, and comprises approximately 460 m competent cream limestone with inter-bedded argillaceous limestone [21]. The type section of this formation was measured in the Asmari anticline, which is located in the east of the study area. This formation shows mechanical brittle behavior and is predominantly fractured by existent fault zones. It has good porosity and permeability for forming a hydrocarbon reservoir.

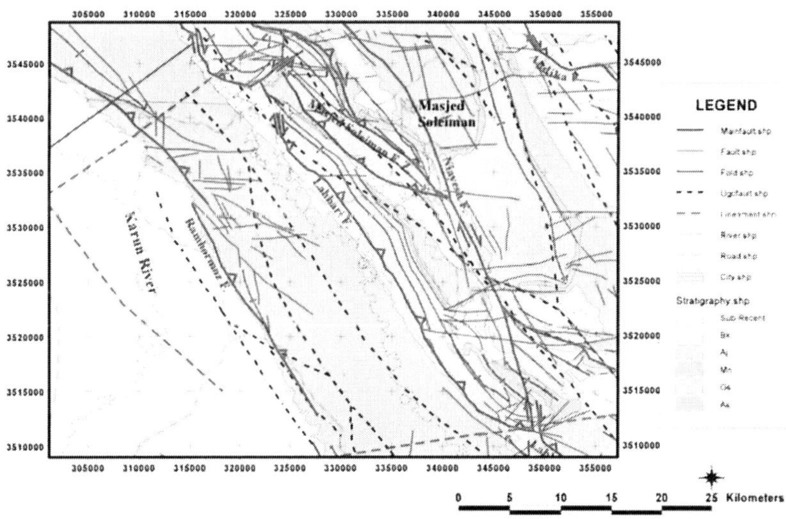

FIGURE 2. Geological map of the Masjed-i-Soleiman area.

2.2. Gachsaran Formation

This formation dates from the Miocene age and comprises a sequence of anhydrite and marl with marly limestone and salt inter-beds (approximately 1,500 m thickness) [21]. Gachsaran formation shows a mechanical ductile behavior and is insignificantly fractured. Hence, it plays an important role as cap rock in the MIS oilfield.

3. METHODOLOGY

The following methodology has been adopted in this research for seismic hazard mapping for the shallow oil fields and its impact on the settlement. The following steps have been attempted in this research: (a) image processing and field investigations for distinguishing outcropped lithology and extracting structural features such as. faults and lineaments, (b) investigation of Recent Fault activities by seismotectonics assessment, (c) digitizing of underground contour map (UGC) and then, preparation of 3D-Sketch showing the relationship between the subsurface and surface structures.

3.1. Structural Geology

In this research, the spatial investigations consist of the interpretation of satellite image, field surveys and measurement of fault characteristics for the creation of the structural map.

Satellite Landsat 7 ETM+ image (NASA 2000) was geometrically corrected using 100 ground control points. The digital image processing techniques, such as filtering techniques using convolution kernel size, have been applied to extract the lineaments features from the image [4]. The key elements for the image interpretation are tone, shape, topography, erosion, vegetation and linearity. False color composite (FCC) 741 (RGB) band composition image was developed using the ENVI software [22]. In order to extract the linear structures, directional-filters were also applied. The extracted lineaments were checked and measured during the field investigations and then the final structural map was produced (Plate 1).

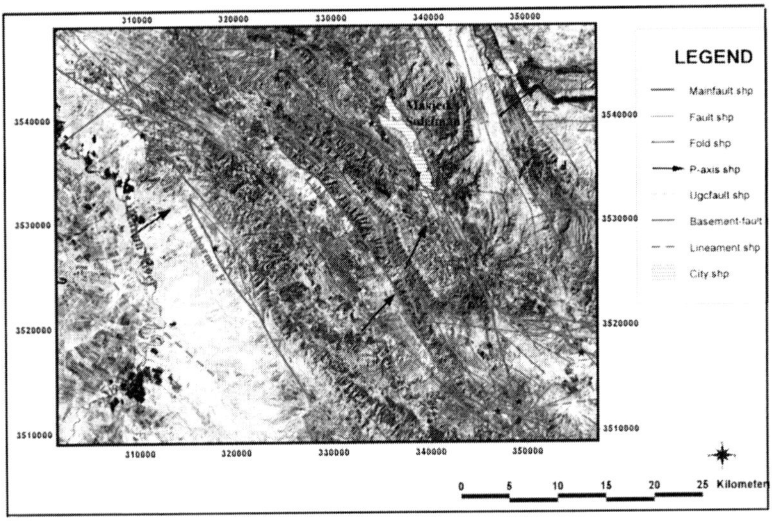

PLATE 1. Structural map (with satellite image base) of the Masjed-i-Soleiman area.

3.2. Seismotectonics

In this research, the emphasis was given to the seismotectonic analysis in order to find out the Quaternary (and recent) fault activities; and investigation of earthquakes. On the basis of seismotectonic studies, four main quaternary Masjed-i-Soleiman, Niayesh, Lahbari and Andika fault zones have been detected (Figure 3), which have the potential for seismic activities. The activity impression, related earthquake epicenters and seismic potential of these fault zones, were also studied.

The seismic parameters of these faults such as magnitudes and horizontal ground accelerations were evaluated. The evaluations of these parameters were done using the deterministic (analytical) method. In this method, the maximum magnitude of probable earthquake is evaluated. In this research, the Nowroozi, Ambraseys and Melville modeling techniques were used for Iran's earthquakes [23,24]. The equations of these models are given below:

Nowroozi model: $Ms = 1.259 + 1.244 \log L$, where L is the fault length (m) (1)

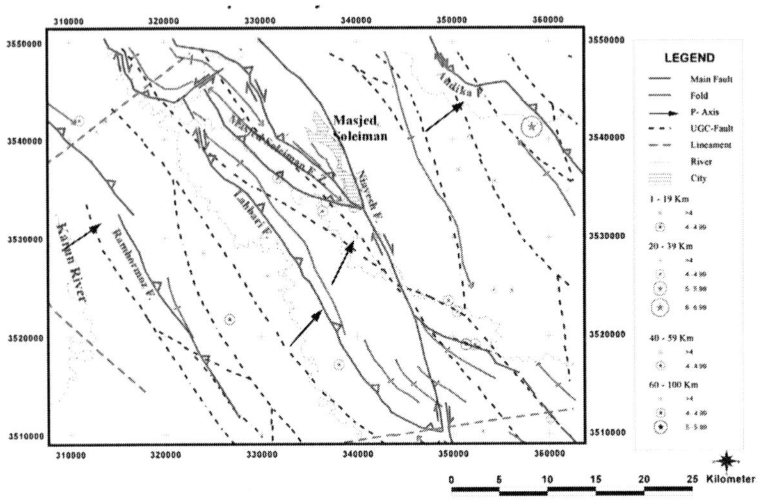

FIGURE 3. Seismotectonic map of the Masjed-i-Soleiman area.

Ambraseys and Melville model: $Ms = 1.429 \log L + 4.699$, where L is the fault length (Km)

(2)

After evaluating the maximum analytical magnitude, the Donovan and McGuire attenuation equations have been used to evaluate the maximum horizontal ground acceleration [14]. These equations are as follows:

Donovan equation: $a = 1079\ e0.5MS\ (R + 25) - 1.32$

(3)

McGuire equation: $a = 472.3\ e0.64MS\ (R + 25) - 1.301$

(4)

where R is the distance in Km, a is the maximum horizontal acceleration (Cm/s2), and e is Neperian base logarithm.

3.3. Underground Investigations

In order to make underground investigations, the underground contour map (UGC) of the top of the Asmari Formation has been digitized [10]. Further, the Iso-potential surface of this UGC was reconstructed. The surface structures (fold axis and main fault zones) and the location of gas leakages (and oil seepages) were overlaid on this Iso-potential surface map (Figure 4a). Using the GIS techniques, a 3D underground schematic sketch was prepared using Arcview 3.2 [25]. This 3D map helped us to visually detect the relationship between the surface structures (such as faults and folds) and the subsurface structures (such as structural contours and UGC-faults which are shown on UGC-map). Moreover, the locational features of these gas leakages are exhibited on the generated 3D-sketch. Figure 4b shows the rotation of this schematic 3D-sketch, the new viewpoints (aspects) and the surface and subsurface structural relationships between MIS anticline, main fault zones and gas leakage locations (Figure 4b).

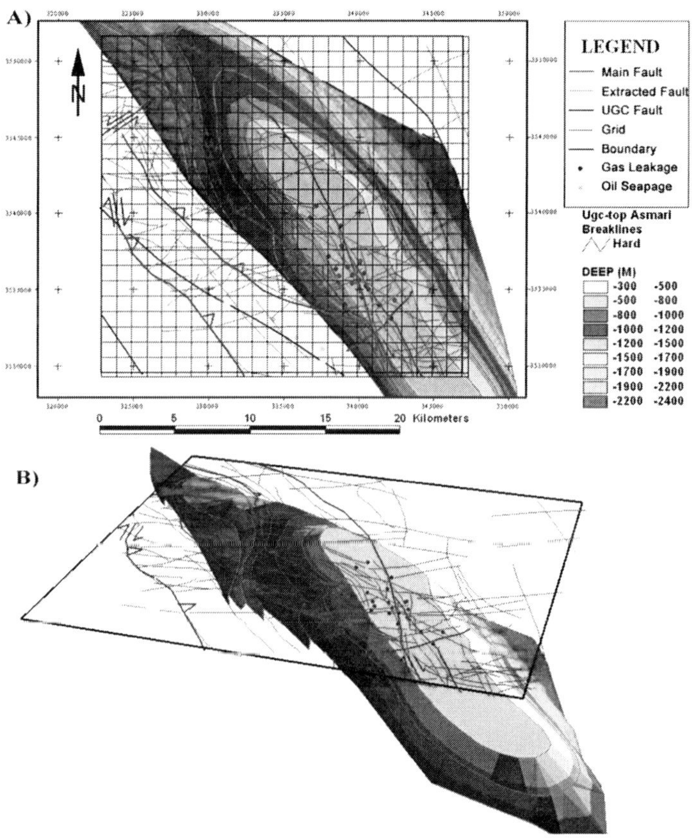

Figure 4. Showing the 3D model of structures-gas leakages relationship (a) Preparation of Iso-potential surface of UGC map and overlaying surface structures [10] (b) Construction of 3D-sketch from iso-potential surface and its rotation.

4. RESULTS

This research paper discusses the oilfield and gas leakages that have occurred during the last decades due to the tectonics and seismic activities. The seismotectonics (or major fault) activities generated numerous lineaments and fractures, which play a major role in transmission of crude oils and H_2S gas from the uppermost crust of the reservoir. Remote sensing techniques have proven to be a powerful tool for detecting lineaments, faults, folds and other structural features. The linearity of the vegetation and rugged topography of the studied area helped us to interpret structural features from the images. This study shows that the directional filtering and liner 2% enhancement techniques increased the interpretation ability for the analyst. The field observation coupled with remote sensing and GIS techniques were very helpful for the interpretation of the lineaments and faults at the surface and near the surface.

4.1. Structural Geology

Four main fault zones were detected in the studied area namely, Masjed-i-Soleiman, Niayesh, Lahbari and Andika. The main characteristics of these fault zones have been studied.

4.1.1. Masjed-i-Soleiman Fault Zone

This fault zone is 22 km in length, and extends from the northwest to southeast of the MIS area. It comprises of several reverse faults showing N115-125/65-NE trends, forming a structural Duplex (Plate 2a) [26]. The kinematical mechanism of this fault zone is reverse faulting with minor dextral strike slip component. The activities of this fault zone created the Masjed-i-Soleiman anticline as a fault related fold (Figure 5).

4.1.2. Niayesh Fault Zone

This dextral strike slip fault zone is 72 km in length and extends from the north of the MIS to the southwest of Haftgel city. This transverse fault zone crosses the MIS residential area (Plate 2b). The main structural attitude of the Niayesh fault is N160-170/75-NE. The kinematical mechanism of this fault zone is dextral strike slip. The major activities of this fault caused truncating and dextral dragging of MIS anticline.

PLATE 2. Field evidence of main fault zones in Masjed-i-Soleiman area (a) MIS fault zone as a duplex; (b) Niayesh fault in residential area of MIS city (c) Lahbari thrust fault zone; (d) Andika fault zone.

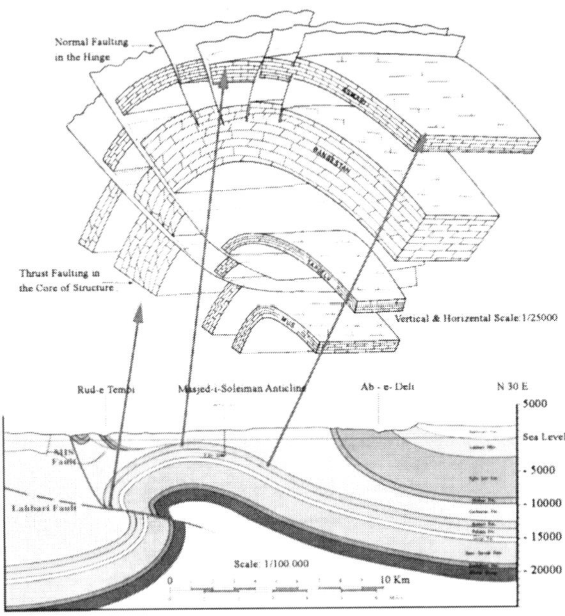

FIGURE 5. Formation of Masjed-i-Soleiman anticline due to activity of MIS fault zone (Modified from Keyhanfar 1995 and NIOC 1967) [10,27].

4.1.3. Lahbari Fault Zone

This reverse fault zone is 90 Km in length and extends from the Batvand area (located in the west-northwest of MIS region) to the northeast of Haftgel city (located 45 km south of MIS region). The main structural attitude of the Lahbari fault is N140/65-NE. The kinematical mechanism of this fault zone is reverse faulting with minor dextral strike slip component. The activities of this fault caused thrusting of Gachsaran Formation (in Miocene age) over Quaternary Bakhtiari conglomerates (Plate 2c).

4.1.4. Andika Fault Zone

This fault zone is 45 km in length and is located approximately 14 km northeast of the MIS city (in northeast corner of study area) (Plate 2d). The main structural attitude of this fault is N145/50-NE. The kinematical mechanism of this fault zone is dextral strike slip with minor reverse component (Plate 2d). The activity of this fault caused dextral dragging of other structures (such as faults and anticlines).

4.2. Seismotectonics

Siesmotectonically, Quaternary fault activities and its possible earthquake investigation during the last decades have shown that the entire aforementioned fault zone plays a major role in the generation of seepages (and leakages) in the study area. The description is given as follows:

4.2.1. Masjed-i-Soleiman Fault Zone

The field activity of this fault zone revealed that the Quaternary activity played a major role in thrusting the Gachsaran Formation over the Tembi Quaternary plain (Plate 1). The recent activity has also tilted and caused truncating of recent alluvium along the old MIS road. The earthquake epicenters that are related to the recent activities of this fault have been depicted in Table 1. On the basis of analytical evaluation of seismic parameters (Table 2), we can say that this fault can generate future earthquakes with maximum magnitude of Ms = 6.23 and maximum horizontal ground acceleration 0.37 g in MIS city.

4.2.2. Lahbari Fault Zone

The activity of this fault caused a thrusting of Miocene evaporites over Quaternary conglomerates (Plate 2c). The earthquake epicenters which are related to the recent activities of this fault are illustrated in Table 1. This fault has the capacity to generate future earthquakes with a maximum magnitude Ms = 7.05 and maximum horizontal ground acceleration 0.32 g in MIS city.

TABLE 1. The correlation of main quaternary fault zones and earthquakes. Ms = maximum magnitude on the Richter scale.

Date	Caused Fault	Long.	Lat.	Depth	Ms
1964/01/12	Niayesh	49.4	31.5	67	5.2
1989/03/01	Niayesh	49.46	31.58	46	4.5
1994/10/27	Niayesh	49.41	31.43	33	4.6
1995/04/18	Niayesh	49.43	31.8	18	4.9
2002/12/30	Niayesh	49.41	31.84	33	4.8
2002/12/30	Niayesh	49.41	31.84	33	4.8
2003/01/01	Niayesh	49.44	31.8	33	4.7
1995/06/07	MIS	49.23	31.99	33	4.3
2001/12/23	MIS	49.29	31.78	33	4.7
2001/12/23	Lahbari	49.24	31.67	33	4.5
2003/12/11	Lahbari	49.22	31.95	33	5

TABLE 2. Evaluation of seismic parameters of the main fault zones in Masjed-i-Soleiman area.

Evaluated Max. Acceleration			Evaluated Max. Magnitude			Distance From MIS (Km)	Mechanism	Length (Km)	Fault Name
Mean (g)	McGuire (1973) Cm/s^2	Donovan (1972) Cm/s^2	Mean	Ambraseys &Melville (1982)	Nowroozi (1985)				
0.37	386.5	347.2	6.23	6.18	6.28	Crossing the city	Reverse	22	MIS
0.55	593.4	485.3	6.92	6.92	6.92	Crossing the city	Dextral Strike slip	72	Niayesh
0.32	354.4	281.3	7.05	7.06	7.04	15	Reverse	90	Lahbari
0.26	283.5	238.1	6.65	6.63	6.67	14	Dextral Strike slip	45	Andika

4.2.3. Andika Fault

The field evidences of this fault zone indicate that the truncating and tilting of young Quaternary (to Sub-Recent) lithified sediments are exhibited in the Andika region. The main earthquake epicenter that is related to the recent activity of this fault zone was reported on 15 July 1929, an earthquake with Ms = 6.25. This fault has the capacity to generate future earthquake with a maximum magnitude Ms = 6.65 and maximum horizontal ground acceleration 0.26 g in MIS city.

5. DISCUSSIONS

MIS city is located in the Dezful embayment zone of the Zagros fold-thrust belt. This seismically active fold-thrust belt has several anticlines representing medium to large scale oilfields. Due to the tectonic settings of the Dezful embayment, many of these oilfields are situated at a shallow depth. The MIS oilfield is situated at a shallow depth, close to historical places such as the Zoroastrian temple, which may be established due to the oil seepages and minor gas leakages.

In this research, four main fault zones namely, Niayesh, Lahbari, Andika and MIS fault zone are recognized in the studied area. These were detected from the satellite image based on image interpretation technique.

The structural studies and controlling of structural cross section revealed that the MIS anticline is formed due to the thrusting movement along the MIS fault zone; hence creating a propagation fold (Figure 5). This reverse movement along fault zone formed the shallow buried anticline and, consequently, a shallow oilfield with 300 to 600 m depth (on the basis of UGC of top of Asmari formation) [9].

The accumulation of hydrocarbons in the competent Asmari Formation of this anticline has created the MIS oilfield. The main reservoir of this oilfield consists of competent cream limestone of Asmari which is overlaid by evaporitic rocks of Gachsaran Formation (as cap rock). MIS city is a high populated residential area and is situated in the center of the shallow oilfield.

The Niayesh and MIS fault zone activities have affected the oilfield. Subsequently, a highly dense fracture zone has developed within the intersection of Niayesh and MIS fault zones. Therefore, the cap rocks are highly fractured in nature.

On the basis of the 3D model and geological interpretation, it can be seen that a high density fracture zone has formed within the intersection between the Niayesh and the MIS fault zones. This is organized over a hinge zone with dense tensile fractures of the MIS buried anticline (Figure 5). The highly fractured reservoir and cap rocks caused several oil seepages and gas leakages from the MIS oilfield in the residential areas at a shallow depth. Moreover, the long-term oil production from this oilfield (from 1908 to present) has formed a thick gas cap (predominantly H_2S) in the upper most part of the hydrocarbon reservoir in the studied area. Unfortunately, in recent years, many oil seepages have been causeddue to the leakages of gas.

The seismotectonic behavior of the region is studied and it shows that four main fault zones are active faults. These fault zones have evidences of Quaternary to Recent activities such as: Tilting, dragging and truncating of lithified Sub-Recent and Quaternary sediments showing that the thrusting of Gachsaran Formation has taken place over the Plio-Pleistocene Bakhtiari Formation. The activities of these fault zones caused the high rate regional seismicity in the Masjed-i-Soleiman residential areas. The strongest earthquake, which recently took place on 15 July 1929, has a recorded magnitude of Ms = 6.25 on the Richter scale (i.e., in Andika area placed in 14 Km of NE of MIS city). The seismotectonic study revealed that re-activation of these active faults might have caused the large earthquakes with maximum magnitude $6.23 < Ms < 7.05$ (on the Richter scale) and maximum horizontal acceleration $0.26 < a < 0.55$ (g) (Table 2). Furthermore, the re-activation of these fault zones in future may generate a large earthquake. This may develop the large scale fracturing of Gachsaran ductile cap rocks and thus, may release a large volume of H_2S gas from the upper most of the reservoir.

6. CONCLUSIONS

The activity of the MIS reverse fault zone has developed in the Dezful embayment zone within the Zagros fold-thrust belt. It has formed the MIS shallow buried anticline and acts as a fault related fold. The accumulation of hydrocarbon in the competent Asmari Formation (overlaid by ductile Gachsaran Formation) established the MIS shallow oilfield with a thick gas cap under MIS residential area. The intersection of the Niayesh and the MIS fault zones has created high fracture zones that have caused gas leakage (and oil seepage) through the thick gas cap of this oilfield.

The activities of main active fault zones could generate destructive earthquakes in future; with geological seismite such as large scale opening of previous fractures and or forming new fractures in the Gachsaran ductile cap rocks. The creation of large scale tensile fractures in the hinge zone of the shallow buried MIS anticline (with reservoir contain thick gas cap) is the main cause of release of large volumes of gas, predominantly H_2S, from the upper most part of this oilfield, posing danger to the inhabitants and damaging the environment. This study concluded that the integration of remote sensing and GIS techniques in conjunction with field observation can be successfully utilized for mapping structural features and also environmental impacts in oilfield and seepages.

ACKNOWLEDGEMENTS

The authors would like to the anonymous reviewers for their very useful comments which helped bring the manuscript to its current form.

REFERENCES

1. Bordenave, M.L.; Burwood, R. Source rock distribution and maturation in the Zagros Orogenic Belt: Provenance of the Asmari and Bangestan reservoir oil accumulations. Org. Geochem. **1990**, 16, 369–387.
2. Colman-Sadd, S.P. Fold development in Zagros simply folded belt, Southwest Iran. AAPG Bull. **1978**, 62, 984–1003.
3. Edgell, H.S. Regional case studies salt tectonics in the Persian Gulf basin. In Salt Tectonics; Geological Society of London: London, UK, 1996; Special Publications No. 100; pp. 129–151.
4. Ali, S.A.; Pirasteh, S. Geological application of Landsat ETM for mapping structural geology and interpretation: aided by remote sensing and GIS. Int. J. Remote Sens. **2004**, 25, 4715–4727.
5. Pirasteh, S.; Woodbridge, K.; Rizvi, S.M. Geo-information technology (GiT) and tectonic signatures: the River Karun and Dez, Zagros Orogen in south-west Iran. Int. J. Remote Sens. **2009**, 30, 389–404.
6. Safari, H.; Pirasteh, S.; Pradhan, B. Uplifting estimation in Zagros Transverse faults Iran: An Application of Geoinformation Technology. Remote Sens. **2009**, 1, 1240–1256.
7. Pradhan, B.; Singh, R.P.; Buchroithner, M.F. Estimation of stress and its use in evaluation of landslide prone regions using remote sensing data. Adv. Space Res. **2006**, 37, 698–709.
8. Youssef, A.M.; Pradhan, B.; Gaber, A.F.D.; Buchroithner, M.F. Geomorphological hazard analysis along the Egyptian Red Sea Coast between Safaga and Quseir. Nat. Hazards Earth Syst. **2009**, 9, 751–766.
9. Pirasteh, S.; Pradhan, B.; Rizvi, S.M. Tectonic process analysis in Zagros Mountain with the aid of drainage networks and topography maps dated 1950–2001 in GIS. Arab. J. Geosci. **2009**.
10. NIOC. The Condition of Existence Hydrocarbon Seepages in Masjed-i-Soleiman City; Internal Report; NIOC: Tehran, Iran, 1995; p. 19.

11. Shishegar, A. Environmental hazards due to oil seepages in Masjed-i-Soleiman city. In Proceeding of First Conference on Environmental Crisis and Collate Techniques in Iran, Tehran, Iran; 2002; pp. 177–186.
12. Flottmann, T.; McClay, K.; Soleimany, B.; Goodarzi, H.; Majedi, M. Basement fault control on upper crustal folding, Dezful Embayment and Zagros Mountains, Iran. In Proceedings of International conference on theory and application of fault related folding in foreland basin, Beijing, China; 2005; p. 17.
13. Hatzfeld, D.; Tatar, M.; Priestley, K.; Ghafory-Ashtiany, M. Seismological constraints on the crustal structure beneath the Zagros Mountain belt (Iran). Geophys J. Int. **2003**, 155, 403–410.
14. McQuarrie, N. Crustal scale geometry of the Zagros fold-thrust belt, Iran. J. Struct. Geol. **2004**, 26, 519–535.
15. Alavi, M. Tectonics of the Zagros orogenic belt of Iran; New data and interpretations. Tectonophysics **1994**, 229, 211–238.
16. Falcon, N.L. Southern Iran: Zagros Mountains. In Mesozoic-Cenozoic Orogenic Belts; Spencer, A.M., Ed.; Scottish Academic Press: Edinburgh, UK, 1974; pp. 199–211.
17. Berberian, M. Master blind thrust faults hidden under the Zagros folds: Active basement tectonics and surface Morpho-tectonics. Tectonics **1995**, 241, 193–224.
18. Inger, S.; Alllen, M.; Hassani, H.; Talebian, M.; Jackson, J. Regional to reservoir scale tectonic evolution of the zagros orogenic belt. In Proceedings of AApg Annual Meeting, Tehran, Iran, 10–13 March, 2002.
19. Pirasteh, S.; Ali, S.A.; Hussaini, S. Lineaments development processes for geomorphological controls in Zagros Mountains, Southwest Iran: An application of geo-information technology. J. Geomatics **2007**, 1, 19–27.
20. Safari, H.; Chitsazan, M. Determination and structural analysis of Izeh fault zone, Zagros, Iran. J. Sci. Shahid Chamran University, Iran **2005**, 12, 57–75. (In Persian).
21. James, G.A.; Wynd, J.G. Stratigraphic nomenclature of Iranian oil consortium agreement area. Amer. Assn. Petrol. Geo Bull. **1965**, 49, 2182–2245.
22. Lillesand, T.M.; Kiefer, R.W. Remote Sensing and Image Interpretation, 4th ed.; John Wiley and Sons: New York, NY, USA, 2000; pp. 20–180.
23. Nowroozi, A. Empirical relations between magnitudes and fault parameters for earthquakes in Iran. Bull. Seismol. Soc. Amer. **1985**, 75, 1327–1338.

24. Ambraseys, N.N.; Melville, C.P. A History of Persian Earthquakes; Cambridge University Press: Cambridge, UK, 1982; p. 67.
25. Aronoff, S. Geographic Information System: A Management Perspective; WDL publication: Ottawa, Canada, 1989; p. 200.
26. Ramsay, J.G.; Hubber, M. The techniques of modern structural geology: Folds and fractures (vol. II). Academic Press: London, UK, 1987; p. 700.
27. Keyhanfar, S. Idealize Block Diagram Showing Thrust Faults in Lower Khami Formation; Internal Report No.3929; NIOC: Tehran, Iran, 1995.

5 CHAPTER

CHANGE DETECTION OF DESERT SAND DUNES: A REMOTE SENSING APPROACH

Surender Varma[1*], Vatsal Shah[2], Biplab Banerjee[1], Krishna Mohan Buddhiraju[1]

[1]*Center of Studies in Resources Engineering, IIT Bombay, Mumbai, India*

[2]*ElectricalEnginering, IIT Bombay, Mumbai, India*

ABSTRACT

Deserts are one of the major landforms on the Earth. While deserts occupy about one-fifth of Earth's land surface, they have been studied to a much lesser extent. All over the world, desert landforms are expanding ever rapidly and more and more human settlements are finding place in desert regions for habitation. Thus, quantifying and monitoring dunes becomes more relevant from a managerial perspective. Analyzing desert areas using satellite imagery is a challenging task due to weak textural differences and nearly homogeneous spectral responses in most parts of the terrain. In this paper, a post-clustering methodology for change detection of desert sand dunes is proposed. Features based on Radon spectrum are used to cluster dunes of various orientations. These clustered boundaries are used to detect if there are any changes occurring in the dune re- gions. In the experiments, remote sensing data covering various dune regions of the world are observed for possible changes in dune orientations. In all the cases, it is seen that there are no major changes in desert dune orientations since three decades.

Keywords: Change Detection; Desert Land Forms; Monitoring Sand Dunes

1. INTRODUCTION

Desert regions are increasingly facing imminent threats and becoming vulnerable to effects of climate changes, urbanization, land degradation, water shortage, etc. [1] . Deserts are extremely important habitat for specialized flora and fauna. Besides, deserts possess extensive amounts of minerals such as lithium, boron, etc. Low surface runoff which is one of the major characteristics of desert regions helps in preserving these minerals on the land. Quantification of dune morphology not only helps for mapping purposes but also provides important informa- tion about the regional geologic history, environmental change, and desertification. Knowledge about changes occurring in dune boundaries and their directions will also contribute to the ability in making prediction about wind patterns. From a human perspective, if the dunes are changing in direction and are towards populated re- gions, then immediate precautionary steps are to be initiated by the respective authorities. Hence, it has become imperative to accurately map and monitor deserts to formulate and implement various management activities. Categorizing dunes using field instruments is generally arduous as they generate sparse data. Quantitative me- thods using remote sensing data are more viable for collecting dune information and monitoring them.

Major studies in respect of desert dynamics are mainly focused in and around vegetation or water bodies and changes occurring with respect to them. Dunes can either be stable or mobile as a function of vegetation and wind power. Unrestrained erosion is generally a feature of bare dunes. Stabilized dunes, on the other hand, do not undergo major changes in textures, and can thus facilitate the growth of vegetation [2] [3] . In the work by [4] , a combination of Digital Elevation Modeling and GPS guided field survey were used to obtain the base data about dunes. NDVI index was often used to monitor the growth vegetation and its relation to surrounding set- tlements. [5] used brightness index, semi-variogram, and other statistical data to analyze the changes in texture orientations. [6] has made use of Principal Component Analysis to detect changes in the images.

While above techniques capture certain aspects of desert environments, they are not specifically devised to- wards orientation of dunes. Analyzing desert areas using satellite imagery is a challenging task due to weak textural differences and nearly homogeneous spectral responses in most parts of the terrain, i.e., categorization of dunes via traditional clustering techniques is a challenging task as dunes with differing orientation character will have similar gray value, NDVI index and also brightness index.

Researchers in remote sensing community have extensively worked on a variety of methodologies for other land covers but fewer methodologies have been proposed specifically for dune regions. In this paper we present a novel way of identifying changes in dune orientations.

Radon Transform

Radon developed the Radon Transform in the early part of 19th century [7] . A line with normal vector in direc- tion θ at a distance of s from the origin satisfies the line equation (1):

$$x\cos\theta + y\sin\theta - s = 0 \tag{1}$$

$$g(s,\theta) = \iint f(x,y)\cdot\delta(x\cos\theta + y\sin\theta - s)\,\mathrm{d}x\mathrm{d}y \tag{2}$$

The mass density $f(x,y)$ evaluated over different viewing directions will give rise to a series of 1D projec- tions parameterized by s. The Dirac delta function δ, is zero for every argument except on s axis. Its integral, $g(s,\theta)$ is the radon transform expressed as given in Equation (2). Maximum angle of Radon spectrum can be used to represent surface orientations of sand dunes in [8] . Radon spectrum also contains additional information which can be used to delineate dunes. This additional information can be captured using statistical features based on Radon Spectra which can help in better clustering of dunes. In a related work by [9] , Radon Transform was used to compute the global orientation of the texture present in the image. In the current paper, initially statistical features computed from Radon spectrum are used to cluster various texture regions. Further a change detection technique has been proposed to monitor the changes occurring in these dune texture boundaries.

The study areas selected belong to various dune regions across the world. Depending on the availability of good quality data, it has been ensured that the images selected correspond to the same time of the year. This acts as the prerequisite for change detection so that the changes occurring are not due to seasonal changes.

2. METHODOLOGY

The main objective of this methodology is to identify regions of change in the multi temporal desert dune da- tasets. A post-clustering based approach was developed for identifying changes in the orientations of dunes. Initially contiguous regions representing dunes with each region having similar dune orientation are generated. The region boundaries are used as basis for finding changes in multi temporal remote sensing data. i.e., on a temporal scale, if the boundaries of corresponding regions match then it is considered that there is "no-change" in dune patterns else we say that "change" has occurred.

The methodology adopted is shown in the flowchart (Figure 1). The input image and the reference image be- longing to different times but corresponding to the same region are first registered to ensure the correctness of corresponding pixels for the change detection algorithm. Histogram matching is later performed on to ensure that the changes observed are not due to radiometric differences. Radon Transform is then applied on the im- agery and then similar to Haralick texture features [10] , three statistical features (Equations (3), (4), and (5)) basing on Radon spectrum have been defined namely Mode, standard deviation, and entropy. These features are seen to be superior over mean, median and correlation features for dune texture delineation. It is to be noted that these statistical quantities quantifying Radon data are extracted at each pixel location using a moving window (block) processing technique.

1) Mode radon angle

$$\theta_{\text{mode}} = \left\{\theta_{\max} : f\left(\theta_{\max}\right) \geq f(\theta), \forall \theta \in [0, \pi]\right\} \tag{3}$$

2) Entropy

$$\theta_{\text{entropy}} = -\sum_{i=0}^{\pi} p(\theta_i)\ln\left[p(\theta_i)\right] \text{ where } p(\theta_i) = \frac{\theta_i f(\theta_i)}{\sum_{i=0}^{\pi} f(\theta_i)} \tag{4}$$

3) Standard Deviation

$$\theta_{sd} = \sqrt{\frac{\sum_{i=0}^{\pi} f(\theta_i)[\theta_{\text{mean}} - \theta_i]^2}{\sum_{i=0}^{\pi} f(\theta_i)}} \tag{5}$$

In Equations (3) to (5), the function f(q) denotes the number of pixels with angle q within the window cen- tered around the pixel in consideration. Mode angle feature shown in Equation (3) is the angle with the highest frequency within that block. Entropy feature signifies the randomness of angular information. In Equation (4), p(q) indicates the fraction of pixels having orientation q in the neighborhood. Equation (5) gives the standard deviation feature signifying the spread of angles with respect to the mean angle.

Three different images of the same size as the original image are created by using the above calculated statistical features for each block. Next, each pixel is assigned the same attribute as that of the majority of pixels within a block. Thresholds are assigned empirically using trial and error basing on their ability to delineate dune patterns. A binary image is thus created for each feature. Once again, these discrete binary feature outputs are transformed into continuous feature output images using a recursive smoothing step. Finally, the texture clustered output is obtained by combining the feature outputs using logical "or" operator.

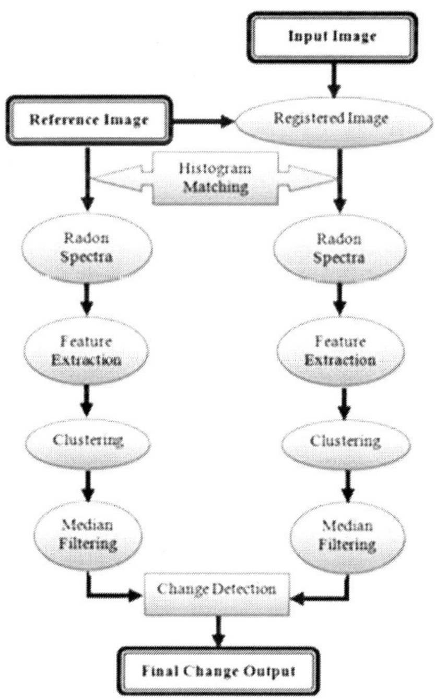

FIGURE 1. Flowchart of the methodology.

The window based processing technique is used in order to get contiguous cluster regions. Window size is dependent on a host of factors such as the image resolution, size of the texture patterns on the dunes, distance between successive dune crests, etc. The window size, thresholds used in clustering and other such parameters used in the clustering algorithm are kept constant for both the reference and the input images.

The clustered regions obtained from the reference and the registered input images are then compared with each other. The changes occurring in the dune orientations are captured by observing the variations in the boundaries of clustered regions for identical dune orientations. These changes can be easily depicted by taking the difference of the individual clustered outputs.

All the operations explained in this section are performed for each type of texture orientation individually. For example, if an image has two major regions of different textural orientations, the image will be scanned successively to identify the regions having differing textural orientations one after another. If there are no changes over the temporal scale then the clustering algorithm should ideally give similar boundaries for each cluster region. On the other hand, changes in dune orientations will correspond to changes in the boundaries of the obtained cluster regions.

3. RESULTS AND DISCUSSION

The datasets used for the experiments are summarized in Table 1. Figure 2 and Figure 3 are the images cor- responding to the Gobi desert of central Asia in China-Mangolia region. Figure 4 and Figure 5 indicate the corresponding clustered outputs obtained using the Clustering algorithm mentioned in Section II. The orange colored region in Figure 4 and Figure 5 encompasses the desert regions with horizontal orientation of sand dunes, while the blue region indicates the vertically oriented sand dunes. Figure 6 indicates the difference out- put which can be obtained using the Change Detection Algorithm in Section 2.

Figure 7 and Figure 8 show the superimposition of corresponding input images and clustered outputs.

This gives a good visualization of the matching between image regions and cluster regions.

For the 2nd dataset (Figure 9 and Figure 10), imagery correspond to Thar Desert region in the north western region of India is taken. Figure 11 and Figure 12 show the corresponding clustered outputs respectively. The orange colored region in indicates the desert regions with diagonally oriented sand dunes, while the blue region indicates the sand dunes with horizontal orientation. Figure 13 indicates the difference output. As next dataset (Figure 14 and Figure 15), we consider another region of Thar Desert. Figure 16 and Figure 17 are the corres- ponding clustered outputs for the third dataset. The orange colored region in Figure 16 and Figure 17 indicates the desert regions with horizontally oriented sand dunes, while the blue region indicates the sand dunes with no particular orientation. Figure 18 indicates the change detection output for third dataset.

TABLE 1. Description of datasets

Figure Number	Region	Date	Sensor and Resolution
Figure 1	Gobi Desert	December 1972	Landsat, 30 m
Figure 2	Gobi Desert	January 2011	Landsat, 30 m
Figure 8	Thar Desert Region 1	March 1999	LISS-III, 23.5 m
Figure 9	Thar Desert Region 1	March 2009	LISS-III, 23.5 m
Figure 13	Thar Desert Region 2	March 1999	LISS-III, 23.5 m
Figure 14	Thar Desert Region 2	March 2009	LISS-III, 23.5 m
Figure 18	Sahara Desert	March 1988	Landsat, 30 m
Figure 19	Sahara Desert	March 2005	Landsat, 30 m

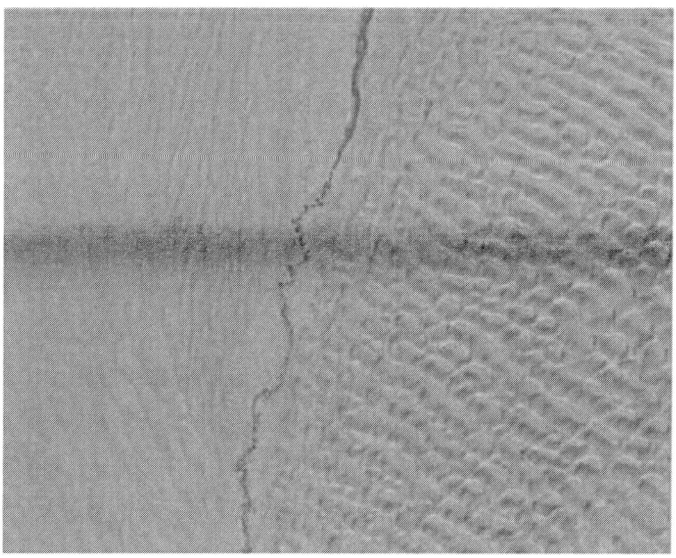

FIGURE 2. Gobi Desert region 1 at time 1.

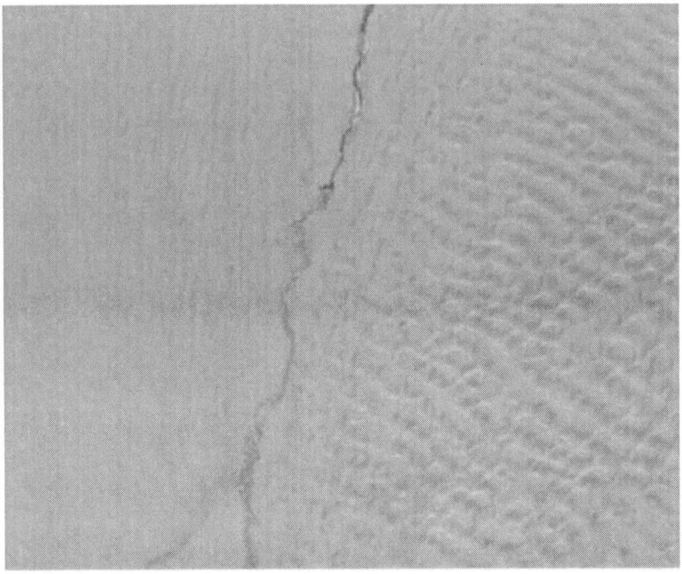

FIGURE 3. Gobi Desert region 1 at time 2.

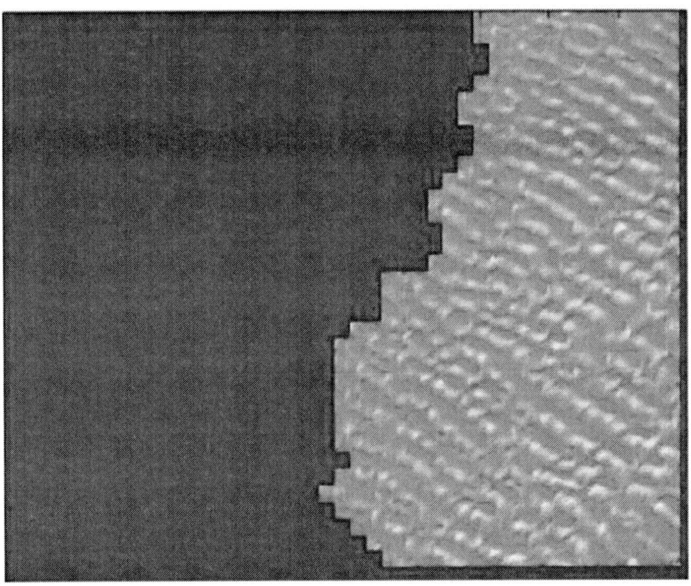

FIGURE 4. Clustered output of image in Figure 1.

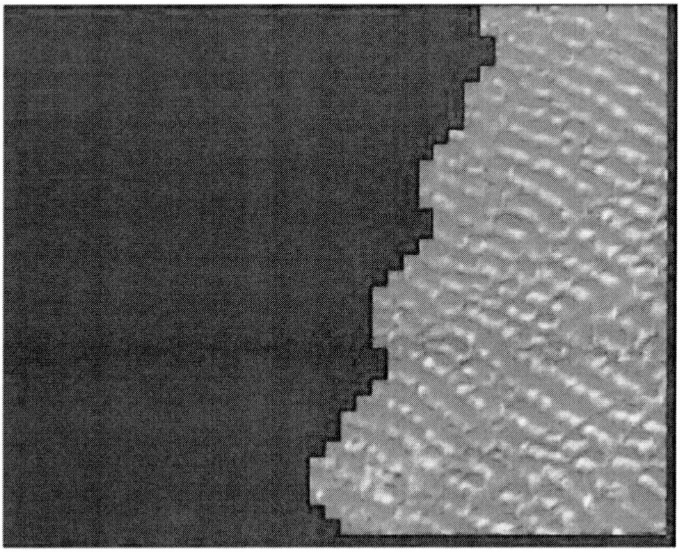

FIGURE 5. Clustered output of image in Figure 2.

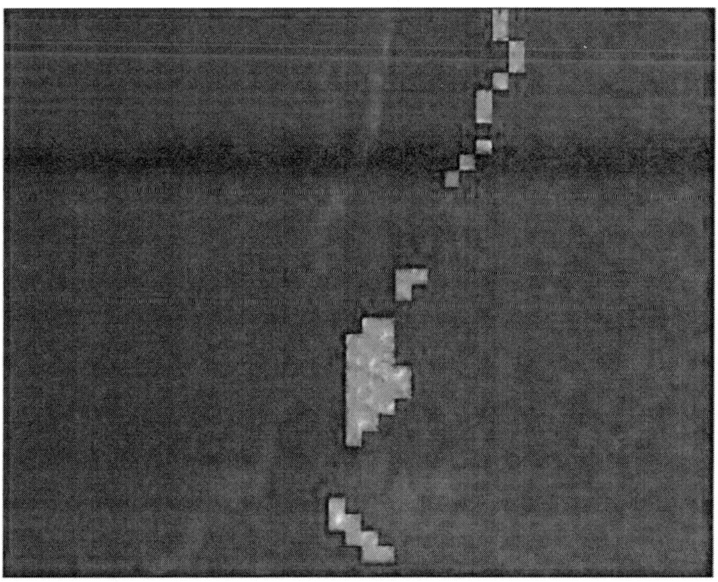

FIGURE 6. Change detection output: difference output of Figure 4 and Figure 5.

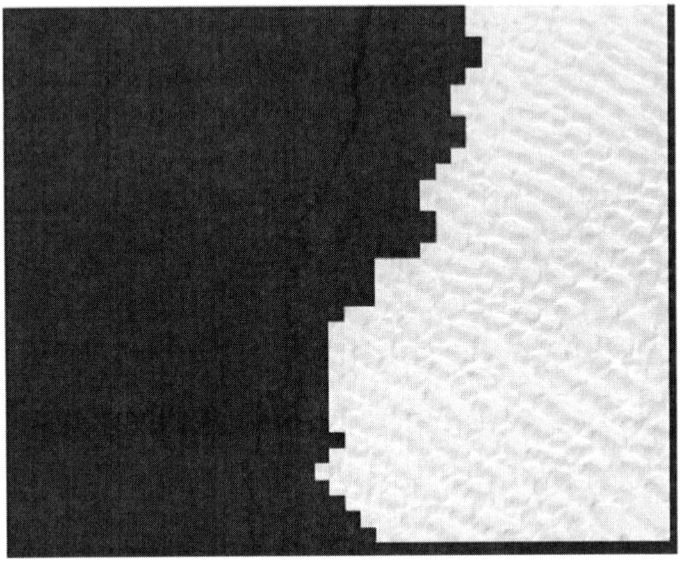

FIGURE 7. Superimposed clustered output of image in Figure 2.

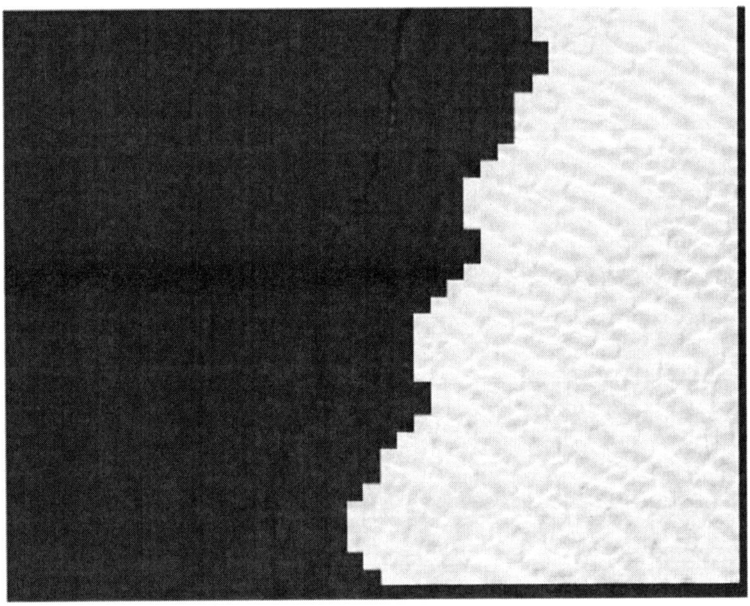

FIGURE 8. Superimposed clustered output of image in Figure 3.

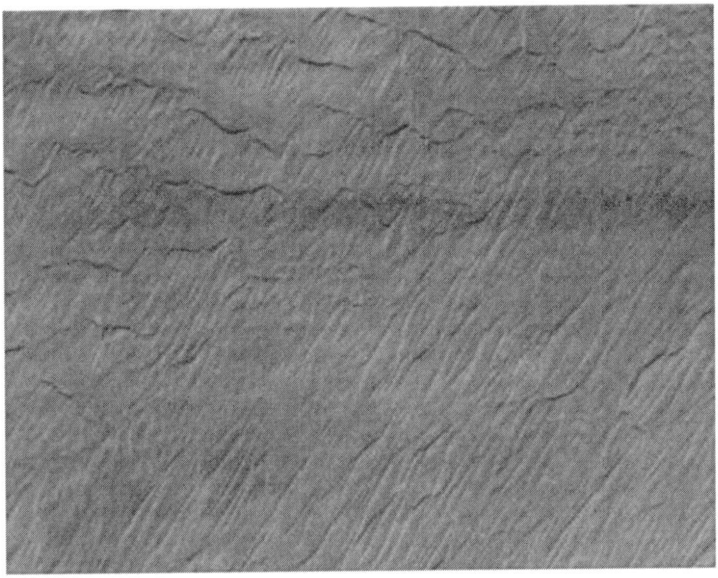

FIGURE 9. Thar Desert region 1 at time 1.

FIGURE 10. Thar Desert region 1 at time 2.

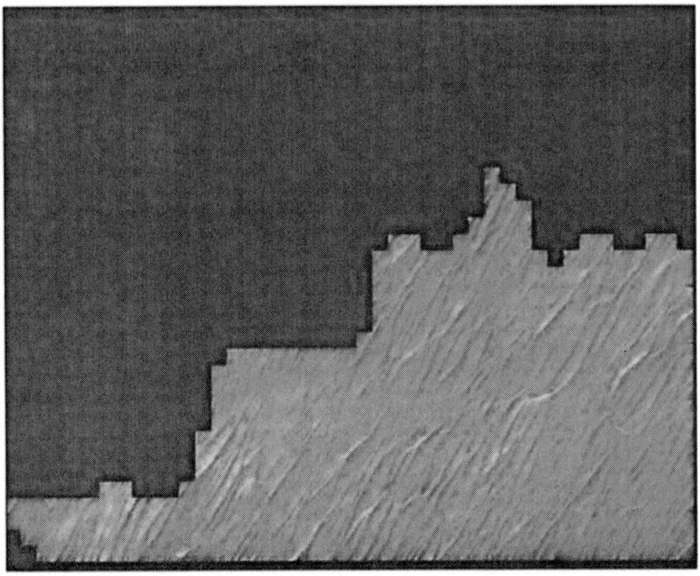

FIGURE 11. Clustered output of image in Figure 9.

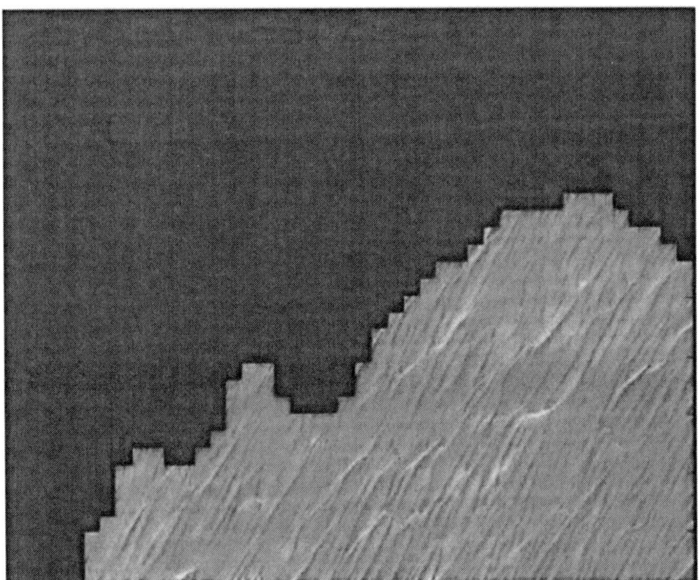

FIGURE 12. Clustered output of image in Figure 10.

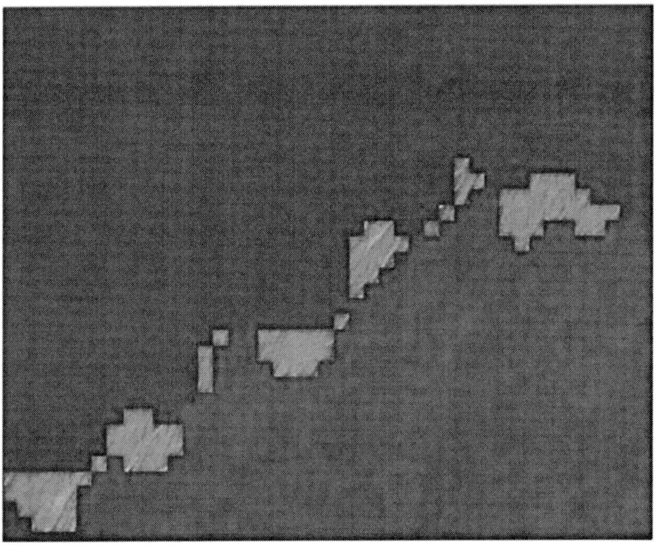

FIGURE 13. Change detection output: difference output of Figure 11 and Figure 12.

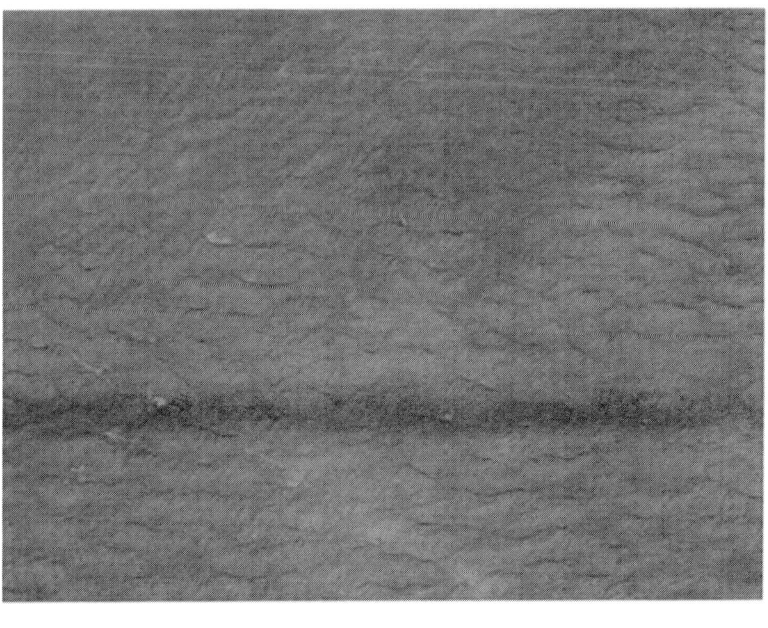

FIGURE 14. Thar Desert region 2 at time 1.

FIGURE 15. Thar Desert region 2 at time 2.

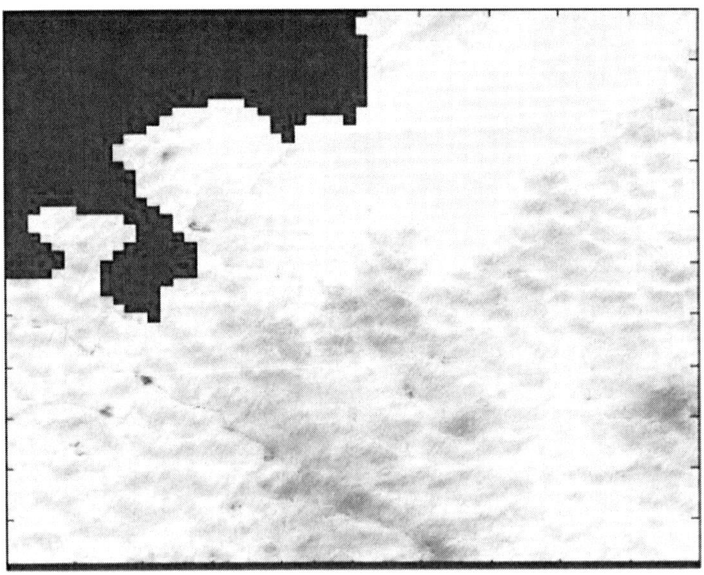

FIGURE 16. Clustered output of image in Figure 14.

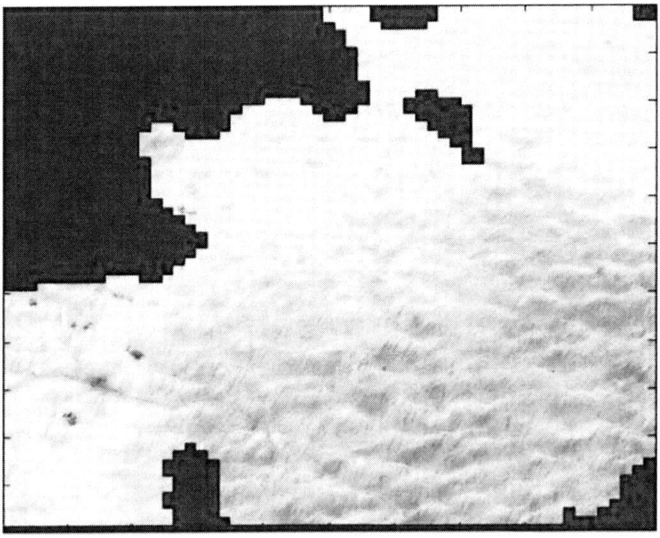

FIGURE 17. Clustered output of image in Figure 15.

Sahara desert region in Yemen is taken as the 4th dataset (Figure 19 and Figure 20) for experimentation. Figure 21 and Figure 22 are the corresponding clustered outputs for dataset 4. The orange colored regions in both these images denote the diagonally oriented sand dunes, while the blue region indicates the absence of dunes with any specific orientation. Figure 23 indicates the change output of dataset 4.

The attribute allocation after Radon transform is performed block wise rather than pixel wise. Block size is dependent on a host of factors such as the image resolution, size of the texture patterns on the dunes, distance between identically oriented dunes, etc. Larger block sizes have thus been used to cluster the dataset1 imagery, while smaller block sizes were used to cluster imagery of 2nd dataset.

Optimal thresholds for correct clustering of dunes are decided by observing the texture patterns and compar- ing them with the generated output regions. It has been ensured that exactly same thresholds were used for clus- tering images corresponding to each image set.

The change outputs obtained by the methodology adopted in case of all the four datasets used indicate that the interiors of the clustered regions are not undergoing any changes. All the minor changesare observed to occur at the boundaries of these clustered regions.

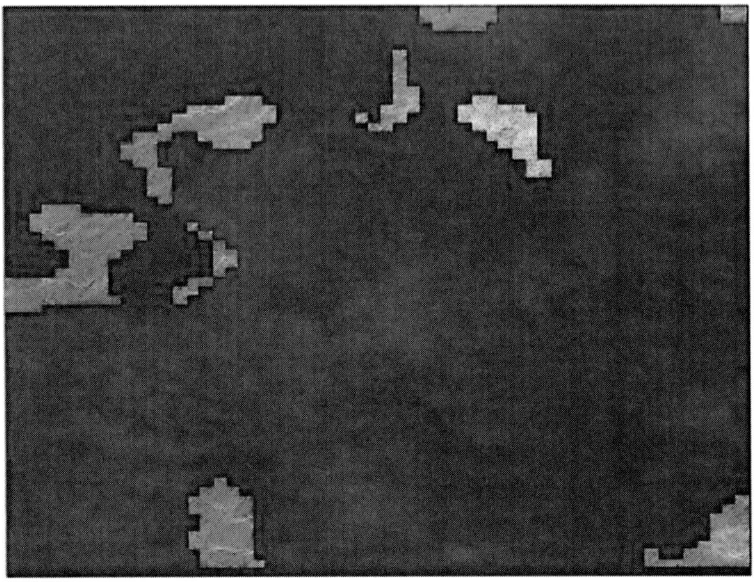

FIGURE 18. Change detection output: difference output of Figure 16 and Figure 17.

FIGURE 19. Sahara Desert region at time 1.

FIGURE 20. Sahara Desert region at time 2.

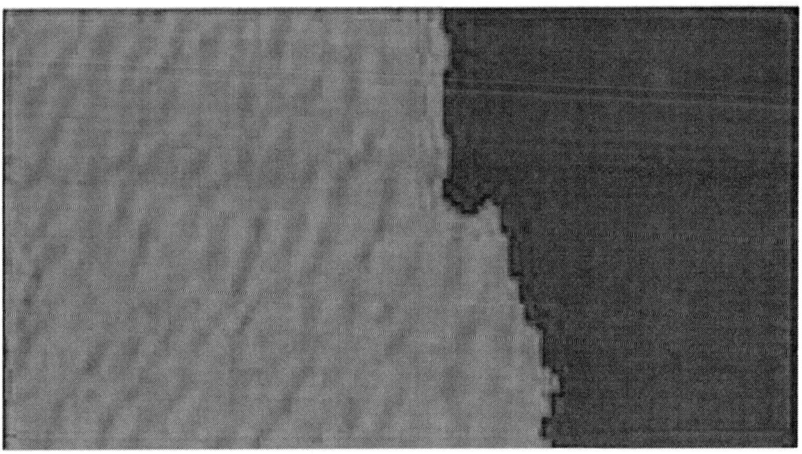

FIGURE 21. Clustered output of image in Figure 19.

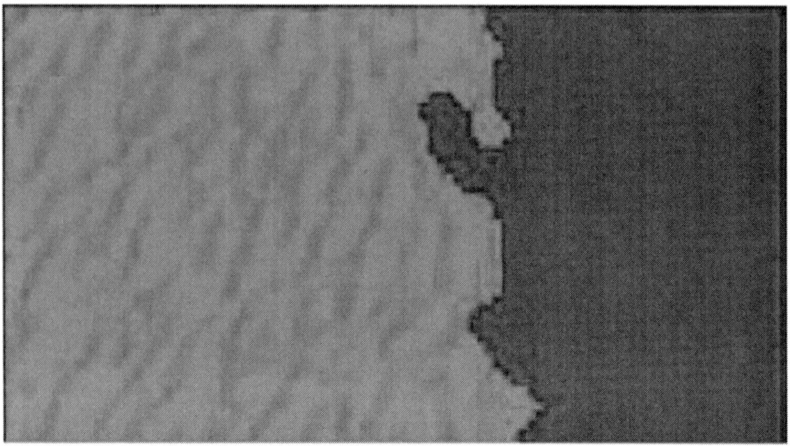

FIGURE 22. Clustered output of image in Figure 20.

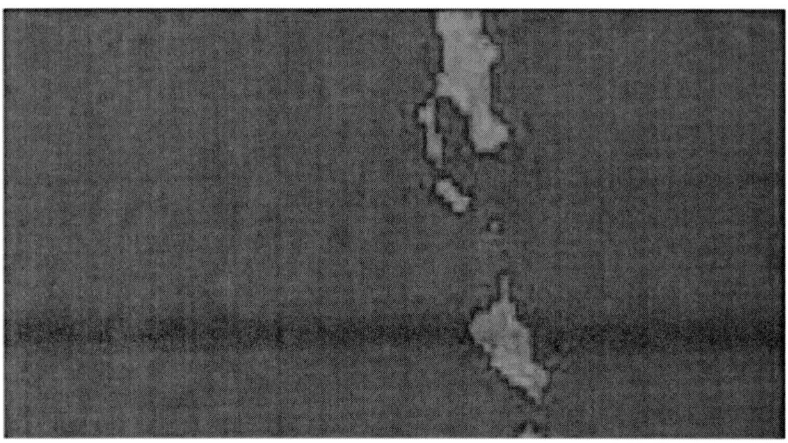

FIGURE 23. Change detection output: difference output of Figure 21 and Figure 22.

4. CONCLUSIONS

In this paper, we have proposed post clustering Methodology to monitor sand dunes based on changes occur- ring in their cluster boundaries. The separation of different textured regions has been achieved with the help of selective statistical Radon features. This knowledge of categorical dune regions and changes occurring over a period of time will aid the administrative setup in taking precautionary measures and thus protect the surround- ing ecosystem.

Change detection of desert sand dunes was performed on desert data sets belonging to three different dune re- gions of the world. The proposed change detection algorithm worked very effectively for the interior parts of imagery but faltered slightly at the borders of these clusters. This is primarily due to the block based processing technique which will enhance the subtle variations at the edges. These Block effect is particularly useful in ge- neralization and smoothing of clustered regions.

It has been observed that essentially the dune regions in all the considered cases remained constant in terms of their orientation structure since around three decades. There are only minor changes occurring at the interface of dune patterns. This indicates the stability of wind direction and other factors affecting dune orientations. The constancy of dune regions with similar textural boundaries over the time periods considered may not necessarily imply the dormancy of sand throughout but it does point out to the similar patterns of wind directions in that time span. Hence it can be inferred that over the years, though the actual sand particles may be migrating, but there is a definite pattern in their direction of migration. While these findings have implications for understanding

of dune geomorphology and the changes occurring in dune directions, they also draw the attention to a wider study of dune orientations and their feedback to associated phenomena both at regional and global scales.

Furthermore, the fact that the proposed block based approach does not require any additional information apart from the one available through satellite imagery suggests that it can be used for a host of applications. It is to be veri- fied if the constancy of dune orientations holds well from a historical perspective. Non-availability of open domain remote sensing data prior to 1972 limits the scope of using the proposed methodology for dune change detection.

REFERENCES

1. Belaid, M.A. (2010) Remote Sensing of Desert Cities in Developing Countries. In: Tarek, R. and Carsten, J., Eds., Remote Sensing of Urban and Suburban Areas, Vol. 10, Springer Publisher, New York, 245-265.
2. Lancaster, N. (1988) Development of Linear Dunes in the Southwestern Kalahari, Southern Africa. Journal of Arid Environments, 14, 233-244.
3. Paisley, E.C.I., Lancaster, N., Gaddis, L.R. and Greeley, R. (1991) Discrimination of Active and Inactive Sand from Remote Sensing: Kelso Dunes, Mojave Desert, California. Remote Sensing of Environment, 37, 153-166. http://dx.doi.org/10.1016/0034-4257(91)90078-K
4. Yang, X. (2011) Visualization Urban Spatial Growth of Desert Cities from Satellite Imagery: A Preliminary Study. Proceedings of the 25th International Cartographic Conference, Paris, 3-8 July 2011, CO-439.
5. Karnieli, A., Gilad, U., Ponzet, M., Svoray, T., Mirzadinov, R. and Fedorina, O. (2008) Assessing Land-Cover Change and Degradation in the Central Asian Deserts Using Satellite Image Processing and Geostatistical Methods. Journal of Arid Environments, 72, 2093-2105.
6. Aldossary, A. (2012) Analysis of Urban Change Detection Techniques in Desert Cities Using Remote Sensing. Master's Thesis, California State University, Northridge.
7. Radon, J. (1917) On the Determination of Functions from Their Integrals along Certain Manifolds. Mathematische- Physikalische, 69, 262-267.
8. SurenderVarma G., Banerjee, B. and Buddhiraju, K.M. (2012) Representation of Desert Sand Dunes by Surface Orientation Using Radon Transform. Proceedings of IGARSS, Munich, 22-27 July 2012, 2733-2736.
9. Jafari-Khouzani, K. and Soltanian-Zadeh, H. (2005) Radon Transform Orientation Estimation for Rotation Invariant Texture Analysis. IEEE Transactions on Pattern Analysis & Machine Intelligence, 27, 1004-1008.
10. Haralick, R.M., Shanmugam, K. and Dinstein, I. (1973) Textural Features for Image Classification. IEEE Transactions on Systems, Man and Cybernetics, 6, 610-621.

6

CHAPTER

HABITAT SUITABILITY MODELING OF ENDANGERED PRIMATES IN NIGERIA: INTEGRATING SATELLITE REMOTE SENSING AND SPATIAL MODELING TECHNIQUES

Alex O. Onojeghuo[1,2*], Alan G. Blackburn[1], Francis Okeke[3], Ajoke R. Onojeghuo[2,4]

[1]*Lancaster Environment Centre, Lancaster University, Lancaster, UK*

[2]*Department of Surveying and Geoinformatics, Nnamdi Azikiwe University, Awka, Nigeria*

[3]*Wildlife Conservation Society, Calabar, Nigeria*

[4]*Centre for Landscape and Climate Research, University of Leicester, Leicester, UK*

ABSTRACT

This paper investigates the impacts of forest cover and spatial structure changes on the forest landscape across Afi-Mbe-Okwangwo protected area of Cross River State, Nigeria and its corresponding implication on two endangered primates (Cross River Gorilla and Nigeria-Cameroon Chimpanzee) habitat using satellite remote sensing and modeling techniques. Using remote sensing change detection analysis, the spatial extent and annual rate of deforestation for the study area was determined as 34,620 hectares and 1.5% respectively (from 2000 to 2014). The protected areas with highest annual deforestation rates were Afi Mountain Wildlife Sanctuary (2.6%) and Mbe Mountains (2.2%), both prominent for gorilla and chimpanzee sightings and nests. Further investigations on changes to the forest landscape structure revealed high levels of forest fragmentation across the study area for the 14-year period investigated. As a means of further understanding effects of forest landscapes changes across the study area, a 14-year forward simulation was performed using the Markov model as to determine the spatial extent of futuristic forest cover changes. The results showed that if this current trend of forest cover change continued, 28,121 hectares of forests would be lost to deforestation in 2028 (approximately 16% of the total landmass of the entire study area). Using Maxent modeling, suitable primate habitats were predicted and the total coverage determined as 30,940 hectares (54.4% situated in CRNP—Okwangwo division, 29.4% in AMWS, 14.3% in Mbe Mountains and 1.9% in ARFR). Further analysis revealed 6468 hectares of predicted primate habitat were affected by deforestation in 2014 (21% of the predicted primate habitats). These results indicate that suitable primate habitats (particularly for gorillas and chimpanzees) are under immense pressure from deforestation and forest fragmentation. This paper presents a cost effective and time saving approach for determining suitable primate habitats and understanding the effects of forest transition on primate habitat suitability.

Keywords: Maxent, Remote Sensing, Primates, GIS, Deforestation, Markov, FRAGSTATS

1. INTRODUCTION

The negative impacts of forest loss and hunting activities across the world's tropical forests are sources of concern, as it affects the continued existence of endangered primates [1] -[3] . The Afi-Mbe-Okwangwo forest landscape is a biodiversity hotspot with known sightings of four Red List endangered primates namely: the Nigeria-Cameroon Chimpanzee Pan Troglodytes ellioti, Cross River Gorilla Gorilla gorilla diehli, Preuss's guenon Cercopithecus preussi and drill Mandrillus leucophaeus [4] -[6] . The area is also known to have suitable roosting sites for migratory birds [4] . However, this region of great ecological importance is faced with a range of human induced threats to the existing wildlife and forest ecosystem at large. Furthermore, the large and growing population of Cross River State depends greatly on subsistence agriculture, which has over the years resulted in high rates of forest loss and over-exploitation of natural resources [1] . This once undisturbed stretch of tropical forests in Cross River State is faced with diverse threats of deforestation in forms of large commercial plantations of plantain, rubber, cashew and pineapples [7] -[9] . In Africa, primates population have been most affected by loss and modification of forest habitats resulting from factors like intense logging, shifting cultivation, bush burning, over harvesting of fuel wood for commercial and household use, creation of grazing land and clearance for plantation farming [7] [10] -[12] . The interaction of high population growth rates and deforestation rates has caused a significant loss in the spatial coverage of remaining tropical forest landscapes across Africa [13] [14] . The study by [4] noted that primates, such as gorillas, are particularly vulnerable given their reliance on tropical forest habitats, slow rates of reproduction and massive body sizes. Hence, a better understanding of environmental factors such as forest cover distribution, pattern of forest landscape dynamics and modeling of species distribution would prove vital in enhancing conservation and proper management of forest protected areas across Africa.

The combined use of satellite remote sensing and ground truth data has shown to be an effective method of understanding changes in forest landscapes across the world [4] [15] -[18] . Using outputs from remote sensing in form of classified images, a better understanding of forest structure changes is possible, providing valuable results that are crucial to formulating strategies which assist in forest conservation and preservation of existing wildlife. A major problem identified in a number of conservation related studies is the negative impact of forest fragmentation on existing wildlife that relies on forested habitats for survival [19] -[21] . The study [20] , assessed levels of forest fragmentation in the conservation of priority landscapes across parts of India using FRAGSTATS analysis. The investigation as to changes in the spatial structure of forest landscape provided a better understanding of forest fragmentation and its impact on protected areas across the study area.

An important component of this study was to perform a forward simulation of forest landscape transition in the study area as a means to create futuristic scenarios based on existing trends of forest cover dynamics. This would provide a direct representation as to the spatial and structural changes of forest landscape in the future, provided the current trend of forest transition continues. With such outputs, it would be particularly useful for wildlife conservation and decision makers involving in the day to day running and management of protected areas across the country. To this end, the Markovian process which had capabilities to model future state of landscapes using the immediate preceding state as a basis for forward prediction [22] was investigated in the study. [23] proposed making use of the cellular automata-Markov Chain model to simulate a counterfactual scenario of evaluating the role of forest conservation and investigate degrees of forest fragmentation protected areas in part of India.

Having a good understanding of wildlife species distribution and identifying factors that influence habitat selection would greatly assist wildlife conservation and management. Within Afi-Mbe-Okwangwo protected area landscape, delineating the current and potential distribution of endangered species, such as the Cross River Gorillas and Chimpanzees primates are critical to conservation management and planning. Part of this study shall investigate the use of presence-only data and other environmental variables for mapping suitable habitats for endangered primates in the study area. The Maximum Entropy modeling (Maxent), allows for suitable habitat distribution through combined use of presence data and environmental variables [24] -[28] . The advantages of Maxent modeling include high levels of relative insensitivity to positional errors of species location data, non- dependence on large datasets for predictive modeling and better performance levels when compared to other predictive modeling techniques [27] . Although we acknowledge the existence of other predictive models, emphasis of this study focuses on use of Maxent model in primate habitat distribution across parts of Nigeria. Since the Maxent modeling technique relies greatly on species presence data, collection of such information should be unbiased in nature and great care must be taken to avoid data duplication [25] .

Taking into consideration the afore-mentioned, this study aimed to investigate impacts of forest cover and spatial structure changes on the forest landscape of Afi-Mbe-Okwangwo protected area in Cross River State, Nigeria particularly its corresponding implication on two endangered primates (Cross River Gorilla and Nigeria-Cameroon Chimpanzee) habitat using satellite remote sensing and modeling techniques. The key objectives of the study were: 1) to examine the changes in forest cover and spatial structure across the study area using satellite remote sensing and ground data; 2) to perform predicted forward simulation of forest cover changes across the study area as a means of projecting possible impacts of deforestation on existing forest landscapes; 3) to model habitat suitability of endangered primates using a combination of species presence data and other key environmental variables; and 4) to estimate the spatial extent and effects of deforestation on predicted primate habitats.

2. MATERIALS AND METHODS

2.1. Study Area

The study area, Afi-Mbe-Okwangwo forest landscape, is composed of four protected areas namely Afi Mountain Wildlife Sanctuary (AMWS), Afi River Forest Reserve (ARFR), Mbe Mountains and Cross River National Park (CRNP) (Okwangwo division) all situated in the northern part of Cross River State (Figure 1). The entire landscape covers an approximate area of 176,835 hectares and is geographically located between latitudes 6°04'19" to 6°28'35" and longitudes 8°49'55" to 9°21'47".

The AMWS is the westernmost protected area that serves as home to estimated 25 - 30 Cross River Gorilla sightings. The sanctuary created in 2000 was established to protect wildlife in approximately 100 km2 of lowland and hill forests in the north-western corner of the 380 km^2 Afi River Forest Reserve (ARFR) [3] . The protected area is currently managed by the Cross River State Forestry Commission with assistance from the environmental NGO Wildlife Conservation Society (WCS). Other NGOs that have been involved region include Fauna and Flora International, Nigerian Conservation Foundation and Pandrillus. The AMWS is dominated by rocky hills with sparse vegetation and large grasslands making it a suitable roost for migratory European barn swallows (Hirundo rustica). [4] estimates 20 million winter birds, particularly swallows, roost in the suitable grasslands of the AMWS each year. The sanctuary is surrounded by sixteen villages and farms which has led to cases of encroachment beyond its boundaries, scattered logging in the lowland areas, occurrence of forest fires during farm clearing by farmers in farming seasons, poaching of endangered wildlife (such as the Cross River Gorilla, chimpanzee etc.) and over exploitation of non-timber forest products such as herbs for commercial and household use [1] [3] [4] [29] .

To the east of the Afi Mountains and adjourning the ARFR is the Mbe Mountains protected area, a forest dominated hilly landscape with sightings of gorillas since 1983 was established a wildlife sanctuary in 2000 [30] . It covers an estimate area of 100 km^2 and has hills rising to elevation of 900 metres. To the northern, western and southern parts of Mbe Mountains are nine villages (namely, Kanyang 1, Kanyang 2, Bamba, Abo Mkpang, Abo Obisu, Abo Ogbagante, Bokalom, Wula 1 and Wula 2) which have led to problems of encroachment by surrounding communities. The Mbe Mountains is an important corridor between AMWS and east of the CRNP (Okwangwo Division). The Mbe Mountains has been managed as a community wildlife sanctuary by the Conservation Association of the Mbe

Mountains (CAMM) established by the earlier mentioned nine communities. The community wildlife sanctuary is managed by CAMM and supported by the CRS Forestry Commission and Wildlife Conservation Society (WCS). As is the case with AMWS, similar challenges from agricultural activeties, poaching, bush burning and fragmentation of existing forest landscapes are key issues of concern in the protected area. [30] confirmed the severity of threats from hunting, farm encroachment, fire and logging as threats to the ecological integrity of the landscape which left unchecked could affect the continued existence of the Cross River Gorilla and other primates present.

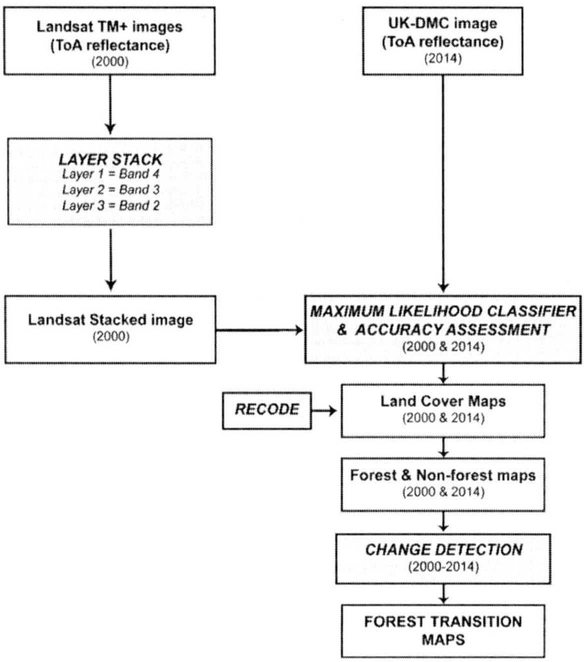

FIGURE 1. Map of study area showing protected areas investigated and an insert map of Nigeria showing Cross Rivers State.

Further to the east of Mbe Mountains, divided by the Okon River is CRNP Okwangwo Division. The CRNP located in south eastern Nigeria consists of two divisions, the Oban division to the south of Cross River state and the Okwangwo division to the north as shown in Figure 1. The Okwangwo division of the CRNP, created in 1991, covers an approximate area of 4000 km^2 and comprises of the former Boshi, Okwangwo, Boshi Extension Forest reserves and some inhabited enclaves (Okwangwo and Okwe 1) [2] (Figure 1). The CRNP Okwangwo division landscape, dominated by mixture of primary and secondary high forests and savanna woodland formations, has a high biodiversity particularly with respect to primates like the Cross River Gorilla [31] . Although the wildlife in the park are in theory under protected status, hunting in the park is widespread and cases of fire in the montane forests on the edge of the Obudu Plateau a reoccurring event [3] . The gorilla habitats in the CRNP (Okwangwo division) are under immense threat from farming encroachment as the population within both enclaves has greatly increased over the years.

2.2. Field Data Collection

The field data on endangered primates (gorillas and chimpanzees) presence across the study area was collected using cyber-tracking technology employed by patrol officers of WCS and CRSFC [32] . The signs of gorilla and chimpanzee presence were based on vocalisation; presence of nests, food remains, dung and direct encounters. The use of cyber tracking based systems was introduced in the Mbe Mountains in 2009 and has since allowed for ease in tracking gorillas across the protected area since [32] . Other sources of endangered primate sightings were obtained from the WCS annual reports available on the official website (http://www.wcsnigeria.org/). The gorilla and chimpanzee presence data from 2011 to 2014 was used for this study. Past studies on gorilla distribution in Cross River observed that combining earlier surveys did not alter the current distribution of the primates, as recent surveys showed that all earlier known areas were still occupied [4] .

2.3. Forest Cover Mapping Using Remote Sensing

The change in the forest landscape across the study area was performed using satellite remote sensing data for two epochs (2000 and 2014). For this study Landsat 7 Enhanced Thematic Mapper (ETM+) with World Reference System path 187, row 056 (10 December 2010) and UK-DMC satellite data (7 January 2014) were used. The Landsat data was downloaded from the United States Geological Survey GLOVIS web-page (http://glovis.usgs.gov/) while the UK-DMC satellite data was supplied by Nigeria Space Agency, NASRA (National Space Research and Development Agency), based on sharing agreement with UK partners (Disaster Monitoring Constellation Company, DMCii). The Landsat image was processed to Level 1 Terrain-corrected, implying it was already ortho-rectified making it a suitable product for direct image-to-image comparison in change detection analysis. Similarly the UK-DMC satellite image was radiometrically and geometrically corrected. Since the images were already radiometrically and geometrically corrected on delivery, the Landsat and UK-DMC images were atmospherically corrected [33] and the digital numbers converted to ToA (i.e. Top of Atmosphere Reflectance) [34] [35] . This conversion process was done to remove all forms of noise present in the satellite images (such as instrument error, errors from view changes during data acquisition and effects of illumination geometry) [34] [36] [37] . A similar approach was implemented on the UK-DMC image used in the study. Before image classification, Landsat bands with same spectral ranges as the UK-DMC image were selected for input during image classification (Figure 2).

For this study, the supervised maximum likelihood classifier (MLC) was applied to each satellite image. A total of six broad classes were used in the study, based on the Intergovernmental Panel on Climate Change land use classification scheme [38] . The six broad classes comprised of forestland, cropland, grassland, wetlands, settlements and other land classes. For purpose of this study theses classes were further re-categorized into two distinct classes namely Forest and Non-forest. The process of image classification and accuracy assessment was performed using independent training and testing data. The training and ground truth data were obtained using a variety of sources namely Google Earth, GPS (global positioning system) data over the study area, historic/recent aerial photographs and visual interpretation of the satellite images.

2.4. Forest Cover Change Analysis

In this study, the IDRISI Selva Land Cover Modeler was used to analyses forest cover change over the study area. Using results of the forest cover maps for 2000 and 2014, forest cover change maps across the protected areas were determined for 2000 to 2014. The forest cover change maps for both time intervals showed deforested, unchanged forest and afforested landscapes classes. In order to calculate, the annual deforestation and forest changes for the two time periods were used.

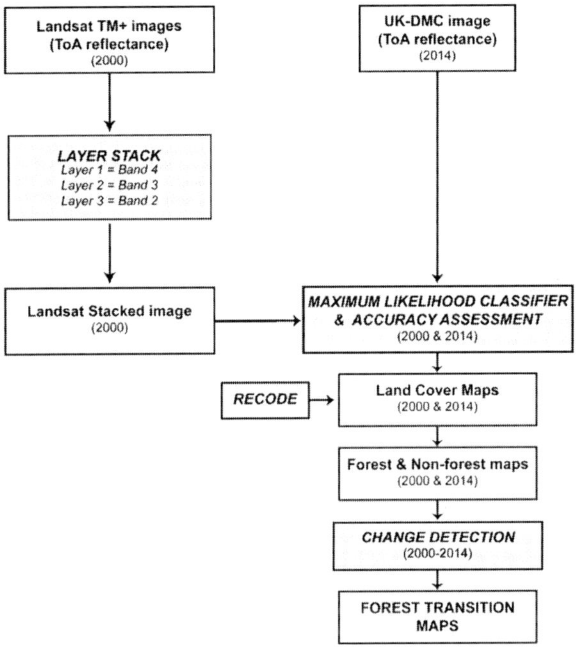

FIGURE 2. Procedure for post classification comparison using Landsat and UK-DMC satellite images.

2.5. Spatial Dynamics of Forest Landscape and Predictive Modeling

The open source software, FRASTATS (version 4.3.1.603), [39] was used to calculate the landscape and class metrics needed to better understand the spatial structure of forest cover classifications. The class metrics used in this study included the number of patches (NP), mean patch area (MPA) and patch density (PD) (which is the number of patches of a patch type divided by the total landscape area).

An important component of this study, with particular emphasis on the management of protected areas and eventual benefits to endangered primate habitats, was predicting the future status forest landscapes across the study area using current trends of forest cover transition. To this end, the Markov Chain Analysis and Cellular Automata Analysis (CA_MARKOV) module in IDRISI Selva was used to model forest cover scenario over a 14-year future projection (i.e. from 2014 to 2028). The Markovian process models the future state of any chosen landscape using the immediate preceding state as the basis for forward prediction [22] . For this study, the forest cover maps for 2000 and 2014 were used to model future predictions of forest cover over a 14-year period. Using the forest cover maps for 2000 and 2014 as inputs, the MARKOV module was used to produce the transition probability matrices and conditional probability images. To implement the CA_MARKOV module, the transition area text file, transition suitability image and 2014 forest map were used to generate a 14-year forward simulation of forest cover. The Cellular Automata iteration was specified as 14, allowing the model to be based on 1-year increments. The default cellular Automata filter of 5 × 5 was used for the study.

2.6. Maximum Entropy Modeling of Endangered Primates in Cross River State

The Maximum Entropy Distribution (Maxent) [26] [27] was used to model the presence of endangered gorilla and chimpanzee primates across the study area. Maxent is a machine learning programme that utilises maximum entropy to estimate species distribution probabilities' using the presence only data and environmental variables (categorical or continuous) [26] . For this study endangered primates' presence data across the study area and environmental variable (continuous and categorical) layers were used as inputs into the modeling process. The open source software Maxent version 3.3.3k (http://www.cs.princeton.edu/~schapire/maxent/) was used to generate estimates of probabilities for primates presence varying from 0 to 1 (where 0 is the lowest and 1 the highest probability respectively). The presence data, Global Positioning System (GPS) positions of gorilla and chimpanzee nests and sightings for 2009 and 2013 were sample layer inputs used in Maxent to model potential distribution of environmental suitability for the endangered primates. Topographical data used in modeling was taken from the Shuttle Radar Topography Mission Digital Elevation Model (SRTM DEM). The SRTM DEM was downloaded from the Global Land Cover Facility (GLCF) website (http://glcfapp.glcf.umd.edu:8080/esdi/).

The modeling of environmental suitability for endangered primates using the presence data of gorilla and chimpanzee sightings and nests and three key environmental variables namely, DEM (for terrain information), Slope map (derived from the DEM) and forest cover map of the study area. The recommended default values for maximum iteration numbers (50), convergence threshold (100,000) and regularization values (10,000) were used. The random test percentage of value used was 30 (meaning that 30% of the presence data was used for validation while 70% for training of the model). For analysis a threshold was specified to help perform binary predictions of habitat suitability conditions with suitable habitats predicted above such threshold and unsuitable indicated as below. In this study a convergence threshold of 0.00001 was used with maximum iterations of 500. The ASCII raster outputs of predicted primate habitats were further analyzed using ESRI ArcGIS software version 10.2 [40] .

3. RESULTS AND DISCUSSIONS

3.1. Classification Accuracy Results

The overall land cover classification accuracy results were approximately 89 percent (2000) and 91 percent (2014) respectively. Table 1 presents the classification accuracy reports for the classifications performed. The classification accuracy results were within VCS REDD Methodologies specification [41] .

3.2. Changes in Forest Cover and Spatial Structure

The classification result shows that total forest cover for the entire study area was 152,509 hectares in 2000 and 129,339 hectares in 2014 (Table 2). The total area of forest cover across the project area landscape declined from 86% in 2000 to 73% in 2014. Overall, the four protected areas experienced a general decline in the spatial cover of forest landscape for the 14-year period investigated (Table 2).

TABLE 1. Accuracy assessment results showing user accuracy (UA), producer accuracy (PA) and overall accuracy (OA) generated for classified images.

Land Use	2000 (ETM+)		2014 (UK-DMC)	
	UA (%)	PA (%)	UA (%)	PA (%)
Forest	87.1	95.7	91.8	88.2
Farmland	88.6	87.8	90.9	86.2
Grassland	93.6	96.7	91.7	94.3
Wetland	86.4	88.4	90	87.8
Settlement	100	54.6	93.3	96.6
Other land	100	100	86.7	100
OA (%)	*89.4*		*90.8*	

The forest cover transition results indicated that overall deforestation for the study area from 2000 to 2014 was 34,620 hectares (Table 3). The total forest landscape that remained unchanged and afforested for the same period were 117,730 hectares and 11,574 hectares respectively. In terms of percentage landmass, 13.3% of the total forest landscape was deforested over the entire study area. Amongst the four protected areas, the most affected by deforestation were AMWS and Mbe Mountains with annual deforestation rates of 2.6% and 2.2%, followed by CRNP (1.1%) and ARFR (1.5%). The AMWS and Mbe Mountains, both known to have gorilla and chimpanzee sightings, are under immense pressure from illegal logging, encroachment farming and bush burning. The recent WCS annual report (2014) on AMWS and Mbe Mountains identified rise in the number of hunting incidents in the area, a major threat to the existence of endangered primate population. Furthermore, the close proximity of surrounding communities to the primate habitats has further heightened the pressures of farming encroachment and hunting activities, (particularly in forms of wire snares across the sanctuary). These factors have greatly contributed to the rise in deforestation and consequent disturbance of the gorilla and chimpanzee habitats across the protected area.

Using spatial metrics measures (NP, PD and MPA) over time, the changes in spatial structure over the 14-year period was investigated for the study area The NPs for the entire study area increased from 1255 in 2000 to 2843 in 2014. Similarly, the PD of the entire study area increased from 0.48 in 2000 to 1.12 in 2014 while MPA decreased from 122 hectares (in 2000) to 46 hectares (in 2014) (Table 2). The implication of these results was that the forest landscape across the study area experienced high levels of forest fragmentation over the 14-year period. Results of spatial structure dynamics for all four protected areas indicated that there were significant rises in NP and PD and a general decline in MPA (particularly for Mbe Mountain which experienced a sharp MPA decline from 4472 hectares in 2000 to 48 hectares in 2014). The effects of large decline in forest cover and increased levels of forest fragmentation over the 14-year period are indications of habitat loss, decreased connectivity of remaining vegetation (which prevents migration and movement of primates) and reduction in forest patch sizes (unsuitable for primate habitation). Other key deforestation drivers that have ravaged the Afi-Mbe- Okwangwo landscape include over-exploitation of non-timber forest products and occurrence of forest fires (a common practice in seasonal farming activities in the lowland forests) [1] [3] [4]. The combined effects of the afore-mentioned drivers of deforestation are serious threats to the conservation of the forest ecosystem, upon which existing primates depend for survival. Hence, the major challenge of the Cross River State Government and Nigeria at large is the establishment of suitable measures to combat or possibly mitigate these effects of deforestation across existing protected areas.

TABLE 2. Results of spatial structure changes for forest landscape in Afi-Mbe-Okwabgwo protected areas from 2000 to 2014.

Zone	Forest Cover (ha)		Number of Patches		Patch Density (no. per 100 ha.)		Mean Patch Area (ha)	
	2000	2014	2000	2014	2000	2014	2000	2014
Project area	152,509	129,339	1255	2843	0.48	1.12	121.52	45.49
AMWS	9715	7418	76	474	0.35	2.18	127.83	15.65
ARFR	27,033	22,315	113	338	0.13	0.39	239.23	66.02
Mbe Mountains	8944	6728	2	141	0.01	0.67	4475.24	47.72
CRNP (Okwangwo)	61,828	53,614	80	496	0.06	0.37	772.85	108.09

TABLE 3. Results of forest cover transition for the Afi-Mbe-Okwangwo forest landscape from 2000 to 2014.

Zone	Deforested (ha)	Unchanged forest (ha)	Afforested (ha)	Deforested (%)	Unchanged forest (%)	Afforested (%)	Annual deforestation rate (%)	Annual rate of change in forest cover (%)
Project area	34,620	117,730	11,574	13.26	45.09	4.46	1.48	0.51
AMWS	3229	6486	933	14.84	29.80	4.29	2.61	0.84
ARFR	5559	21,474	830	6.43	24.85	0.96	1.48	0.59
Mbe Mountains	2311	6634	94	10.92	31.35	0.45	2.19	0.88
CRNP (Okwangwo)	9623	52,196	1388	7.17	38.89	1.03	1.06	0.44

To this end, the Nigerian Government is actively involved in the United Nations REDD+ (Reducing Emissions from Deforestation and Forest Degradation) programme aimed at preserving existing forest landscape across the country, particularly forest dependent communities [42] . As part of the REDD+ programme in Nigeria, portions of Afi-Mbe-Okwangwo protected areas have been assigned as one of the three main REDD pilot sites for Cross River State. The implementation of the REDD+ programme across the state would facilitate sustainable management of forest landscape and its resources whist promoting wildlife conservation. Other approaches include creating public awareness on the need to refrain from killing endangered primates and avoidance of human activities such as agricultural encroachment, forest fires, illegal logging and over-exploitation of non-timber forest products. Figure 3 shows the forest cover (2000 and 2014) and forest transition (2000-2014) maps of the entire study area.

3.3. Forest Cover Forward Simulation

Figure 4(a) shows the spatial extent of forest cover across the entire study area, including the four protected areas from 2000 to 2028. The results show that for the entire study area, forest landscape experienced a gradual decline over the 28 year period. The spatial extent of forest decline from 2000 to 2014 and 2014 to 2028 were 23,170 hectares and 41,338 hectares respectively. The results of forest landscape decline were representative of approximately 15% and 27% of the earlier forest cover in 2000. Figure 4(c) and Figure 4(d) show the 2014 forest cover map and 2028 forward simulated forest cover map of the study area. At a glance there's an obvious decline in the spatial coverage of forest cover in 2028.

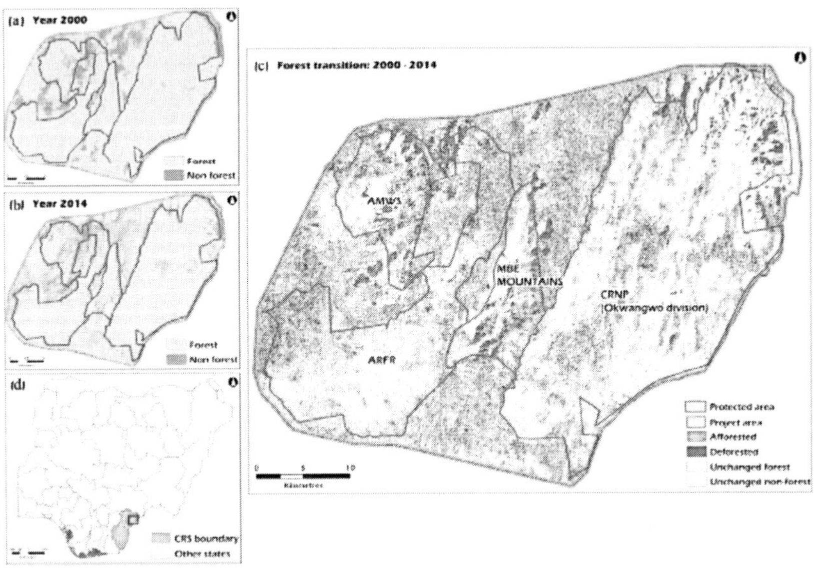

FIGURE 3. Map of study area showing: (a) Forest cover for 2000; (b) Forest cover for 2014; (c) Forest transition for 2000 to 2014 indicating deforested, unchanged forests, afforested and unchanged non forest landscapes across Afi-Mbe-Okwangwo protected areas in Cross River State; and (d) Insert map of Nigeria showing the study area.

The results of forest cover change analysis for the entire study area based on the 2014 and 2028 predicted forest maps showed that 28,121 hectares of forest cover would be deforested. The spatial extent of unchanged forest and afforested landscape between 2014 and 2028 would be 101,183 and 9988 hectares respectively. These predicted results are based on the current trend of forest cover change driven by current forest management practices and pressures from human activities (such as farming encroachment, bush burning and illegal logging). Figure 4(c) shows forest landscape transition highlighting deforested, unchanged and afforested landscapes (for 2000 to 2014 and 2014 to 2028) across the protected areas. Overall, the results show that forest cover transition across the study area is prone to high levels of deforestation particularly outside and within the protected areas (Figure 4(d)). Given the current trend of forest transition, measures should be put in place to mitigate associated threats to endangered primates and human population.

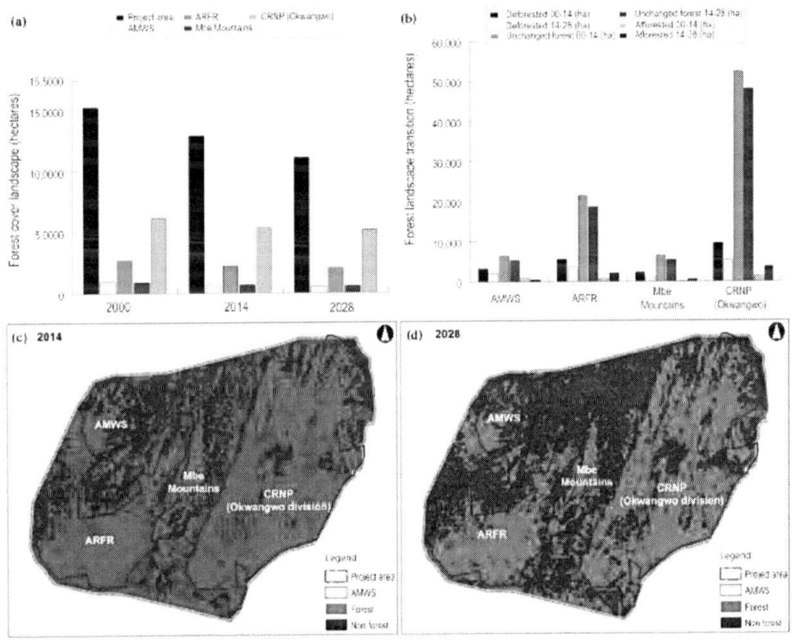

FIGURE 4. Results of forest transition analysis showing (a) Graph of forest cover landscape for 2000, 2014 and 2028 across study area; (b) Graph of forest transition across study area for 2000-2014 and 2014-2028, indicating deforested, unchanged forest and afforested landscape; (c) Forest cover map of 2014; and (d) 2028 predicted forest cover map generated using Markov module.

3.4. Habitat Suitability Modeling of Endangered Primates Using Maxent

The omission and predicted area plots of the endangered primates (Figure 5) show how testing and training omissions and predicted areas vary with the choice of cumulative threshold. Figure 5(a) and Figure 5(b) showed that the omission on test samples (for both primates) was a very good match to the predicted omission rate. Figure 5(c) and Figure 5(d) showed the receiver operating curve (ROC) for both training and test data used in the modeling process. The red (training) line shows the "fit" of the model to the training data while the blue (testing) indicated the fit of the model to the testing data (note, this is the real test of the models predictive power). Given that the blue line in graph 5(c) and graph 5(d) were towards the top left of the graph, it was an indication that the model was better at predicting the presence contained in the test sample of the data [43] . The accuracy of the predictive modeling for both primates was determined by the test area under the ROC (AUC).

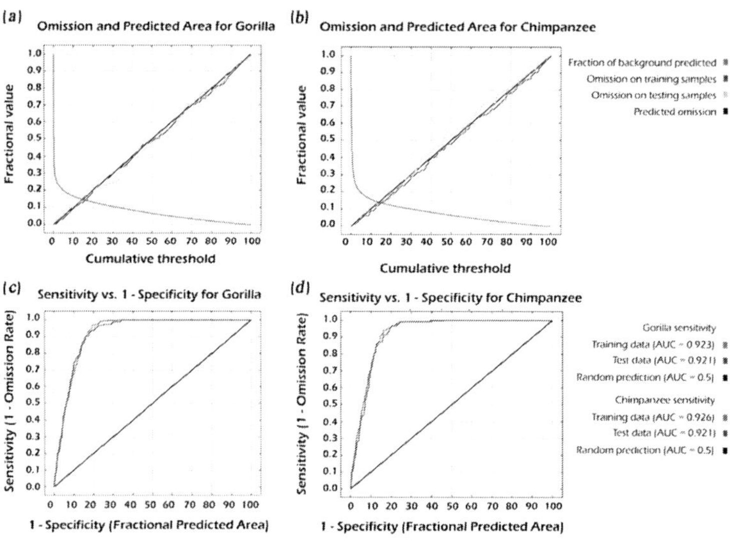

FIGURE 5. Results of statistical analysis of Maxent modeling for endangered species (Cross River Gorilla and Nigeria-Cameroon Chimpanzee) habitat suitability results of forest transition analysis showing (a) Graph of forest cover landscape for 2000, 2014 and 2028 across study area; (b) Graph of forest transition across study area for 2000-2014 and 2014-2028, indicating deforested, unchanged forest and afforested landscape; (c) Forest cover map of 2014; and (d) 2028 predicted forest cover map generated using Markov module.

The AUC and standard deviation for the modeled prediction of gorillas and chimpanzees habitats were 0.921 (standard deviation 0.006) and 0.921 (standard deviation 0.007) respectively. The gorilla suitability habitat modeling was performed using 359 presence records for training and 153 for testing. For the chimpanzee habitat suitability modeling 217 presence records was used for training and 93 for testing. The environmental variables used in the model included the study area DEM, classified forest map and slope map datasets.

During the training phase of species prediction modeling Maxent software keeps track of environmental variables that contribute to fitting the model. Consistent with ecological knowledge the model predicted that Cross River Gorilla and chimpanzees are found mostly in forested areas (particularly tropical high and montane forests) in mountainous locations across the site. Figure 6(a) and Figure 6(b) show the predicted habitat suitability map outputs of endangered gorilla and chimpanzee primates generated using presence data and key environmental variables in Maxent modeling software. The most important variables in order of importance for both primates as indicated by the Maxent model were DEM, Slope derived map and Forest map of study area.

3.5. Effect of Deforestation on Predicted Primates Habitat

The raster datasets showing predicted primate habitats generated using Maxent model were imported into ArcGIS software for further analysis and visualization. The raster outputs were reclassified to binary images showing only predicted primate habitats with possible probability of occurrence (i.e. 0.4 - 1.0) (Figure 6(c)). The binary images were digitized and compared with current and historic primate presence data. The results indicated that the total coverage of predicted suitable habitat for endangered primates over the entire study area was 30,940 hectares (54.4% situated in CRNP—Okwangwo division, 29.4% in AMWS, 14.3% in Mbe Mountains and 1.9% in ARFR). The spatial coverage of modeled primate habitats for both gorillas and chimpanzees for AMWS, ARFR, Mbe Mountains and CRNP (Okwangwo division) were approximately 9094 hectares, 573 hectares, 4428 hectares and 16,846 hectares respectively. These results would prove vital in the effective management and development of conservation plans for endangered primate habitats across the study area required by wildlife conservation organizations and the State government.

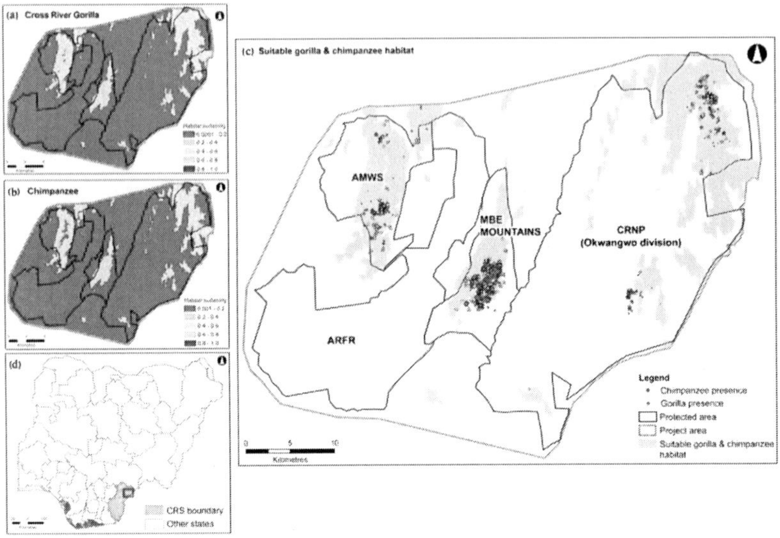

Figure 6. Map of study area showing Maxent predicted habitat suitability maps for (a) Cross River Gorilla and (b) Nigeria-Cameroun Chimpanzee; (c) Combined suitable gorilla and chimpanzee habitats and primate presence data from 2011 to 2014 across protected areas in the Afi-Mbe-Okwangwo forest landscape and (d) Insert map of Nigeria showing location of study area.

Following the modeling of suitable gorilla and chimpanzee habitats, the authors investigated the effects of deforestation trends (2000-2014) on the spatial coverage of predicted primate habitats across the study area. Using spatial analysis techniques in ArcGIS, the forest transition map (Figure 3(c)) and predicted primate habitat map (Figure 6(c)) were reclassified to binary images, each having two classes (deforestation map with 1 = deforested landscape & 0 = others landscape and primate habitat map with 1 = primate habitats & 0 = others habitats). Using the afore-mentioned binary raster, a map showing deforestation hotspots across the predicted primate habitat were generated (Figure 6(c)). The results indicated that approximately 6468 hectares of predicted primate habitats were affected by deforestation (21% of the predicted forested primate habitat) in 2014. The effects of deforestation and forest fragmentation on the predicted primate habitats reflects the poor management of forest landscape in the protected areas across the study area, justifying the urgent need for existing management practices across the protected areas to be reviewed

3.6. Implication of Forest Transition and Human Activities on Primates Habitat and Conservation

The high levels of forest fragmentation across the four protected areas investigated in this study are largely dependent on the current prevailing agricultural practices across the region. The presence of intensified pressure from agricultural encroachment by surrounding villages and communities are key catalysts to the disturbance of existing gorilla and chimpanzee habitats. Over the years the combined effects of forest fragmentation and deforestation has resulted in extensive restriction of primates to isolated and highly fragmented forest patches. This pattern of forest transition combined with uncontrolled hunting in protected areas such as the CRNP (Okwangwo division) has resulted in rapid loss of endangered primates (particularly gorillas and chimpanzees) and in drastic cases extinction of some wildlife species such as the grey-cheeked mangabey (Lophocebus albigena) and crowned guenon (Cercopithecus pogonias) [2] [44] . The problems of hunting and agricultural encroachment are further heightened by the presence of large enclaves in CRNP (Okwangwo division). Similar occurrences are observed in AWFS, ARFR and Mbe Mountains.

In the Mbe Mountains, a key catalyst asides from the above mentioned factors, is the presence of the Ikom—Obudu highway which traverses across Ago Ogbagante, Kayang-1, Kayang-2, Wula-1 and Wula-2 communities. Results of past studies [2] [4] and current interviews have noted a distinct pattern in forest clearance along the Ikom—Obudu highway for agricultural practice by surrounding communities.

The devastating impact of deforestation and other human related activities on remaining forest landscape and wildlife (in particular endangered primates) justifies the need for Cross River State (CRS) Government (in collaboration with the National Government) to put in place mitigative measures and policies/laws that would protect all wildlife and forests. In this regard, the CRS government on February 2014 established a mobile court under the CRSFC charged with the responsibility of prosecuting any individual found guilty of killing or hunting any animal (or endangered species) as listed in the CRS Forestry Commission Law No. 3 of 2010 [45] . Another avenue explored by CRS is the implementation of the UN-REDD+ programme. In 2010 CRS Government demonstrated its commitment to the REDD+ programme by enacting a deforestation moratorium, which is still in force till date, banning all forms of logging in the state [46] [47] .

4. CONCLUSIONS

The study has demonstrated the use of remote sensing data and techniques combined with various modeling approaches to understand effects of forest cover transitions on the spatial structure and prediction of suitable primate habitats. The results show that suitable habitats for gorillas and chimpanzees are under immense pressure from deforestation and high levels of forest fragmentation. Forest change analysis results showed that the overall extent of deforestation across the study area (2000 to 2014) was 34,620 hectares with annual deforestation rate of 1.5%. The protected area with the highest annual deforestation rates were AMWS (2.6%) and Mbe Mountains (2.2%), both prominent for gorilla and chimpanzee sightings and nests. The key drivers of deforestation across the study area were identified as illegal logging, over exploitation of non-timber forest products, hunting, farm encroachment, and forest fires. Further investigations as to changes in forest landscape structure revealed presence of high forest fragmentation levels, factors that had adverse negative effects on endangered primate and other wildlife habitat. As a means of further understanding the effects of the current forest cover transition trends, a forward simulation of forest landscape was performed for a 14-year forward prediction. The results showed that if the current trend of forest cover changes continued, a total of 28,121 hectares of forest cover would be deforested in the 2028 (approximately 16% of the total landmass of the entire study area). Using Maxent, suitable primate habitats were modeled and the total coverage of predicted habitat was estimated to be 30,940 hectares (54.4% situated in CRNP—Okwangwo division, 29.4% in AMWS, 14.3% in Mbe Mountains and 1.9% in ARFR). Further analysis as to the extent of deforestation in the predicted primates habitat revealed that 6468 hectares of forest cover had been deforested in 2014 (an estimated 21% of the predicted primate habitats). These results show that suitable primate habitats (particularly for gorillas and chimpanzees) are under immense pressure from deforestation and forest fragmentation.

This study has demonstrated that using remote sensing analysis combined with predictive modeling tools (such as Markovian and Maxent models) provides better understanding of forest transition and corresponding effects on primate habitats across parts of Africa. The approach presented in this study is a cost effective and time saving method of accurately predicting endangered primate habitats and understanding the dynamics of forest change and structure in the area. The information derived using this approach would assist conservation organizations, wildlife managers and the Government of Nigeria in making decisions and implementing policies that would help protect the forests and conserve endangered primates and other wildlife that are forest dependent for their existence.

In conclusion, the continued conservation of forests across African countries is paramount, as gorillas and chimpanzees alike depend on forest for survival. Hence, the only way to ensure their survival is to fight for the conservation of remaining forests. Such efforts will require support from government of African nations, Conservation Organizations, World Donor

Agencies and most important the assistance of communities where such forests and endangered primates are found. The place of financial independence of forest protected areas can be promoted through ecotourism and infrastructure development [48] . Furthermore, implementing simple non-cap- ital intensive and sustainable projects on a local scale can promote self-sufficiency across forest dependent communities in Nigeria.

ACKNOWLEDGEMENTS

The authors would like to appreciate Wildlife Conservation Society (Cross River) for providing the relevant field data on endangered primates across the study area and National Space Research and Development Agency (NASRDA) for making available the UK-DMC satellite image of Cross River State for image analysis.

REFERENCES

1. Al-Quraishi, A.M.F., Guang, D.H. and Jian, G.C. (2004) Land Degradation Detection, Mapping, and Monitoring in the Northwestern Part of Hebei Province, China, Using RS and GIS Technologies. Map Asia Conference 2004, Beijing, 26-29 August 2004.
2. Angelsen, A. and Kaimowitz, D. (1999) Rethinking the Causes of Deforestation: Lessons from Economic Models. The World Bank Research Observer, 14, 73-98.
3. Baldwin, R.A. (2009) Use of Maximum Entropy Modeling in Wildlife Research. Entropy, 11, 854-66.
4. Bergl, R.A., Brenda, J.B, Nsubuga, A. and Vigilant, L. (2008) Effects of Habitat Fragmentation, Population Size and Demographic History on Genetic Diversity: The Cross River Gorilla in a Comparative Context. American Journal of Primatology, 70, 848-59.
5. Bergl, R.A., Ymke, W., Aaron, N., Dunn, A., Inaoyom, I., Jacqueline, L. Sunderland-Groves and Oates, J.F. (2012) Remote Sensing Analysis Reveals Habitat, Dispersal Corridors and Expanded Distribution for the Critically Endangered Cross River Gorilla Gorilla gorilla diehli. Oryx, 46, 278-89.

6. Boyd, D.S., Foody, G.M. and Ripple, W.J. (2002) Evaluation of Approaches for Forest Cover Estimation in the Pacific Northwest, USA, Using Remote Sensing. Applied Geography, 22, 375-92.
7. Chander, G., Markham, B.L. and Helder, D.L (2009) Summary of Current Radiometric Calibration Coefficients for Landsat MSS, TM, ETM+, and EO-1 ALI Sensors. Remote sensing of Environment, 113, 893-903.
8. Cowlishaw, G. and Dunbar, R.I.M (2000) Primate Conservation Biology. University of Chicago Press, Chicago.
9. Elith, J., Graham, C.H., Anderson, R.P., Dudík, M., Ferrier, S., Guisan, A., Hijmans, R.J., Huettmann, F., Leathwick, J.R., Lehmann, A., Li, J., Lohmann, L.G., Loiselle, B.A., Manion, G., Moritz, C., Nakamura, M., Nakazawa, Y., Mc Overton, J., Peterson, A.T. and Phillips, S.J. (2006) Novel Methods Improve Prediction of Species' Distributions from Occurrence Data. Ecography, 29, 129-51.
10. ESRI (2014) ArcGIS Desktop: Release 10.2. Environmental Systems Research Institute. http://www.esri.com/
11. FAO (2010) Global Forest Resources Assessment 2010. Food and Agriculture Organization of the United Nations, Rome.
12. Geist, H.J. and Lambin, E.F. (2001) What Drives Tropical Deforestation. LUCC Report Series No. 4, 116 p.
13. Harcourt, A.H., Stewart, K.J. and Inahoro, I.M. (1989) Gorilla Quest in Nigeria. Oryx, 23, 7-13.
14. Harcourt, A.H. (1996) Is the Gorilla a Threatened Species? How Should We Judge? Biological Conservation, 75, 165-176.
15. Harcourt, A.H. and Doherty, D.A. (2005) Species-Area Relationships of Primates in Tropical Forest Fragments: A Global Analysis. Journal of Applied Ecology, 42, 630-637.
16. Huang, C., Wylie, B., Yang, L., Homer, C. and Zylstra, G. (2002) Derivation of a Tasselled Cap Transformation Based on Landsat 7 At-Satellite Reflectance. International Journal of Remote Sensing, 23, 1741-1748.
17. Imong, I., Eban, J. and Mengnjo, C. (2014) Using Technology to Save Gorillas in the Mbe Mountains. Gorilla Journal, 48, 15-17.
18. Iqbal, M. (2012) An Introduction to Solar Radiation. Elsevier, Amsterdam.
19. Irish, R.R. (2000) Landsat 7 Science Data User's Handbook. National Aeronautics and Space Administration, Report 430-15-01-003-0.
20. Ite, U.E. and Adams, W.M. (1998) Forest Conversion, Conservation and Forestry in Cross River State, Nigeria. Applied Geography, 18, 301-314.

21. Kehinde, A.L., Akande, J.A. and Ntabe, E. (2009) Forest Stakeholder's Awareness of Reduced Impact Logging (RIL) in Nigeria and Cameroon. Journal of Horticulture and Forestry, 1, 176-181.
22. Lu, D., Mausel, P., Brondizio, E. and Moran, E. (2002) Assessment of Atmospheric Correction Methods for Landsat TM Data Applicable to Amazon Basin LBA Research. International Journal of Remote Sensing, 23, 2651-2671.
23. Mambo, J. and Archer, E. (2007) An Assessment of Land Degradation in the Save Catchment of Zimbabwe. Area, 39, 380-391.
24. McGarigal, K., Cushman, S.A., Neel, M.C. and Ene, E. (2002) FRAGSTATS: Spatial Pattern Analysis Program for Categorical Maps.
25. Mengnjo, C. (2015) Wildlife Conservation Society Annual Report 2014—Cross River National Park (Okwangwo Divison). Wildlife Conservation Society. www.wcsnigeria.org
26. Midha, N. and Mathur, P.K. (2010) Assessment of Forest Fragmentation in the Conservation Priority Dudhwa Landscape, India Using FRAGSTATS Computed Class Level Metrics. Journal of the Indian Society of Remote Sensing, 38, 487-500.
27. Mondal, P. and Southworth, J. (2010) Evaluation of Conservation Interventions Using a Cellular Automata-Markov Model. Forest Ecology and Management, 260, 1716-1725.
28. Morgan, B.J., Adeleke, A., Bassey, T., Bergl, R., Dunn, A., Fotso, R., Gadsby, E., Gonder, K., Greengrass, E., Koulagna, D.K., Mbah, G., Nicholas, A., Oates, J., Omeni, F., Saidu, Y., Sommer, V., Sunderland-Groves, J., Tiebou, J. and Williamson, E. (2011) Regional Action Plan for the Conservation of the Nigeria-Cameroon Chimpanzee (Pan troglodytes ellioti). IUCN/SSC Primate Specialist Group and Zoological Society of San Diego.
29. Myers, N., Mittermeier, R.A., Mittermeier, C.G., Da Fonseca, G.A.B. and Kent, J. (2000) Biodiversity Hotspots for Conservation Priorities. Nature, 403, 853-858.
30. Oates, J.F., Bergl, R.A. and Linder, J.M. (2004) Africa's Gulf of Guinea Forests: Biodiversity Patterns and Conservation Priorities. Advances in Applied Biodiversity Science, Vol. 6, Conservation International, Washington DC.
31. Oates, J.F., White, D., Gadsby, E.L. and Bisong, P.O. (1990) Conservation of Gorillas and Other Species. In: Caldecott, J.O., Oates, J.F. and Ruitenbeek, H.J., Eds., Cross River National Park (Okwangwo Division):

Plan for Developing the Park and Its Support Zone, Appendix 1, WWF-UK, Goldalming.
32. Oates, J.F., Sunderland-Groves, J., Bergl, R., Dunn, A., Nicholas, A., Takang, E., Omeni, F., Imong, I., Fotso, R. and Nkembi, L. (2007) Regional Action Plan for the Conservation of the Cross River Gorilla (Gorilla gorilla diehli). IUCN/SSC Primate Specialist Group and Conservation International, Washington DC.
33. Oyebo, M., Bisong, F. and Morakinyo, T. (2015) A Preliminary Assessment of the Context for REDD in Nigeria. http://www.gcftaskforce.org/meeting/documents/Nigeria_National_Update_GCF_2011.pdf
34. Phillips, S.J., Anderson, R.P. and Schapire, R.E. (2006) Maximum Entropy Modeling of Species Geographic Distributions. Ecological Modelling, 190, 231-259.
35. Phillips, S.J., Dudík, M. and Schapire, R.E. (2004) A Maximum Entropy Approach to Species Distribution Modeling. Proceedings of the 21st International Conference on Machine Learning, Banff, 4-8 July 2004, 83.
36. Plumptre, A.J., McNeilage, A., Hall, J.S. and Williamson, E.A. (2002) The Current Status of Gorillas and Threats to Their Existence at the Beginning of a New Millennium. In: Taylor, A.B. and Goldsmith, M.L., Eds., Gorilla Biology: A Multidisciplinary Perspective, Cambridge University Press, Cambridge, 414-431.
37. Radespiel, U. and Bruford, M.W. (2014) Fragmentation Genetics of Rainforest Animals: Insights from Recent Studies. Conservation Genetics, 15, 245-260.
38. Ravilious, C., Kapos, V., Osti, M., Bertzky, M., Bayliss, J.L., Dahiru, S. and Dickson, B. (2010) Carbon, Biodiversity and Ecosystem Services: Exploring Co-Benefits. Nigeria: Preliminary Results. UNEP-WCMC, Cambridge, UK.
39. Saatchi, S., Buermann, W., ter Steege, H., Mori, S. and Smith, T.B. (2008) Modeling Distribution of Amazonian Tree Species and Diversity Using Remote Sensing Measurements. Remote Sensing of Environment, 112, 2000-2017.
40. Schoneveld, G.C. (2014) The Politics of the Forest Frontier: Negotiating between Conservation, Development, and Indigenous Rights in Cross River State, Nigeria. Land Use Policy, 38, 147-162.

41. Shoch, D., Eaton, J. and Settelmyer, S. (2011) Project Developer's Guidebook to VCS REDD Methodologies. Conservation International, Arlington.
42. Smith, P., Bustamante, M., Ahammad, H., Clark, H., Dong, H., Elsiddig, E.A., Haberl, H., Harper, R., House, J., Jafari, M., Masera, O., Mbow, C., Ravindranath, N.H., Rice, C.W., Robledo Abad, C., Romanovskaya, A., Sperling, F. and Tubiello, F. (2014) Agriculture, Forestry and Other Land Use (AFOLU). In: Edenhofer, O., Pichs-Madruga, R., Sokona, Y., Farahani, E., Kadner, S., Seyboth, K., Adler, A., et al., Eds., Climate Change 2014: Mitigation of Climate Change. Contribution of Working Group III to the Fifth Assessment Report of the Intergovernmental Panel on Climate Change, Cambridge University Press, Cambridge, United Kingdom and New York, 811-922.
43. Fielding, A.H. and Bell, J.F. (1997) A Review of Methods for the Assessment of Prediction Errors in Conservation Presence/Absence Models. Environmental Conservation, 24, 38-49.
44. Souza, C., Firestone, L., Silva, L.M. and Roberts, D. (2003) Mapping Forest Degradation in the Eastern Amazon from SPOT 4 through Spectral Mixture Models. Remote Sensing of Environment, 87, 494-506.
45. CRS (2014) Cross River State of Nigeria Gazette. C.R.M. Court, Ed., No. 1, Ministry of Information, Calabar.
46. Suter, J. and Oates, J. (2000) Sanctuary in Nigeria for Possible Fourth Subspecies of Gorilla. Oryx, 34, 71-71.
47. Takada, T., Miyamoto, A. and Hasegawa, S.F. (2010) Derivation of a Yearly Transition Probability Matrix for Land-Use Dynamics and Its Applications. Landscape Ecology, 25, 561-572.
48. Ezebilo, E.E. and Mattsson, L. (2010) Socio-Economic Benefits of Protected Areas as Perceived by Local People around Cross River National Park, Nigeria. Forest Policy and Economics, 12, 189-193.

7

CHAPTER

STATISTICAL CHARACTERIZATION OF BARE SOIL SURFACE MICRORELIEF

Edwige Vannier[1], Odile Taconet[1], Richard Dusséaux[1] and Olivier Chimi-Chiadjeu[1]

[1] Université de Versailles Saint-Quentin en Yvelines, LATMOS/IPSL/CNRS, Guyancourt, France

1. INTRODUCTION

Because the soil surface occurs at the boundary between the atmosphere and the pedosphere, it plays an important role for geomorphologic processes. Roughness of soil surface is a key parameter to understand soil properties and physical processes related to substrate movement, water infiltration or runoff, and soil erosion. It has been noted by many authors that most of the soil surface and water interaction processes have characteristic lengths in millimeter scales. Soil irregularities at small scale, such as aggregates, clods and interrill depressions, influence water outflow and infiltration rate. They undergo rapid changes caused by farming implements, followed by a slow evolution due to rainfall events. Another objective of soil surface roughness study is investigating the effects of different tillage implements on soil physical properties (friability, compaction, fragmentation and water content) to obtain an optimal crop emergence. Seedbed preparation focuses on the creation of fine aggregates and the size distribution of aggregates and clods produced by tillage operations is frequently measured.

Active microwave remote sensing allows potential monitoring of soil surface roughness or moisture retrieving at field scale using space-based Synthetic Aperture Radars (SAR) with high spatial resolution (metric or decametric). The scattering of microwaves depends on several surface characteristics as well as on imagery configuration. The SAR signal is very sensitive to soil surface irregularities and structures (clod arrangement, furrows) and moisture content in the first few centimeters of soil (depending on the radar wavelength). In order to link the remote sensing observations to scattering physical models as well as for modelling purpose, key features of the soil microtopography should be characterized. However, this characterization is not fully understood and some dispersion of roughness parameters can be observed in the same field according to the methodology used. It seems also, that when describing surface roughness as a whole, some information related to structured elements of the microtopography is lost.

2. DATA SAR – ELECTROMAGNETIC MODELS AND SOIL MODELLING: POSITION OF PROBLEM

Synthetic Aperture Radars allow the study of agricultural bare soils by measuring the backscattered coefficient. The backscattered signal depends on the surface roughness, on the electrical permittivity which is closely linked to the soil moisture and texture. It varies depending on the frequency and polarization of the transmitted electromagnetic wave and on the incidence angle [1-4]. The electromagnetic modeling is a valuable tool for radar data inversion to intend to characterize the roughness and / or the humidity of agricultural soils. The electromagnetic models based on the Maxwell's equations and the boundary conditions can be classified into two categories: analytical models [5] and numerical models [6-9].

The analytical models are based on physical approximations which reduce the applicability domain. But, these models have an obvious interest: the electromagnetic field scattered by the surface and the coherent and incoherent intensities are given by analytical formulas arousing physical interpretations. The first-order small-perturbation method (SPM1) is only valid for surfaces with small roughness versus the wavelength [10] and the Kirchhoff approximation (KA) is applicable to surfaces with long correlation length [11-12]. The small slope approximation (SSA1) has an extended domain of applicability which includes the domains of the two previous approaches [13]. It is likewise for the integral equation model (IEM) which became the most quoted and implemented rough surface scattering model in the field of radar remote sensing for earth observation [14]. These four analytical models assume that the soil can be represented by a stationary random process. The stationarity is not clearly established for agricultural soils, especially in the presence of very marked furrows [15]. These models require the knowledge of the autocorrelation function. In addition, the models KA, SSA and IEM require the one-point and two-point height probability distributions. The calculations are typically conducted with the Gaussian probability density function. The Gaussian character is observed for seedbeds with furrows slightly marked but to our knowledge, it is not the case for ploughed soils [9, 16]. Furthermore, tilled agricultural soils are anisotropic surfaces and we have to take into account the quasi-periodicity of the surface in the estimation of the coherent and incoherent components of the backscattered signal [17]. Fractal and multi-fractal approaches for describing agricultural soils were also implemented. Some works allowed extending electromagnetic analytical models to the fractal character of soils [18-19]. So, the remote sensing studies using these analytical models consider a global description of the scattering surface and do not take into account their structuring objects such as soil clods, holes and aggregates [20]. Then, for the analysis of radar data, it is essential to extend the analytical models by taking into account these structuring objects which needs the understanding of their statistical properties and their influence on the autocorrelation function.

The second class of electromagnetic models relies on numerical methods for solving Maxwell's equations and boundary conditions [6-8]. These models are so called exact if no physical approximation is made. These models do not provide an analytical solution of the problem and require high computational times. This is a major drawback for the inversion of radar data. Advanced numerical methods have been developed over the past two decades to reduce computational time [21]. For these numerical methods, the scattered intensities and the backscatter coefficient are estimated over results of a finite set of surface realizations. The estimation error depends on the number and size of surfaces. These numerical models associated with the Monte Carlo method rely on surface generating algorithms. Agricultural bare soils have structuring objects contributing to the backscattered signal. Except for a few references [22], the surface generators are based on a global approach and do not take into account the objects characterizing agricultural soils. To improve the analysis of radar data, it makes sense to move forward in the statistical description of the structuring objects, to develop new surface generating algorithm in accordance with the statistical properties of these objects and to interface them with numerical electromagnetic models.

3. STATISTICAL AND SPATIAL ANALYSIS OF SOIL ROUGHNESS

Several types of surface roughness may be recognized from micro relief variations to larger level variations representing the landscape. At millimetre or centimetre-scales, soil surface roughness mainly results from the breakdown of the superficial soil layer by tillage operations. Depending on the soil properties, in particular texture, organic matter content and moisture state of the soil material, and on the tillage tool, different aggregate sizes will be produced (figure 1).

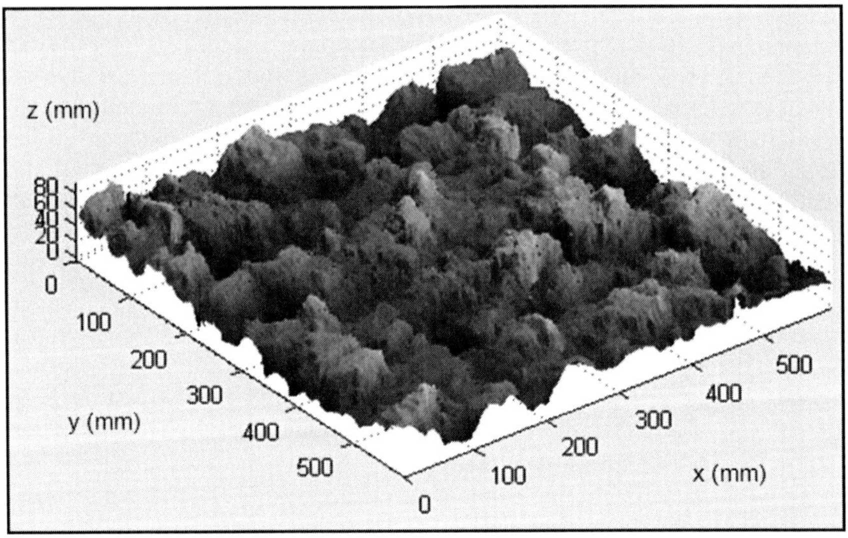

FIGURE 1. Microtopography of a seedbed surface

3.1. Background

Two approaches are commonly investigated for statistical and spatial analysis of soil roughness at these scales. In the first one, the surface roughness is characterized as a whole by its autocorrelation function model or by its variogram and is summarized by statistical indices: root mean square of heights, correlation length, roughness exponent or Hurst coefficient, parameter Z_s, tortuosity or other specific indices [9, 15, 23-31]. Fractal and multifractal description are investigated too [32-34]. In the second one, local and morphological aspects of the surface are brought into focus such as aggregates, clods and mound-and-depression patterns [20, 28, 35-43].

Direct ground field measurement of surface roughness can be performed by either contact techniques [44-46] or non-contact techniques such as close range photogrammetry [15, 26, 47-50] or laser scanning [32, 51-53]. Some authors made comparative studies [30, 54-56] and a review of the measurement techniques can be found in [3]. Although close range photogrammetry and laser scanning enable retrieving a 3D digital elevation model (DEM) of the soil surface, many authors still carry on characterizing soil surface roughness by extracting 1D profiles on the 3D measurement [50, 57-58]. This can be explained by the fact that 1D statistical indices are widely spread and serve as reference, since measurement of 1D profiles was developed in the first place. Nevertheless, it traduces also a lack of microrelief modelling. In the field of soil moisture retrieval from SAR imagery, Lievens and Verhoest, [59], are led to propose methods to circumvent surface roughness parameterization problems.

Blaes and Defourny [30] and Dusséaux et al. [9] suggest using a bidimensional autocorrelation function to characterize bare soil agricultural surfaces. Bidimensional approach is more appropriate to the measurement of roughness anisotropy related to multiscale processes (row structured field, clod arrangement). In order to improve statistical characterization of bare soil surface microrelief, we highly focus on 3D DEM interpretation study. The proposed approach is detecting and localizing clods and big aggregates by segmentation methods, modelling them by a simplified geometric form, estimating the statistical distribution of the modelling structuring object parameters and then generate numerical surfaces by setting structuring objects onto a substrate according to statistical laws.

3.2. 3D-DEM interpretation

Millimetric DEMs of the soil surface were seldom used for detecting and localizing big aggregates and clods. In 3D, an individual clod (or big aggregate) is seen as a local bump standing out with slightly sharp boundaries (figure 2). It is a recognizable structure presenting rather medium or high level values, with a high gradient on the edges, but inside color level values are non-homogenous. Two newly built methods have been presented [41- 42, 60].

Addressing the problem of detecting and localizing clods requires also an evaluation methodology. We propose several qualification tools based on the knowledge of a ground truth serving as reference. The detection performance can be evaluated by estimating the sensitivity *sen* and specificity *spe*, defined as follows:

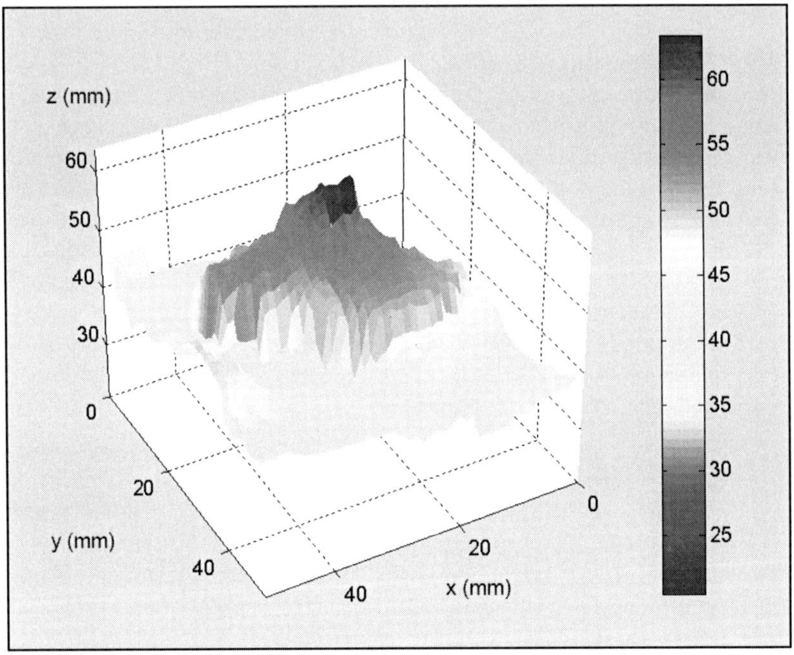

FIGURE 2. Clod representation in 3D

There is usually a compromise between sensitivity and specificity for a detection method. Setting of the method parameters can be made by constructing the receiver operating curve (ROC), eventually introducing weight reflecting the cost of a bad decision [61].

where I_i is the set of pixels located inside the identified boundaries of the clod *i*, I_{ti} the set of pixels located inside the true boundaries of the clod *i*, and [Math Processing Error] the cardinal of I_i. O_v is null either when a true clod is not identified or when a false positive has been detected. The ideal case corresponds to an overlap rate equal to 1 for all the clods. This ideal case leads to a cumulative distribution function (cdf) equal to 0 for an overlap rate strictly lower than 1 and equal to 1 for an overlap rate upper or equal to 1 [60].

Now, in [41], clod identification is obtained by merging information on local maxima enhanced at different scales by wavelet transform. Clods are successfully identified by their summit and two diameters on several kinds of soil surfaces. Boundary points are estimated by thresholding the height slope. The detection performance was evaluated with the help of a soil scientist on a controlled surface made in the laboratory as well as on real seedbed and ploughed surfaces, made by tillage operations in an agricultural field. A minimum size, set here to 7 mm of diameter depending on the DEM horizontal resolution of 1mm, is required for aggregates, in order to prevent a high false positive rate. Indeed, one has to be able to count the aggregates. The identifications of the clods were in good agreement, with an overall sensitivity of 84% and a specificity of 94%. A limitation of this method is that it does not provide the clod boundaries. An example of clods identification on a part of seedbed surface is provided on figure 3. We can see that most of the mounds have been identified. A small one, with center coordinates located near (50, 15) has not been detected.

In order to get the clods boundaries, a contour-based method has been proposed by Taconet et al. [42]. The clod boundary is determined by hierarchy of closed elevation contours going through the pixels of highest gradient values. This second method gives satisfactory results with a clod identification sensitivity of 75% and specificity of 95%. Figure 4 displays the clods localization obtained in the same part of seedbed as on figure 3.

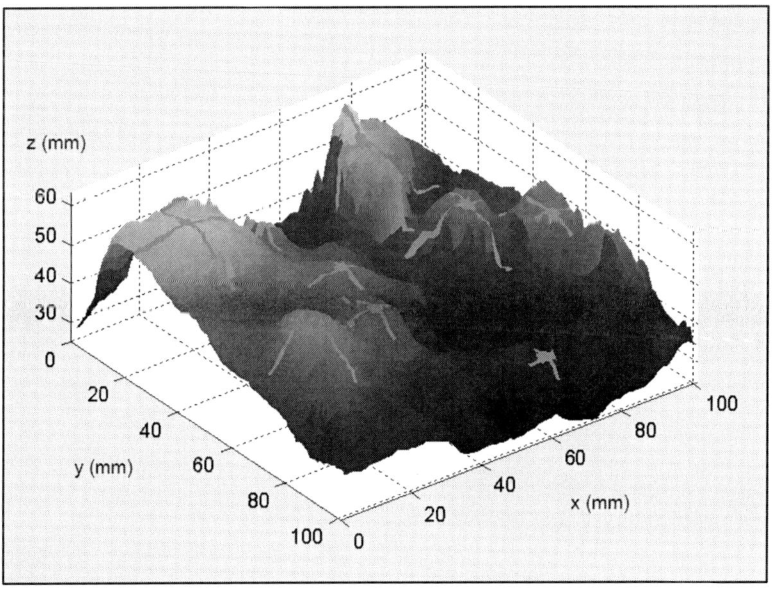

FIGURE 3. Automatic localization of clods by wavelet-based method

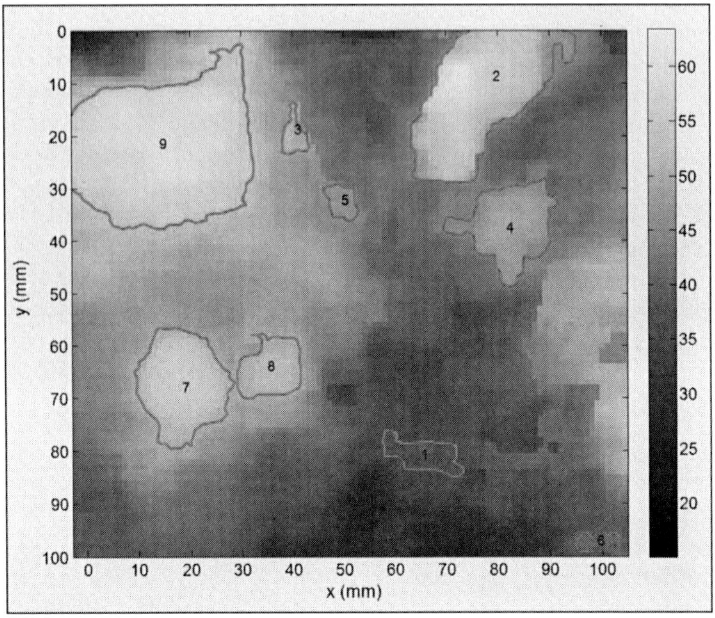

FIGURE 4. Automatic localization of clods by contour-based method

There is a good agreement with the identification of clods by wavelet-based method. The small clod that was not detected is now identified (clod number 5 in figure 4), but two clods are not detected anymore, illustrating the lower sensitivity of the contour-based method.

The contour-based method has been tested until now, on freshly tilled seedbed and artificial soils which both contain mostly distinct clods and aggregates. Furthermore, defining clod boundary as an elevation contour requires that the substrate of clods has little variation, which is not the case of ploughed surfaces. Indeed, the automatically retrieved clod boundaries are generally underestimated and lay at a constant height, which is not very realistic.

Another approach is developing a more classical method of image segmentation based on mathematical morphology, the watershed-based method [62]. Such a method has been introduced in [63]. It relies on a transformation of the DEM image in order to produce a pseudo elevation image of greater dynamic to enhance the clods and on mixing information on heights and gradients so that the local minima correspond to clods bases. Figure 5 shows the cdf of overlap rates estimated for the contour-based and the watershed-based methods on a seedbed DEM.

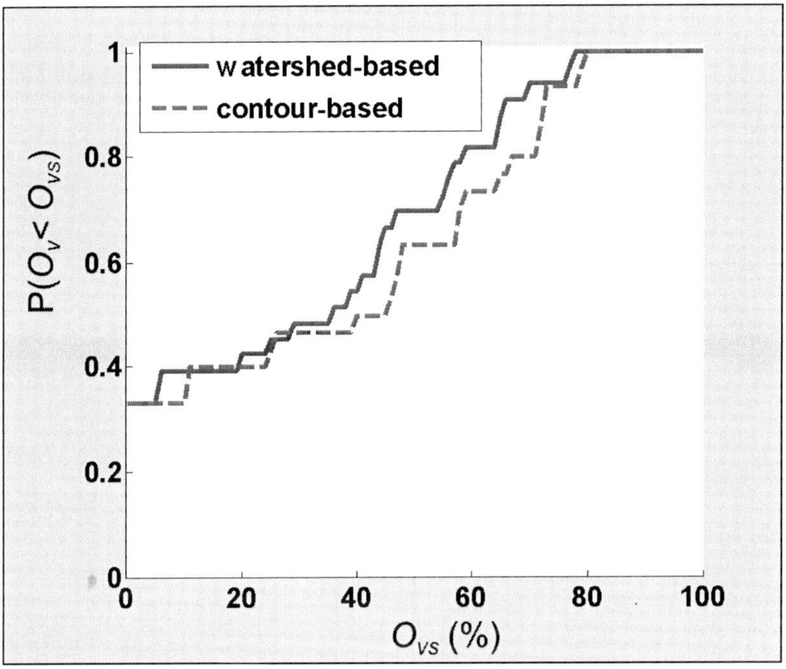

FIGURE 5. Cdf of overlap rates for watershed- and contour-based segmentation methods

The cdfs have the same value at the origin, showing that both methods have the same rates of false detected and not detected clods. The clods localization is all the better than the cdf is low. The watershed-based method cdf being higher than the contour-based one, it shows that, for this surface, the classical image segmentation method does not perform better than the newly built method. However, it does not have the drawback of horizontal clods bases. Thus identifying clods and big aggregates on a millimeter DEM remains a complex problem. Also, a specific approach of contour moving based on the minimization of a cost function related to clods detection properties has been introduced [64]. Based on the characterization of clods, as recognizable structures presenting rather medium or high level values, with a high gradient on the edges and low height values on the boundary, we have defined four criteria, linked to heights and gradients again, that characterize the clod boundary. Contrary to [42], the hypothesis of a horizontal base does not exist anymore. The minimization is performed by a meta-heuristic inspired from the simulated annealing algorithm [65]. Let us notice that each identification method can be used to initialize the contour moving algorithm, however, until now, the results are better if the initial contours lay inside the reference boundaries. The overlap rates can be enhanced by up to 10%.

4. MODELLING OF SOIL CHARACTERISTICS

As clods and aggregates are irregular shaped objects, a way to define relevant indices in size and shape is representing them by a simple approached form which matches closely their horizontal and vertical extents. In numerical generation of soil surfaces, clods and aggregates are usually modelled by half circles [22] or half spheres [38] or part of spheroids [66] put onto a substrate which may be a plane surface, an exponentially correlated random surface, or a surface representing furrows. In [66], the spheroids are equal-sized and regularly placed on a square grid. In [38], an envelope-Boolean surface generation model was introduced, where half spheres were set onto a plane according to the desired statistic. Two artificial surfaces were manually created by distributing aggregates randomly on a plane. Then structuring elements were generated by a Poisson point process reproducing the occurrence of aggregates size classes. It was suggested that the use of other primary function than half-sphere, like ellipsoids, could improve the performance of the surface generation process. In [22], the soil surface is described as a substrate, modelled by a 1D profile of Gaussian distribution of heights, with an exponential autocorrelation, and clods, modelled by half circles, randomly and independently set onto the substrate. Size of clods and distance between them obey to statistical laws that are not estimated from observations.

The choice of a modelling shape is dependent on the type of soil considered in an experiment. In [67], soil compaction extent is described by a half-ellipse. In [20], it is proposed to fit an ellipse to clod horizontal boundary contour and a half cosine function to the height shape. A semi-ellipsoid can also be suitable to fit the height shape of clods and aggregates. Mathematical frame of these modelling is presented hereafter.

4.1. Equivalent Ellipse Modelling Clod Base Contour

The equivalent ellipse denotes the best fitting ellipse of a plane closed contour. Let O be the barycentre of the considered clod, R = (Oxy) its barycentric reference and Re = (OXY) the inertia principal reference. Estimation of the three shape parameters (semi-major axis length *a*, semi-minor axis length *b* and orientation angle [Math Processing Error], (i.e. the angle between the major axis supported by (OX) and the horizontal axis supported by (Ox)) is based on the method of geometric moments [68]. The central second order moments in the barycentric reference, where the summation extends over all the pixels located within the clod boundary contour in the barycentric reference, where the summation extends over all the pixels located within the clod boundary contour. Using these geometric moments, the length of the semi-axes (a, b) and the orientation angle.

4.2. Semi-Arch Of Cosine Modelling Clod Height Shape

The height of the semi-arch of cosine is estimated by least squares minimization of the sum of squared residuals:

4.3. Half Ellipsoid Modelling Clod Height Shape

a and b semi axes of the equivalent ellipse and h, height of the half ellipsoid, that can be estimated by identifying the sum of squares of clod heights and the sum of squares of model heights, in order to keep the second order moment. An example of clod modelling is illustrated on figure 6.

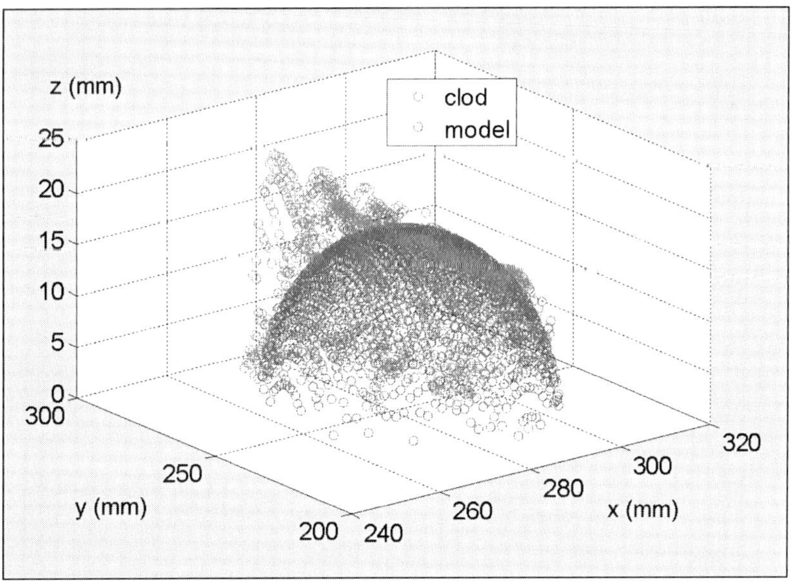

FIGURE 6. Clod modelling by a half ellipsoid

Extracting the identified clods and setting them onto a plane gives a surface with mainly the same autocorrelation function as the same surface where clods are replaced by their half ellipsoid model (see figure 7). This illustrates the goodness of fit, when modelling clods and big aggregates by half ellipsoids. Let us notice that the finite length of surfaces causes oscillations in the autocorrelation functions.

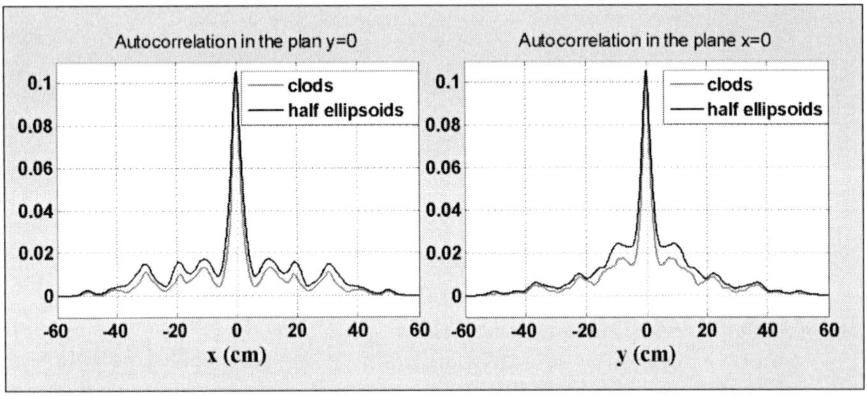

FIGURE 7. Autocorrelation functions of clods set onto a plane and fitted

4.4. Generation Process

Analysis of DEM plots of a soil surface allows deriving the statistical distributions characterizing structuring object parameters. As an example, when analyzing a seedbed surface containing 350 clods, we find that the centers of gravity G and the orientation angles [Math Processing Error] of big aggregates and clods can be assumed uniformly distributed. With the ellipsoid model, the variables a, b and h are not independent. Therefore intermediate variables have to be introduced, such as horizontal and vertical compression factors. Using Pearson's chi-squared tests of independence, we found a plausible independence of a and f_h and of a and f_v. Then, b and h would be derived from f_h and f_v using equations (12) and (13).

Generating realistic numerical surfaces is expected for soil science studies as well as for numerical electromagnetic codes used in remote sensing studies. In order to generate a cloddy surface, several parameters have to be estimated:

- Number of objects (according to the desired density) N_o;
- Probability density functions (pdf) of a, f_h and f_v, θ and G;
- Shape of the substrate, where to put the structuring objects onto.

Then, the proposed generation process is composed of three main steps (see figure 8).

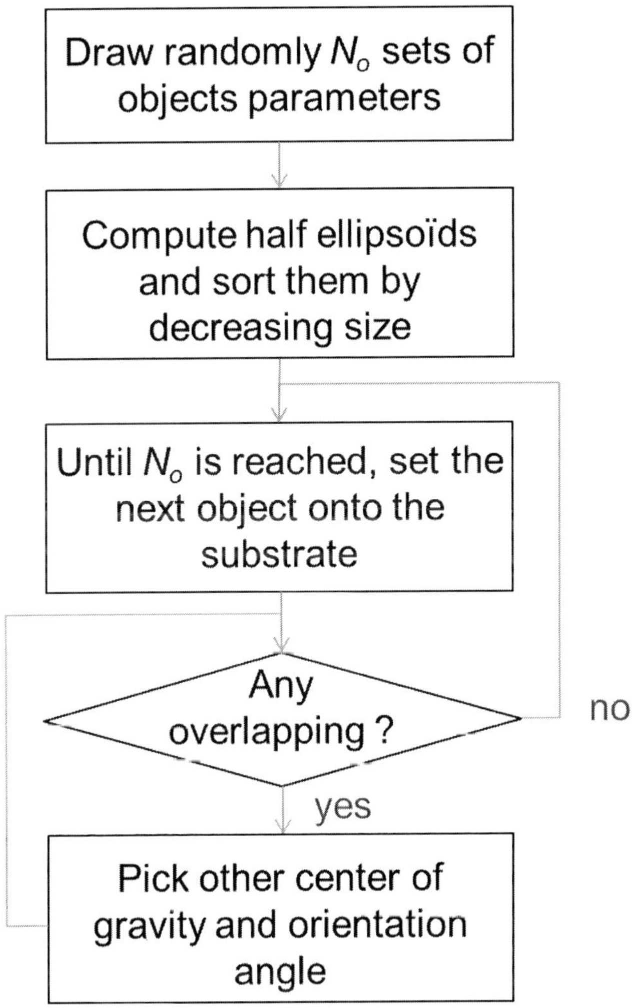

FIGURE 8. Generation process flow chart

First step is drawing randomly N_o sets of objects parameters. Variables a, f_h and f_v as well as θ and G are drawn according to their pdf. Variables b and h are deduced from f_h and f_v using equations (10) and (11). Second step is computing half ellipsoids and sorting them by decreasing size. Last step is setting each structuring object onto the chosen substrate only if it does not overlap with a former object; drawing another couple of center of gravity and orientation angle otherwise.

5. RESULTS AND DISCUSSION

This generation process has been applied, until now, only on a natural freshly tilled seedbed surface. We estimated Gamma pdf for a and Bêta pdf for f_h and f_v. A correction has to be made for a, since the minimum diameter of our aggregates is 7 mm and not zero. Therefore, the computing is performed with the translated variable $a - a_{min}$, where a_{min} designates the minimum value of a. Figures 9 to 11 show the cdf of observed and modelled major semi-axis a and compression factors f_h and f_v. The small difference between the model and the data in each case, less than 3%, confirms the theoretical distributions. The data modeling is also validated by Pearson's chi-squared tests. Figure 12 shows a generated surface obtained by placing half ellipsoids onto a plane.

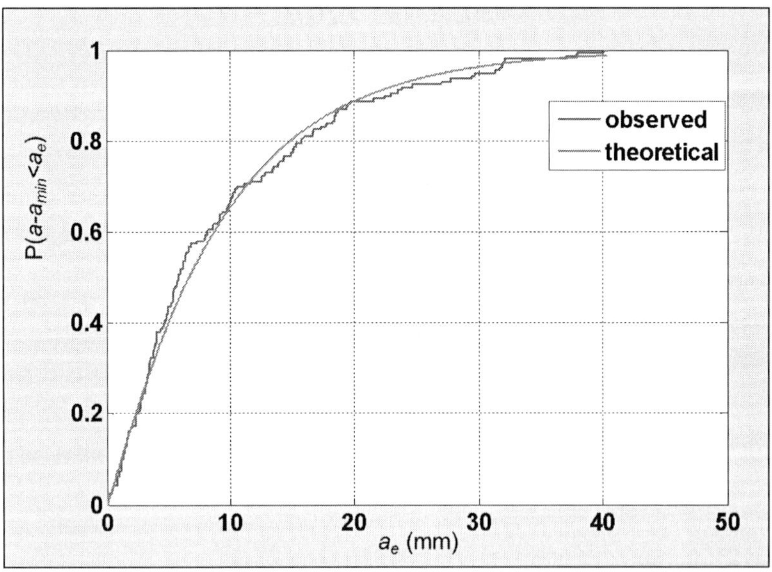

FIGURE 9. Cdf of half ellipsoid major semi-axis a fitted by Gamma function

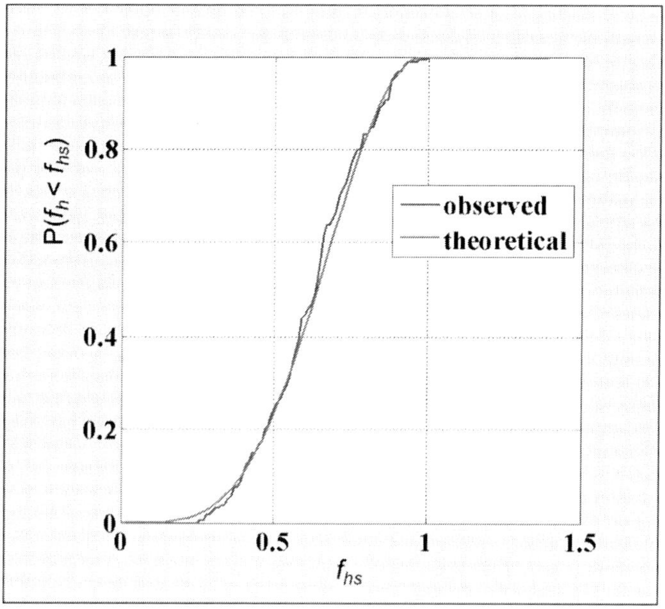

FIGURE 10. Cdf of horizontal compression factor f_h fitted by Bêta function

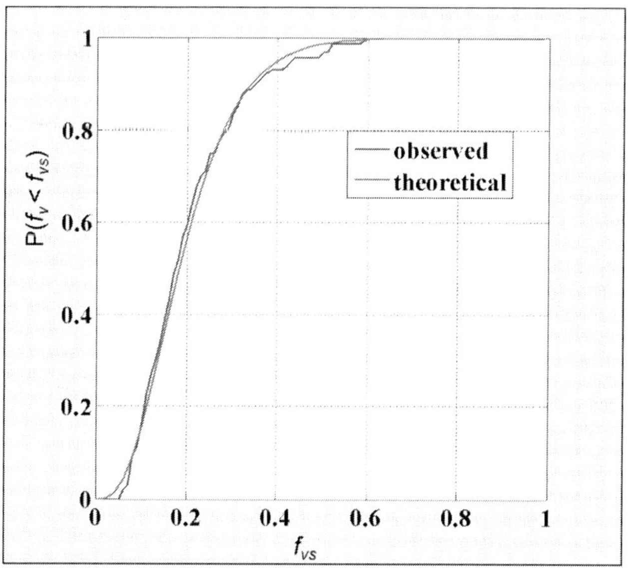

FIGURE 11. Cdf of vertical compression factor f_v fitted by Bêta function

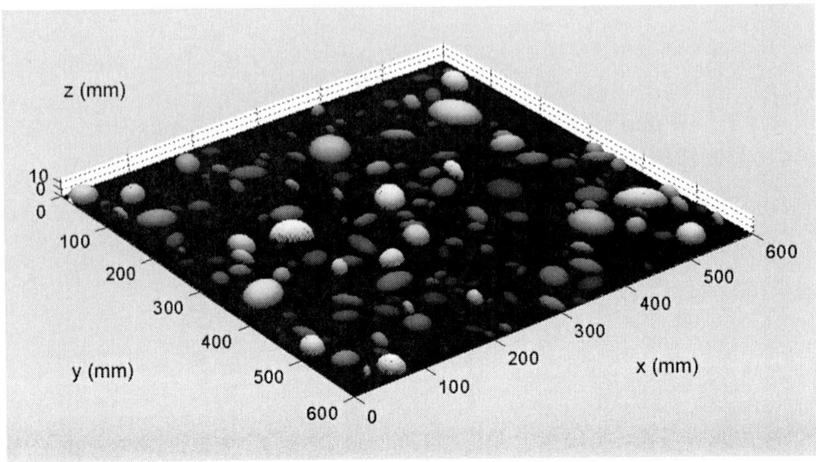

FIGURE 12. Generated numerical surface

The generated soil surface on figure 12 is a simplified surface representing only some irregularities of the soil. Aggregates and clods cover about 30% of a surface. A more complex model would include also the holes or depressions and an appropriate substrate. The segmentation methods presented in this chapter have the potential to detect also the holes. Characterization of a more realistic substrate than a plane is a complex problem that has not been solved until now.

In order to validate the exposed generation process, the autocorrelation function of generated surface was compared to that of the surface obtained by setting the half ellipsoid modelling the automatically identified clods and aggregates of the seedbed surface under study onto a plane. Both autocorrelation functions can be seen on figure 13.

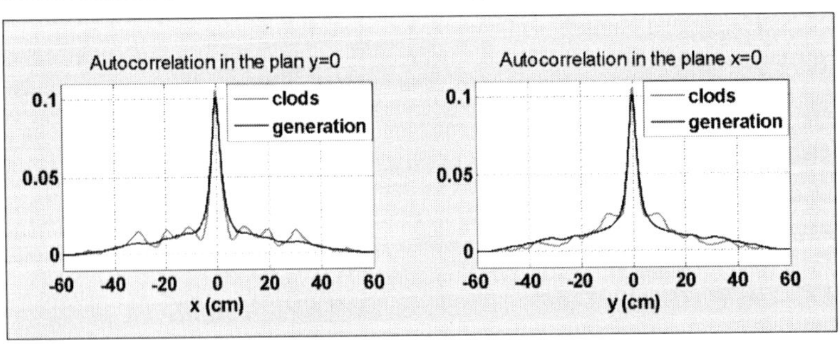

FIGURE 13. Autocorrelation functions of half ellipsoids set onto a plane by fitting detected clods or by generation process

One can see that the autocorrelation function of generated surface reproduces satisfactorily the shape of the autocorrelation function of the modeled clods set onto a plane, which is encouraging for future work. The black curve is a mean autocorrelation function, estimated on 15 realizations of generated surface. It has thus no more oscillations.

Recently, some authors have recalled the need to characterize the soil structure [69] or to define appropriate parameterization of the microtopography [70]. Addressing the arrangement of aggregates and clods contributes to these objectives. The proposed approach for modelling soil characteristics and generating soil surface is robust, as shown by the results. It relies on a geometric model of clods, which gives a physical basis to the characterization of surface microrelief. It represents also a progress in modelling compared to preceding works [22, 38, 66]. Future work should take into account soil depressions in relation to aggregates and clods.

6. CONCLUSION

Taking into account the structured elements when studying roughness effects is more and more brought into focus. The soil irregularities at millimeter or centimeter scales have an impact on hydrological processes, as they influence water outflow and infiltration rate. They have also an influence on remote sensing studies, by producing scattering and shadowing effects. The proposed approaches in this chapter are detecting and characterizing some of the soil surface irregularities that are clods and big aggregates. Little studies were dedicated to this topic. For seedbed-type surfaces, a contour-based identification method enables to automatically retrieve the clods and big aggregates on a 3D DEM with a sensitivity of 75% and a specificity of 95%. Then modelling clods by semi-ellipsoïds is suitable and allows reproducing the same autocorrelation function. The cdf of the semi-ellipsoïds parameters were fitted with relative errors less than 3%. Numerical soil surfaces were successfully generated by placing semi-ellipsoïds onto a plane, according to the statistical distributions. These recent works were applied to a limited number of soils and should be extended to more soils. Interest of a good surface roughness description is allowing generating realistic numerical surfaces. Such surfaces are then very useful for soil science studies as well as for electromagnetics codes used in remote sensing.

REFERENCES

1. Holah N., Baghdadi N., Zribi M., Bruand A., King C. Potential of ASAR/ENVISAT for the characterization of soil surface parameters over bare agricultural fields. Remote Sens. Environ. 2005, 96: 78-86.
2. Rahman M.M., Moran S.M., Thoma D.P., Bryant R., Sano E.E., Holifield Collins C.D, Skirvin S., Kershner C., B.J. Orr, A derivation of roughness correlation length for parameterizing radar backscatter models. Int. J. Remote Sens. 2007, 28: 1907-1920.
3. Verhoest N.E.C., Lievens H., Wagner W., Alvarez-Mozos J., Moran M.S., Mattia F. On the soil roughness parameterization problem in soil moisture retrieval of bare surfaces from Synthetic Aperture radar. Sensors. 2008, 8: 4213-4248.
4. Lievens H., Verhoest N.E.C., De Keyser E., Vernieuwe H., Matgen P., Alvarez-Mozos J., De Baets B. Effective roughness modelling as a tool for soil moisture retrieval from C- and L-band SAR. Hydrology Earth System Sciences. 2009, 15: 151-162.
5. Elfouhaily T.M., Guérin C.A. A critical survey of approximate scattering wave theories from random rough surfaces. Waves Random Media. 2004, 14: R1-10.
6. Warnick K.F., Chew, W.C. Numerical simulation methods for rough surface scattering. Waves Random Media. 2001, 11: R1-R30.
7. Saillard M., Sentennac, A. Rigorous solutions for electromagnetic scattering from rough surfaces. Waves Random Media. 2001, 11: R103-R137.
8. Dusséaux, R., Aït Braham K., Granet G. Implementation and validation of the curvilinear coordinate method for the scattering of electromagnetic waves from two-dimensional dielectric random rough surfaces. Waves Random Complex Media. 2008, 18: 551-570.
9. Dusséaux R., Vannier E., Taconet O., Granet G. Study of backscatter signature for seedbed surface evolution under rainfall – Influence of radar precision. Progress in Electromagnetics Research 2012 ; 125: 415-437.
10. Thorsos E.I., Jackson D.R., The validity of the perturbation approximation for rough surface scattering using a Gaussian roughness spectrum. J. Acoust. Soc. Am. 1989, 86: 261-277.

11. Beckmann P., Spizzichino A. The Scattering of Electromagnetic Waves from Rough Surfaces. 1963. Pergamon, New-York.
12. Thorsos, E.I.The validity of the Kirchhoff approximation for rough surface scattering using a Gaussian roughness spectrum, J. Acoust. Soc. Am. A. 1988, 82: 78-92.
13. Voronovich G. Wave Scattering from Rough Surfaces. 1994. Springer, Berlin.
14. Fung A. K. Microwave scattering and emission models and their applications. 1994. Artech House, Boston.
15. Taconet O., Ciarletti V. Estimating soil roughness indices on a ridge-and-furrow surface using stereo photogrammetric. Soil and Tillage Research 2007; 93: 64-76.
16. Chimi O., Caractérisation probabiliste et synthèse de surfaces agricoles par objets structurants à partir d'images haute résolution. PhD thesis. Univerty of Versailles Saint-Quentin en Yvelines. 2012
17. Mattia F. Coherent and incoherent scattering from tilled soil surfaces. Waves Random Complex Media. 2011, 21: 278-300.
18. Franceschetti G., Migliaccio M., Riccio D. An electromagnetic fractal-based model for the study of fading. Radio Sci. 1996, 31: 1749–1759.
19. Guérin C.A., Holschneider M., Saillard M. Electromagnetic scattering from multi-scale rough surfaces. Wave Random Media 1997: 331-349.
20. Taconet O., Dusséaux R., Vannier E., Chimi-Chiadjeu O. Statistical description of seedbed cloddiness by structuring objects using digital elevation models. Computers and Geosciences 2013; 60: 117-125.
21. Tsang L., Kong J.A., Ding K.H., Ao C.O. Scattering of electromagnetic waves – Numerical Simulations. Wiley-Interscience, New York: 2001.
22. Zribi M., Le Morvan A., Dechambre M., Baghdadi N. Numerical backscattering analysis for rough surfaces including a cloddy structure. IEEE Transaction Geosciences Remote Sensing 2010; 48 (5): 2367-2374.
23. Currence H. D., Lovely W. G. The analysis of soil surface roughness. Transaction ASAE 1970; 13: 710-714.
24. Römkens M.J.M., Wang, J.Y. Effect of tillage on surface roughness. Transactions of the American Society of Agricultural Engineers 1986; 29: 429-433.
25. Bertuzzi P., Rauws G., Courault D. Testing roughness indices to estimate soil roughness changes due to simulated rainfall. Soil and Tillage Research 1990; 17: 87-99.

26. Helming K., Roth C. H., Wolf R., Diestel H. Characterization of rainfall-microrelief interactions with runoff using parameters derived from digital elevation model (DEMs). Soil Technology 1993; 6: 273-286.
27. Takken I., Govers G. Hydraulics of interrill overland flow on rough, bare soil surfaces. Earth Surface Processes and Landforms 2000; 25: 1387-1402.
28. Darboux F., Davy P., Gascuel-Odoux C., Huang C. Evolution of soil surface roughness and flowpath connectivity in overland flow experiments. Catena 2001; 46: 125-139.
29. Zribi M., Dechambre M. A new empirical model to retrieve soil moisture and roughness from C-band radar data. Remote sensing of environnement 2002; 84: 42-52.
30. Blaes X., Defourny P. Characterizing bidimensional roughness of agricultural soil surfaces for SAR modelling. IEEE Transactions on geoscience and remote sensing 2008; 46: 4050-4061.
31. Croft H., Anderson K., Kuhn N.J. Reflectance anisotropy for measuring soil surface roughness of multiple soil types. Catena 2012; 93: 87-96.
32. Huang C., Bradford J.M. Application of a laser scanner to quantify soil microtopography. Soil Science Society of America Journal 1992; 56: 14-21.
33. Vidal Vázquez E., Vivas Miranda J.G., Paz González A. Characterizing anisotropy and heterogeneity of soil surface microtopography using fractal models. Ecological Modelling 2005 ; 182: 337-353.
34. Vidal Vázquez E., García Moreno R., Miranda J. G. V., Díaz M.C., Saa Requejo A., Paz Ferreiro J., Tarquis A.M. Assessing soil surface roughness decay during simulated rainfall by multifractal analysis. Nonlinear Processes in Geophysics 2008; 15: 457-468.
35. Moldenhauer W.C. Influence of rainfall energy on soil loss and infiltration rates: II. Effect of clod size distribution. Soil Science Society of America Journal 1970; 34: 673–677.
36. Ambassa-Kiki R., Lal R. Surface clod size distribution as a factor influencing soil erosion under simulated rain. Soil and Tillage Research 1992; 22: 311-322.
37. Sandri R., Anken T., Hilfiker T., Sartori L., Bollhalder H. Comparison of methods for determining cloddiness in seedbed preparation. Soil and Tillage Research 1998; 45: 75–90.

38. Kamphorst E.C., Chadœuf J., Jetten V., Guérif J. Generating 3D soil surfaces from 2D height measurements to determine depression storage. Catena 2005; 62: 189-205.
39. Arvidsson J., Bölenius E. Effects of soil water content during primary tillage - laser measurements of soil surface changes. Soil and Tillage Research 2006; 90: 222–229.
40. Keller T., Arvidsson J., Dexter A.R. Soil structures produced by tillage as affected by soil water content and the physical quality of soil. Soil Tillage Research 2007; 92: 45-52.
41. Vannier E., Ciarletti V., Darboux F. Wavelet-based detection of clods on a soil surface. Computers and Geosciences 2009 ; 35: 2259-2267.
42. Taconet O., Vannier E., Le Hégarat-Mascle S. A boundary-based approach for clods identification and characterization on a soil surface. Soil Tillage Research 2010 ; 109: 123-132.
43. Borselli L., Torri D. Soil roughness, slope and surface storage relationship for impervious areas. Journal of Hydrology 2010; 393: 389-400.
44. Merrill D. Comments on the chain method for measuring soil surface roughness: use of the chain set. Soil Science Society American Journal 1998; 62: 1147-1149.
45. Planchon O., Esteves M., Silvera N., Lapetite J. Microrelief induced by tillage : measurement and modelling of surface storage capacity. Catena 2001; 46: 141-157.
46. Guzha A.C. Effects of tillage on soil microrelief, surface depression storage and soil water storage. Soil and Tillage Research 2004; 76: 105-114.
47. Brasington J., Smart R.M.A. Close range digital photogrammetric analysis of experimental drainage basin evolution. Earth Surface Processes and Landforms 2003; 28: 231-247.
48. Rieke-Zapp D., Nearing M.A. Digital close range photogrammetry for measurement of soil erosion. Photogrammetric Record 2005; 20 (109): 69-87.
49. Abd Elbasit M.A.M., Anyoji H., Yasuda H., Yamamoto S. Potential of low cost close-range photogrammetry system in soil microtopography quantification. Hydrological Processes 2009; 23: 1408–1417.
50. Bretar F., Arab-Sedze M., Champion J., Pierrot-Deseilligny M., Heggy E., Jacquemoud S. An advanced photogrammetric method to measure surface roughness : application to volcanic terrains in the Piton de la

Fournaise, Reunion Island. Remote Sensing of Environment 2013 ; 135: 1-11.

51. Helming K., Römkens M.J.M., Prasad S.N. Surface roughness related processes of runoff and soil loss: a flume study. Soil Science Society of America Journal 1998; 62: 243-250.
52. Darboux F., Huang C. An instantaneous-profile laser scanner to measure soil surface microtopography. Soil Science Society America Journal 2003; 67: 92-99.
53. Zhixiong L., Nan C., Perdok U.D., Hoogmoed W.B. Characterization of soil profile roughness. Biosystem Engineering 2005; 91: 369-377.
54. Frede H.-G., Gäth S. Soil surface roughness as the result of aggregate size distribution 1. Report: measuring and evaluation method. Pflanzenernaehr. Bodenk. 1995; 158: 31-55.
55. Jester W., Klik A. Soil surface roughness measurement- methods, applicability and surface representation. Catena 2005; 64: 174-192.
56. Aguilar M.A., Aguilar F.J., Negreiros J. Off-the-shelf laser scanning and close-range digital photogrammetry for measuring agricultural soils microrelief. Biosystem Engineering 2009; 103: 504-517.
57. Marzahn P., Ludwig R. On the derivation of soil surface roughness from multi parametric PolSAR data and ist potential for hydrological modelling: Hydrology Earth System Sciences 2009; 13: 381-394.
58. Vulfson L., Genis A., Blumberg D.G., Sprintsin M., Kotlyar A., Freilikher V., Ben-Asher J. Retrieval of surface roughness parameters of bare soil from the radar satellite data. Journal of Arid Environments 2012; 87: 77-84.
59. Lievens H., Verhoest N.E.C. Spatial and temporal soil moisture estimation from RADARSAT-2 imagery over Flevoland, the Netherlands. Journal of Hydrology 2012; 456-457: 44-56.
60. Chimi-Chiadjeu O., Vannier E., Dusséaux R., Taconet O. Influence of gradient estimation on clod identification on seedbed digital elevation model. Environmental & Engineering Geoscience 2011; 17: 337-352.
61. Fawcett T. ROC Graphs: Notes and Practical Considerations for Researchers. Technical Report. HP Laboratories, Palo Alto 2004, USA.
62. Beucher S., Meyer F. The morphological approach to segmentation: the watershed transformation. In: E. R. Dougherty (ed) Mathematical Morphology in Image Processing. 1993. p433–481.
63. Chimi-Chiadjeu O., Vannier E., Dusséaux R., Le Hégarat-Mascle S., Taconet O. Segmentation of elevation images based on a mathematical

morphology approach for agricultural clod detection: proceedings of 5th international congress on image and signal processing CISP 2012, Chongqing, China, 16-18 October 2012, 701-705.

64. Chimi-Chiadjeu O., Vannier E., Dusséaux R., Le Hégarat-Mascle S., Taconet O. Using simulated annealing algorithm to move clod boundaries on seedbed digital elevation model. Computers & Geosciences 2013 ; 57: 68-76.
65. Kim Y.H., Rana S., Wise S. Exploring multiple viewshed analysis using terrain features and optimisation techniques. Computer & Geosciences 2004; 30 : 1019–1032.
66. Cierniewski J., Verbrugghe M., Marlewski A. Effects of farming works on soil surface bidirectional reflectance measurements and modelling. International Journal of Remote Sensing 2002 ; 23 (6): 1075-1094.
67. Richard G., Boizard H., Roger-Estrade J., Boiffin J., Guérif J. Field study of soil compaction due to traffic in northen France: pore space and morphological analysis of the compacted zones. Soil and Tillage Research 1999; 51: 151-160.
68. Feral L., Mesnard F., Sauvageot H., Castanet L., Lemorton J. Rain cells shape and orientation distribution in south-west of France. Physics and Chemistry of the Earth, Part B: Hydrology, Oceans and Atmosphere 2000; 25 (10): 1073-1078.
69. Roger-Estrade J., Richard G., Caneill J., Boizard H., Coquet Y., Defossez P., Manichon H. Morphological caracterisation of soil structure in tilled fields: from a diagnosis method to the modeling of structural changes over time. Soil and Tillage Research 2004; 79: 33-49.
70. Martin Y., Valeo C., Tait M. Centimetre-scale digital representations of terrain and impacts on depression storage and runoff. Catena 2008; 75: 223-233.

8
CHAPTER

SOIL SALINITY MAPPING AND MONITORING IN ARID AND SEMI-ARID REGIONS USING REMOTE SENSING TECHNOLOGY: A REVIEW

Amal Allbed, Lalit Kumar

School of Environmental and Rural Science, University of New England, Armidale, Australia

ABSTRACT

Soil salinity is a serious environmental problem especially in arid and semiarid areas. It either occurs naturally or is human-induced. High levels of soil salinity negatively affect crop growth and productivity leading land degradation ultimately. Thus, it is important to monitor and map soil salinity at an early stage to enact effective soil reclamation program that helps lessen or prevent future increase in soil salinity. Remote sensing has outperformed the traditional method for assessing soil salinity offering more informative and professional rapid assessment techniques for monitoring and mapping soil salinity. Soil salinity can be identified from remote sensing data obtained by different sensors by way of direct indicators that refer to salt features that are visible at the soil surface as well as indirect indicators such as the presence of halophytic plant and assessing the performance level of salt-tolerant crops. The purposes of this paper are to 1) discuss some soil salinity indicators; 2) review the satellite sensors and methods used for remote monitoring, detecting and mapping of soil salinity, particularly in arid and semi-arid regions; 3) review various spectral vegetation and salinity indices that have been developed and proposed for soil salinity detection and mapping, with an emphasis on soil salinity mapping and assessment in arid and semi-arid regions; and 4) highlight the most important issues limiting the use of remote sensing for soil salinity mapping, particularly in arid and semi-arid regions.

Keywords: Soil Salinity; Remote Sensing; Halophytic Plant; Salinity Index

1. INTRODUCTION

According to the US Salinity Staff Laboratory, soils with conductivity of the saturation extract (EC) > 4 deciSiemens per meter (dS/m) at 25°C, Exchangeable Sodium Percentage (ESP) < 15 and pH (soil reaction) < 8.5 are referred to saline soils [1]. Salt in the soil mostly derives from the weathering of rocks and primary minerals, which formed in situ or transported by water or wind [2]. Other causes of soil salinity are topography, irrigation and dryland salinity, which occur due to forest clearance, overgrazing, and cutting bushes that cause water-tables to rise and bring saline groundwater close to the land surface. Thus, soil salinity categories are either primary salinity which is naturally occurring or secondary salinity which is human-induced.

Soil salinity is a prevalent environmental hazard in arid and semiarid regions around the world [3]. The United Nations Food and Agriculture Organization (FAO) has estimated that saline soil covered 397 million hectares of the total land area of the world [4]. Africa, Asia, Australia, Europe, Latin America, Near East and North America are the most affected areas [4]. According to the most recent estimates from a survey of farmers, about 2 million hectares and 20,000 farms across Australia alone showed some signs of salinity [5]. A recent study conducted by McFarlane et al. [6] has estimated that, in Western Australia alone, secondary salinity covers about 10% of the land, and could reach 23% if action is not taken to stop its spread.

Soil salinity adversely affects plant growth, crop production, soil and water quality, and it eventually results in soil erosion and land degradation [7-9]. Soil salinity impacts are not limited only to the environment but also extend to the economy. For instance, the economic losses due to secondary salinisation in Batinah region in Oman have been estimated at US$ 1604

ha^{-1} (28%) when the salinity increases from low to medium level and US$ 4352 ha^{-1} (76%) if it jumps from low to high level [10].

Basically, soil salinity is a dynamic process with severe consequences for the soil, hydrological, climatic, geochemical, agricultural, social, and economic aspects. Therefore, for greater development and implementation of sufficient soil reclamation programs and preventing any further salinization to sustain agricultural lands and natural ecosystems, information on the spatial extent, nature and distribution of soil salinity is becoming very essential. Thus, timely detection of soil salinity, monitoring and assessment of its severity level and extent become very important in its beginning at local and regional scales.

Conventionally, soil salinity has been measured by collecting in situ soil samples and analyzing those samples in the laboratory to determine their solute concentrations or electrical conductivity. However, these methods are time-consuming and costly since dense sampling is required to adequately characterize the spatial variability of an area [11-14]. Ghabour & Daels [15] agreed that detection soil salinity traditionally is time consuming, but remote sensing data and techniques offer more efficiently and economically rapid tools and techniques for monitoring and mapping soil salinity.

Remote sensing data and techniques have been progressively applied to monitor and map soil salinity since 1960s when black-and-white and color aerial photographs are used to delineate salt-affected soils [16]. Multispectral data such as Landsat, Satellite Pour l'Observation de la Terre (SPOT), IKONOS, QuickBird and the Indian Remote Sensing (IRS) series of satellites, as well as hyperspectral data such as EO-1 Hyperion and HyMap, have been found to be useful in detecting, mapping, and monitoring soil salinity [17-23]. Generally, remote sensing uses the electromagnetic energy reflected from targets to obtain information about the Earth's surface with different levels of detail. So based on this concept, the spectral reflectance of the salt features at the soil surface has been widely studied using remote sensing and used as a direct indicator for soil salinity detection and mapping. However, when the soil moisture is high or the crust salt is invisible on the soil surface or mixed with other soil constituents, this direct approach becomes complicated and may yield unreliable results since these factors influence the soil spectral reflectance. But, the present scattered vegetation or halophytes on the soil surfaces can serve as a sign of the salinity problem, making it possible to indirectly detect and map areas that are affected by soil salinity using the reflectance from vegetation. Normally, unhealthy vegetation has a lower photosynthetic activity, causing increased visible reflectance and the reduced near-infrared reflectance (NIR) from the vegetation [24]. This pattern has been found in various plants subjected to salinity stress [25]. Therefore, based on this finding, several vegetation indices (VIs) such as Normalized Differential Vegetation Index (NDVI) and Soil Adjusted Vegetation Index (SAVI) have been used as indirect indicators assess and map soil salinity. Similarly, a number of researchers have developed different salinity indices to detect and map soil salinity such as Normalized Difference Salinity Index (NDSI) and Salinity Index (SI).

This review concentrates on the problem of soil salinity and discusses some soil salinity indicators and how remote sensing data and technologies are used for monitoring and mapping soil salinity. Additionally, it will discuss the most current vegetation and salinity indices used for soil salinity detecting and mapping, and it highlights some of the limitations and problems of using remote sensing for monitoring and mapping this hazard with an emphasis on soil salinity mapping and monitoring techniques for arid and semi-arid regions.

2. SOIL SALINITY AND REMOTE SENSING

2.1. Soil Salinity Symptoms

Soil salinity can be detected directly from remotely sensed data through salt features that are visible at the soil surface, such as bare soil with white salt crusts on the surface [19,26] or indirectly from indicators such as the presence of halophytic plant, the performance level of salt-tolerant crops [27-30].

2.1.1. Salt Features at the Soil Surface

The dynamic processes at the surface of saline soil limit the monitoring and assessment of the salinization process because they influence the spectral, spatial and temporal behavior of the salt features [31]. Via the physicochemical properties of soil such as soil moisture content, organic matter, soil texture, types of clay color and surface roughness soil spectral reflectance is determined [32-36]. Due to salinity these soil properties change which affect the spectral reflectance of features that occur at the soil surface, including salt crusts and efflorescence besides variations in surface texture and structure [37,38]. For example, Schmid et al. [37]found that crusted saline soil reflects strongly in the visible and near-infrared (NIR) bands; moreover, Rao Singh and Sirohi [39] noted that a crusted saline soil surface is generally smoother than a non-saline surface and exhibits high reflectance in the visible and NIR bands, which has been confirmed by Rao et al. [40]. On the other hand, Metternicht and Zinck [41] found that the reflectance in the visible and NIR bands is highly affected by both the crust color and surface roughness factors.

Despite the effects of salt features on the soil surface on the spectral reflectance, they have been considered good direct indicators of soil salinity. For example, Fernandez-Buces et al. [42] used surface features to predict soil salinity. They found that the correlation coefficient between surface colors, EC and the sodium adsorption ratio (SAR) were statistically significant, which suggested that efflorescence color is a promising surface indicator with which to estimate soil salinity.

2.1.2. Presence of Halophytic Plants

Halophytic plants (salt-tolerant plants) are plants that tolerate high salt concentrations of the soil and can be grown on a salt affected land [43]. Although halophytic plants are common in saline areas, not all have been found to be good remote sensing indicators of soil salinity.For instance, Metternicht [44] found the spectral reflectance curve with high absorption in the visible range and high reflectance in the NIR range of halophyte Chenopodiaceae in Bolivia, to be equivalent to that of chlorophyll-rich vegetation. In contrast, due to lower chlorophyll content the spectral reflectance curve of C. dactylon, also a halophyte, increased continuously in the visible and NIR bands. This study concluded that halophytic plants were promise indicator to distinguish saline areas from non-affected ones.

2.1.3. Crop Performance

The performance of some crops that can be grown on saline soils, such as alfalfa, barley, and cotton, reflect the severity of soil salinity. Cotton is largely cultivated on irrigated land, is therefore considered an ideal indirect indicator for soil salinity, so it has been used as salinity indicators in a variety of studies [31]. For example, based on the high correlations between the Normalized Difference Vegetation Index (NDVI) values of cotton, sugarcane crops and the EC, Wiegand et al. [45,46] successfully assessed the severity and extent of soil salinity in terms of the economic impact on crop production and also distinguished saline soils from non-affected soils. This strong relationship most likely exists only where salinity is the major factor that causes crop yield variability; lands that suffer from soil salinity are likely to have other factors that affect yields as much or more than salinity, such as high or low temperatures, topography and land management. Therefore, deduction of the relationship between a number of saline fields and an entire landscape is likely to result in large errors, so the possible use of this indicator for determining the level of soil salinity must be checked carefully. To overcome this issue, some researchers have proposed using average crop production over a series of years to mask out the noise from non-soil factors that differ from year to year. For example, Lobell et al. [47,48] warn of the use of inter-annual changes on crop yield as an indirect indicator of soil salinity at regional scales because yield mapping that is limited to only one year does not always give a reliable estimator of soil salinity, particularly when strongly saline soils are scarce. Thus, through the use of a 6-year temporal series of satellite images of yield, they obtained a strong correlation between yield losses and soil salinity; they highlighted that yield loss in agricultural regions could be primarily due to several factors, including soil salinity.

3. MAPPING AND MONITORING SOIL SALINITY USING REMOTE SENSING DATA

Advantages of using remote sensing technology include saving time, wide coverage (satellite remote sensing provides the only source when data is required over large areas or regions), are faster than ground methods, and facilitate long term monitoring. These techniques provide multispectral image with resolutions that can be ranged from medium to high, as well as hyperspctral image. These remotely sensed data have been successfully used for monitoring and mapping soil salinity for decades with mixed results. Many researchers have used different techniques to monitor and map soil salinity using remote sensing data, as discussed below.

3.1. Multispectral Satellite Sensors for Mapping and Monitoring Soil Salinity

Extensive research using satellite imagery for mapping and monitoring soil salinity has been conducted over the last three decades, mostly with multispectral sensors. These include Landsat Thematic Mapper (TM), Landsat Multispectral Scanner System (MSS), Landsat Enhanced Thematic Mapper Plus (ETM+), SPOT, Advanced Spaceborne Thermal Emission and Reflection Radiometer (Terra-ASTER), Linear imaging self-scanning sensor (LISS-III) and IKONOS [23,49,50]. For example, in the United States of America (USA), Elnaggar and Noller [51] used Landsat TM imagery integrated with decision-tree analysis (DTA) to map soil salinity in central Malheur County. They found that there was a significant relationship between EC values and reflectance in Landsat bands 1, 2, 3 and 4 as well as the Brightness (BI) and Wetness (WI) indices. Maximum likelihood supervised classification was used to classify the image into nonsaline soils (EC $<$ 4 dS/m) and saline soils, with accuracies of 97% and 60%

respectively, whereas DTA predicted five classes of soil salinity with an overall accuracy of approximately 99%. Their results indicated that the use of Landsat TM imagery effectively identified bare soils that were characterized by high spectral reflectance due to a high salt content on the surface, and the approach of integrating DTA with remote sensing data was more accurate and effective compared to using remote sensing analysis alone.

Many researchers, including Katawatin and Kotrapat [52], Mehrjardi [53] and Yu et al. [54] have investigated the utility and effectiveness of ETM+ data for soil salinity mapping and monitoring. For example, in Thailand, Katawatin and Kotrapat [52] investigated the use of Landsat-7 ETM+ with different combinations of three sources of ancillary data (topography, geology, and underground water quality) for soil salinity mapping. A maximum likelihood classification method was employed in this study. Their results showed that the use of Landsat ETM+ data bands 4, 5 and 7 in combination with all three types of ancillary data yielded the most accurate soil salinity map, with 83.6% overall accuracy. Additionally, Douaoui et al. [55], Farifteh et al. [56] and Eldeiry and Garcia[57] agreed that an integrated approach using remote sensing techniques in addition to ancillary data such as field data, topography and spatial models geophysical surveys can improve the development of high quality soil salinity maps.

Using multispectral sensors for soil salinity research has also been studied by Goossens et al. [58]. Their study examined and compared the accuracy of Landsat TM, MSS, and SPOT XS imagery for soil salinity mapping. They found that Landsat TM was optimal for soil salinity mapping. Another comparative assessment of the suitability of multisensor data for soil salinity studies was conducted in Pakistan by Ahmed and Andrianasolo[59]. They compared the performance of Landsat TM and SPOT XS in mapping salinity at a semi-detailed level. Their results were completely opposite to that of Goossens et al. [58]. They found that the SPOT XS data were more helpful than Landsat TM as it provided finer details of various thematic variables.

Thermal band has proven to be a useful tool in soil salinity studies. It has a key role in differentiating saline soils, especially in areas of bare soil or sparse and similar vegetation. For example, Verma et al. [50] demonstrated that the addition of the thermal band of Landsat TM to the visible-NIR bands helped overcome spectral similarity issues with saline soils. Furthermore, thermal band was used to discriminate saltand sodium-affected soils by Metternicht and Zinck [41]. They found that the incorporation of the thermal band allowed for better salt and sodium detection. Furthermore, in a case study in Iran, AlaviPanah and Goossens [60] found that the addition of the thermal band to the best Landsat TM visible-NIR band combination had great potential for separating saline soil from gypsiferous soil. This study confirmed the result obtained by Goossens et al. [61], who reported the key role of the TM thermal band in separating gypsiferous soil from saline soils. In China, Huang et al. [62] used Terra ASTER imagery to identified saline areas dominated by sodium chlorides and sodium sulfates. Their results showed a good correlation between surface salt concentrations and band 1 of the ASTER sensor, followed by bands 2 and 3.

It cannot be denied that medium or low spatial resolution of the satellite images can limit the mapping and detection of saline regions, particularly when the affected areas are smaller than the pixel size. Thus, high resolution multispectral sensors with pixel size of less than 5 m are becoming an essential for soil salinity studies [23]. However, only limited attempts have been made to identify and map soil salinity problems using fine spatial resolution (0.6 m - 4 m) images that are available from IKONOS (4 bands) and Quickbird (4 bands) satellites, as well as WorldView-2, which has 8 multispectral bands at 1.84 m spatial resolution and one panchromatic band at 0.5 m spatial resolution [63]. This is most likely due to the higher cost of this higher-resolution imagery and these sensors being more recently developed systems. Elhaddad and Garcia [64] have used IKONOS satellite imagery and crop reflectance to identify the severity level of soil salinity and its effect on crop yield in Arkansas River Basin, Colorado. Image enhancement was used to separate the crop condition into several classes, and supervised classification was applied to delineate the different levels of soil salinity. They concluded that the superior effectiveness of their approach was primarily due to the use of high spatial resolution imagery. On the other hand, Dwivedi et al. [23] conducted a comparative study on the performance of IKONOS imagery and imagery from the IRS-ID LISS-III sensor for mapping salt-affected soils. Different image classification and transformation techniques were used in their study, and an overall accuracy of 92.4% was gained when using IKONOS data compared to an overall accuracy of 78.4% and 84.3% obtained when using the IRS-ID LISS-III multispectral sensor, which indicates the great potential of high spatial resolution IKONOS images for soil salinity mapping and detection. In South Australia, Setia et al. [22] detected and mapped soil salinity in an agricultural area using QuickBird imagery. They found that 99% of the variation in spectral values occurred in bands 2 and 4. Furthermore, they found that by dividing the image into hundreds of paddocks (small fields) and performing an unsupervised classification using a paddock-by-paddock approach, which is a procedure where each individual paddock is classified and mapped separately, they were able to map soil salinity severity levels more accurately. Thus, they concluded that a paddock-by-paddock classification approach for QuickBird imagery is a promising method for detecting degrees of salinity severity at a farm level. However, in spite of these promising results, it should be noted that it would extremely difficult to assess the validity and reliability of this approach at regional scales because it is a time-consuming and labor-intensive procedure. Thus, further investigations will help to elucidate whether and how a paddock-by-paddock classification approach will work at regional scales.

3.2. Limitations of Multispectral Satellite Sensors in Soil Salinity Mapping and Monitoring

Multispectral satellite sensors have been and still are the preferred method for mapping and monitoring soil salinity. This is primarily due to the low cost of the imagery (e.g., Landsat, SPOT) and the ability to map extreme surface expressions of salinity. Nevertheless, multispectral data has limited diagnostic capability because of its coarse spatial and spectral resolutions [65]. For example, Furby et al. [66] and Howari [67] reported that direct mapping of soil salinity with multispectral imagery had major limitations that arise, especially where there are no salt features on the soil surface and where saline soils are dominated by halophyte plants. Additionally, Furby et al. [66] stated that multispectral satellite sensors caused confusing reflectance, as they found that non-saline soils were confused with bare, extremely saline areas. In the case of Landsat imagery, Fraser and Joseph [68] reported that the spectral resolution of Landsat was insufficient due to the difference between the spectra of saline land and waterlogged land not being sufficient to allow spectral separation, as well as the variable spectral response of saline soil. Additionally, Hick & Russell [69] state that the ability of discriminating plant species and plant health conditions is challenging with Landsat imagery due to the absence of narrow bands in the range of 700 - 730 nm, 730 - 760 nm and 900 - 1100 nm.

3.3. Hyperspectral Remote Sensing Data

Most of the studies discussed above have attempted to map and monitor soil salinity using different multispectral satellite sensors. However, the use of such sensors is restrictive, as their spectral resolution influences the quality and quantity of the information they provide. The development of airborne and satellite-based hyperspectral sensors has overcome some of the spatial and spectral limitations of multispectral satellite imagery for monitoring and mapping soil salinity, both regionally and locally. Hyperspectral sensors offer a large number of spectral bands with high spatial resolution that allow the discrimination of halophyte plants from non-halophyte plants as well as the identification of surface salt features in more detail than the multispectral sensors [70,71]. Taylor et al. [72] demonstrated the possibility of using airborne hyperspectral data to map salinity in the soil. They described the use of visible-NIR and shortwave infrared (SWIR) hyperspectral data that were collected with airborne Geoscan sensor to map soil salinity at Pyramid Hill, Victoria, Australia. The found that differentiation of salt-affected soils based on the mapping of halophytic plants simply achieved via the employing of the principal component analysis of Geoscan imagery. The potential of the HyMap airborne hyperspectral sensor, which captures images within a spectral range of 450 - 2500 nm in 128 bands, for soil salinity studies has been tested by Dehaan and Taylor [73,74]. They concluded that HyMap has considerable potential for mapping saline areas that characterize the variety of salinity levels and scattered halophyte plants. Likewise, Farifteh et al. [75] measured reflectance spectra from multiple sources (experimental, field, and airborne datasets) to predict salt concentrations with Partial Least Squares Regression (PLSR) and Artificial Neural Networks (ANN) and maped soil salinity using airborne hyperspectral data acquired with HyMap. Their results indicated that both PLSR and ANN enabled good mapping of soil salinity.

After the year 2000, two experimental hyperspectral sensors were launched, CHRIS and Hyperion [76]. Compared to airborne hyperspectral remote sensing, very limited numbers of studies on soil salinity have been conducted with these space-borne hyperspectral data. In Turkey, soil salinity was assessed by Satir et al. [77] using CHRIS PROBA sensor imagery and the spectral characteristics of indicator crops. Their study showed a strong correlation between the spectral wavebands of CHRIS PROBA and the reflected signals of cotton and wheat for detecting soil salinity, which confirmed that the condition of cotton and wheat crops is a good indirect indicator of soil salinity. On the other hand, a study by Dutkiewicz [71] evaluated the performance of Hyperion imagery for mapping surface symptoms of dryland salinity using mixture-tuned matched filtering in southern Australia. She found that the hyperspectral imagery wasunable to distinguish halophytic samphire vegetation at slight or moderate levels of salinity; however, it could be used to map high to very high and extremely high salinity. In China, Weng et al. [18] have investigated the potential of data from the spaceborne Earth Observing 1 (EO-1) Hyperion sensor for the prediction of soil salinity. Partial Least-Squares Regression (PLSR) and Stepwise Linear Regression (SWR) were used as prediction models. Their results indicated that the PLSR model produced more accurate estimations of soil salt content than SWR and could overcome the difficulties that were caused by high dimensionality and strong correlation among input variables as well as noisy data. Thus, they concluded that PLSR was a promising approach for the quantitative mapping of soil salinity with

Hyperion data over a large area. These results suggest that further studies need to be conducted to examine prediction models based on nonlinear regression methods, and the spectral noise of hyperspectral data should be taken into consideration to increase the accuracy of soil salinity mapping. In addition, a more recent study was conducted in Iran by Hamzeh et al. [78] to investigate the ability of Hyperion spaceborne hyperspectral data for mapping salinity stress in sugarcane fields. Different classifications such as Support Vector Machine (SVM), Spectral Angle Mapper (SAM), Minimum Distance (MD) and Maximum Likelihood (ML) were used with different band combinations to classify soil salinity into three classes (low, moderate and high salinity). Their results indicated that SVM classification using all bands as input data yielded a salinity map with good accuracy, with an overall accuracy and kappa coefficient of 78.7% and 0.68, respectively.

4. USING VEGETATION AND SOIL INDICES IN SOIL SALINITY STUDIES

4.1. Vegetation and Soil Indices

As mentioned previously, halophytic plants grow naturally in saline soil, and can be adapted to high soil salinity. Therefore, vegetation has been used as an indirect indicator to predict and map soil salinity. Accordingly, numerous researchers have conducted studies on the mapping and delineation of soil salinity using different Spectral Vegetation Indices (SVI). Among the vegetation indices, NDVI, SAVI, Ratio Vegetation Index (RVI) and Tasseled Cap Transformation that consisted of the Soil Brightness Index (SBI), the Green Vegetation Index (GVI), and the Wetness Index (WI) have been used in soil salinity studies [26,29,30,57,79-81].

Due to absorption in the visible range and high reflectance in the NIR range of the electromagnetic spectrum, the NDVI (Table 1) has been widely used to map soil salinity by monitoring halophytic plants [42,51,79]. The difference in reflectance between the visible and NIR bands is divided by the sum of the two bands' reflectance (Table 1). This normalizes differences in the amount of incoming light and produces a number from −1 to 1; the range of actual values is approximately 0.1 for bare soils to 0.9 for healthy vegetation [82]. In Mexico, FernandezBuces et al. [42] found a significant correlation between NDVI, EC and SAR. Moreover, Pérez González et al. [83] have correlated the NDVI of halophytic vegetation with the spatial variability of the chemical and physical properties of a transect to identify saline hydromorphic soils. Their results showed the NDVI to be very proper in detecting halophytic plant and relating it to saline soils. Additionally,

Bannari et al. [84] have stated that because plant growth declines due to soil salinity, salt stress could be predicted using the NDVI. However, researchers such as Metternicht and Zinck [76] and Zhang, et al. [29] argue that detecting soil salinity using the NDVI is challenging because the presence of vegetation could cause spectral confusion with the reflectance properties of salt and also because the NDVI is considered an unreliable indicator, as it is also correlated to other yield variables such as chlorophyll content, biomass and leaf area. Liu and Huete [85] have developed a modification of the NDVI to reduce the atmospheric and canopy background noise, the enhanced soil and atmosphere resistant vegetation index (EVI) (Table 1). A comparison study of the efficiency of the EVI and NDVI calculated from Multiyear Moderate Resolution Imaging Spectroradiometer (MODIS) imagery for assessing soil salinity in the Red River Valley, United States has been conducted by Lobell et al. [48]. They found that the EVI is a more reliable indicator of salinity than the NDVI. The use of the RVI (Table 1) to measure the spectral reflectance of soybean canopy and elephant grass under different salinity and irrigation treatments was investigated by [80] and [86]. In both studies, the results showed that the canopy spectral reflectance in the NIR region was reduced as salinity level increase.

Furthermore, SAVI (Table 1) was developed by Huete [87] to eliminate soil-induced variation and for use in areas where soil backgrounds differed and the low canopy cover was present. Depending on the crop density, the L factor generally ranges from 0 for higher densities and 1 for lower densities [87]. The resulting SAVI values in the classified image are either positive, negative or zero. A positive SAVI value indicates that there is a decrease in the vegetation, while a negative value indicates an increase in vegetation. A zero value indicates no change in vegetation. The effectiveness of this index for soil salinity detection and mapping has been studied by several researchers, including Zhang et al. [29], Alhammadi [88], Koshal [89], Elnaggar & Noller [51] and Masoud & Koike [90]. For example, in the United Arab Emirates (UAE) Alhammadi and Glenn [27] used the SAVI index for detecting date palm health under soil salinity. They found that the SAVI values decreased with increasing soil salinity; for instance, the SAVI value was 0.155 at the lowest salinity level of 6900 parts per million (ppm), whereas the value decreased to 0.104 at the very high salinity of 41,000 ppm. These results showed the potential of using the date palm, which is a halophytic plant, as an indirect indicator of soil salinity as well as the effectiveness of the SAVI in detecting plant stress related to severe salinity and thus permitting the identification and mapping of saline areas indirectly. Recently, Zhang et al. [29] have proposed four Soil-adjusted Salinity Indices (SASIs) through the most sensitive bands in a SAVI form. For halophyte plants, SASIs produced better results compared to other selected vegetation indices such as the NDVI and SAVI. These results indicate that in highly saline areas that are covered mainly by halophyte plants, SASIs would give superior results, whereas VIs such as NDVI and SAVI would only be proper to assess salinity in low saline areas covered by salt-sensitive plants.

Table 1. Vegetation and soil salinity indices that have been proposed and used for soil salinity monitoring and mapping.

	Indices	Equation	References
1	Normalized Differential Vegetation Index	$NDVI = (NIR - R)/(NIR + R)$	[82]
2	Enhanced Vegetation Index	$EVI = 2.5(NIR - R)/(NIR + 6R - 7.5BLUE + 1)$	[85]
3	Soil Adjusted Vegetation Index	$SAVI = (NIR - R)/(NIR + R + L) \times (1 + L)$	[87]
4	Ratio Vegetation Index	$RVI = NIR/R$	[91]
5	Normalized Differential Salinity Index	$NDSI = (R - NIR)/(R + NIR)$	
6	Brightness Index	$BI = \sqrt{(R^2 + NIR^2)}$	[92]
7	Salinity Index	$SI = \sqrt{BLUE \times R}$	
8	Salinity Index	$SI1 = \sqrt{G \times R}$	
9	Salinity Index	$SI2 = \sqrt{G^2 + R^2 + NIR^2}$	[55]
10	Salinity Index	$SI3 = \sqrt{G^2 + R^2}$	
11	Salinity Index	$SI\text{-}1 = ALI9/ALI10$	
12	Salinity Index	$SI\text{-}2 = (ALI6 - ALI9)/(ALI6 - ALI9)$	
13	Salinity Index	$SI\text{-}3 = (ALI9 - ALI10)/(ALI9 + ALI10)$	[84]
14	Soil Salinity and Sodicity Indices	$SSSI\text{-}1 = (ALI9 - ALI10)$	
15	Soil Salinity and Sodicity Indices	$SSSI\text{-}2 = (ALI9 \times ALI10 - ALI10 \times ALI10)/ALI9$	
16	Salinity Index	$S_1 = Blue/R$	
17	Salinity Index	$S_2 = (Blue - R)/(Blue + R)$	
19	Salinity Index	$S_3 = (G \times R)/Blue$	[93]
20	Salinity Index	$S_4 = \sqrt{Blue \times R}$	
21	Salinity Index	$S_5 = (Blue \times R)/G$	
22	Salinity Index	$S_6 = (R \times NIR)/G$	

Similarly, various spectral salinity indices have been developed for salt mineral detection and mapping. Douaoui et al. [55] have proposed three salinity indices (Table 1) produced from SPOT XS imagery to detect and map soil salinity hazards in a semi-arid environment in Algeria. They found that those indices were strongly correlated with measured values, but considerably underestimated the salinity of areas with high levels of surface salt. Besides, Khan et al. [92] have proposed three spectral salinity indices: the Brightness index (BI), Normalized Difference Salinity Index (NDSI) and Salinity Index (SI) (Table 1) from the LISS-II sensor of the IRS-1B satellite to assess hydrosalinized land degradation in Pakistan. Among these indices, they found that NDSI yielded the most acceptable results in identifying different salt classes. Another study conducted by Vidal et al. [94] and

Vincent et al. [95] looked at salinity by differenting vegetated from non-vegetated areas using NDVI; then the BI was computed to identify the moisture and salinity status of fallow land and deserted fields. Furthermore, three different salinity indices, SI-1, SI-2 and SI-3 (Table 1) from the EO-1 ALI spectral bands, have been proposed by Bannari et al. [84] to discriminate slight and moderate soil salinity and sodicity in Morocco. Although the results showed that SI-3 had the highest correlation (46.9%), the result from this index was not adequate to provide precise information. Therefore, they devised another two Soils Salinity and Sodicity Indices (SSSI) (Table 1). Their results indicated that these SSSI indices were likely to increase the identification accuracy in areas with low and medium salinity because they offered the most significant correlation (52.9%) with the ground EC measurement. In Pakistan, Abbas and Khan [93] have suggested an integrated approach based on the spatial analysis of both ground and satellite data to assess soil salinity. Remotely sensed data-based salinity indices and a Principal Components Analysis (PCA) were developed to detect soil salinity. Their result showed that out of the six salinity indices (Table 1) S3 produced the most promising result compared to ground measurements. Moreover, they concluded that PCA and salinity indices are promising techniques for soil salinity prediction based on satellite images.

Looking at these vegetation and soil salinity indices in the literature, a number of results stand out. Utilizing vegetation indices in the assessment and mapping of soil salinity in areas of densely vegetated soils will yield promising results, whereas on bare soils, the identification of salt based on vegetation indices will not work. Thus, soil salinity indices will be the appropriate method in the case of bare soils or soils with very low scattered vegetation cover, providing super results. These observations are in agreement with Bouaziz et al. [96] and Fan et al. [97]. Bouaziz et al. [96] found that vegetation indices such as SAVI, NDVI and EVI had a low correlation with EC due to an insufficient density of vegetation cover, whereas soil salinity indices indices exhibited higher correlations with EC. Additionally, Fan et al. [97] found that NDVI values had a significant negative relationship with soil salinity in soils covered by vegetation, whereas this relationship was not clear on bare soil.

5. ISSUES IN MAPPING SOIL SALINITY IN ARID AND SEMI-ARID REGIONS USING REMOTE SENSING

The spatial, temporal and vertical variability in the soil profile are the limiting factors in the assessment and mapping of soil salinity using remote sensing data because the spectral reflectance is unable to provide information on the whole soil profile, as it only observes the soil surface [98,99]. Moreover, the surface characteristics in many cases may not be representative of the deeper soil profile [100]. However, combining remote sensing data with geophysical surveys and simulation models can be an alternative option [56].

The direct detection of soil salinity becomes applicable and much easier for bare soils and/or whenever salt-related symptoms (e.g., crusts) and scattered vegetation are visible on the surface; in these cases, information such as salt types and quantity as well as the crust thickness can be provided based on the soil surface spectral signatures [67,101]. Yet, the direct deduction of soil salinity in arid and semi-arid regions that are characterized by dense vegetation cover would be difficult, as the vegetation will cause spectral mixing [102,103].

On the other hand, different studies such as Tashi et al. [104], Fernandez-Buces, et al. [42] have successfully used indirect indicator to map soil salinity through monitoring the vegetation condition by using NDVI. Generally, these methods assume that soil salinity is the only stressor decreasing and damaging the crop condition, whereas other factors such as inappropriate soil management and water quality are neglected. Moreover, reductions in vegetation growth and vigor could be related to a lack of necessary nutrients rather than the occurrence of salt. Additionally, different plants generally grow in different levels of salinity, so that the NDVI is considered an uncertain indicator for soil salinity monitoring and mapping. Besides, the existence of halophytic plants may confuse soil salinity detection based on the NDVI due to mixing with the spectral signature of salt, which then will lead to classification errors [98]. Hence, to overcome this issue and remove classification errors to some degree, the SAVI index and other indices and enhancement models have helped to separate soil and vegetation signals [105].

Additionally, low spectral resolution satellite images limit the direct detection and mapping of soil salinity, primarily due to their inability to detect particular absorption bands of some salt types and the frequent occurrence of problems with mixedspectral signatures that come from a variety of surface components [74,106]. Nevertheless, the advance of hyperspectral sensors has enabled spectral features associated to the characteristic absorption bands of salt minerals to be mapped with more detail [107]. The above shortcomings indicate that detecting and mapping soil salinity in arid and semi-arid regions using remote sensing is challenging. This is primarily attributed to the weakness of the spectral signals from saline soils compared to the noise caused by other factors. Obviously, there is no agreed-on best approach to this technology for monitoring and mapping soil salinity, as many researchers have used and applied different tools and techniques to map and monitor saline soils with varying degrees of success. For example, in the case of vegetation indices, some researchers like Elnaggar & Noller [51] have found that vegetation indices (NDVI, SAVI, and GVI) had a weak correlation with the EC measurements, which suggests that halophytes could not be used to identify salt-affected soils under vegetation cover. Sethi et al. [108] in a study in India, found the same, whereas Pérez González et al. [83] found the NDVI to be very useful in detecting halophytic plants and relating it to saline soils. Despite these varying results of using halophytic plants as indirect indicators for soil salinity detection and mapping, it is ineffective to monitor and map soil salinity through non-halophyte plants, as they cannot live in highly saline areas. Thus, taking into consideration the spectral reflectance of halophytic plants is necessary for soil salinity detection and mapping, particularly in highly saline areas.

6. CONCLUSIONS

Soil salinity, either naturally occurring or human-induced, is a serious global environmental problem, especially in arid and semi-arid regions. This is a complex dynamic process with serious consequences for the soil environment as well as, geochemical, hydrological, climatic, agricultural, and economic impacts. Being a severe environmental hazard, the frequent detection of soil salinity and assessment of its extent and severity at an early stage become very important at both local and regional scales. Traditionally, soil salinity was assessed via collecting in situ soil samples and analyzing those samples in the laboratory. Undertaking this method, especially over a large area, is expensive and time consuming. Remote sensing represents a good alternative for monitoring and mapping changes in soil salinity.

Remote sensing data have been used extensively to identify and map saline areas, and the potential of remote sensing for assessing and mapping soil salinity is enormous. Multispectral satellite sensors are the preferred method for mapping and monitoring soil salinity, largely due to the low cost of such imagery and the ability to map extreme surface expressions of salinity. However, multispectral data have limited capabilities due to their spatial and spectral resolution. Hyperspectral imagery, with its fine spatial and spectral resolutions, allows soil salinity mapping in greater detail and represents another alternative.

Surface reflectance is highly affected by soil's moisture content, salt content, color, and surface roughness. High salt concentrations can be identified through the existence of characteristic vegetation types and growth patterns or by the salt efflorescence and crust that are present on bare soils. Similar to vegetation indices, researchers have developed different salinity indices to detect and map soil salinity. As discussed, these indices have been applied with varying degree of success. Field sites differ in terms of levels of salinity and the amount of vegetation cover; hence, a single selected index may not perform best in all cases. Each site needs to be assessed regarding the strengths and weaknesses of the proposed indices before appropriate remote sensingbased indices are used for soil salinity mapping and assessing.

REFERENCES

1. L. Richards, "Diagnosis and Improvement of Saline and alkali Soils," Soil Science, Vol. 78, No. 2, 1954, p. 154.
2. D. P. Shrestha and A. Farshad, "Chapter 13. Mapping Salinity Hazard: An Integrated Application of Remote Sensing and Modeling-Based Techniques," In: G. Metternicht and J. A. Zinck, Eds., Remote Sensing of Soil Salinization: Impact on Land Management, CRC Press, Boca Raton, 2008, p. 257.
3. D. Hillel, "Salinity Management for Sustainable Irrigation: Integrating Science, Environment, and Economics," World Bank Publications, Washington DC, 2000.
4. P. Koohafkan, "Water and Cereals in Drylands," The Food and Agriculture Organization of the United Nations and Earthscan, Rome, 2012.
5. ABS Australian Bureau of Statistics, "Salinity," Canberra, 2002. http://www.abs.gov.au/ausstats/abs@.nsf/Lookup/by%20Subject/1370.0~2010~Chapter~Salinity%20(6.2.4.4)
6. D. J. McFarlane, et al., "The Extent and Potential Area of Salt-Affected Land in Western Australia Estimated Using Remote Sensing and Digital Terrain Models," Engineering Salinity Solutions: 1st National Salinity Engineering Conference 2004, Barton, 2004, pp. 55-60.
7. J. K. Zhu, "Plant Salt Tolerance," Trends in Plant Science, Vol. 6, No. 2, 2001, pp. 66-71.
8. D. Corwin and S. Lesch, "Application of Soil Electrical Conductivity to Precision Agriculture," Agronomy Journal, Vol. 95, No. 3, 2003, pp. 455-471.

9. J. Rhoades and J. Loveday, "Salinity in Irrigated Agriculture," In: B. A. Stewart and D. R. Nielsen, Eds., American Society of Civil Engineers, Irrigation of Agricultural Crops, Vol. 30, Monograph, American Society of Agronomists, Madison, 1990, pp. 1089-1142.
10. A. Naifer, et al., "Economic Impact of Salinity: The Case of Al-Batinah in Oman," International Journal of Agricultural Research, Vol. 6, No. 2, 2011, pp. 134-142.
11. P. Brunner, et al., "Generating Soil Electrical Conductivity Maps at Regional Level by Integrating Measurements on the Ground and Remote Sensing Data," International Journal of Remote Sensing, Vol. 28, No. 15, 2007, pp. 3341-3361.
12. D. Dent and A. Young, "Soil Survey and Land Evaluation," George Allen & Unwin, Crows Nest, 1981.
13. R. Dehaan and G. Taylor, "Field-Derived Spectra of Salinized Soils and Vegetation as Indicators of IrrigationInduced Soil Salinization," Remote Sensing of Environment, Vol. 80, No. 3, 2002, pp. 406-417.
14. M. R. D. Nanni, "Spectral Reflectance Methodology in Comparison to Traditional Soil Analysis," Soil Science Society of America Journal, Vol. 70, No. 2, 2006, pp. 393-407.
15. T. Ghabour and L. Daels, "Mapping and Monitoring of Soil Salinity of ISSN," Egyptian Journal of Soil Science, Vol. 33, No. 4, 1993, pp. 355-370.
16. P. Dale, et al., "Classification of Reflectance on Colour Infrared Aerial Photographs and Sub-Tropical Salt-Marsh Vegetation Types," International Journal of Remote Sensing, Vol. 7, No. 12, 1986, pp. 1783-1788.
17. J. Farifteh, "Imaging Spectroscopy of Salt-Affected Soils: Model-Based Integrated Method," International Institute Geo-Information Science and Earth Observation (ITC) and Utrecht University, Utrecht, 2007.
18. Y. Weng, et al., "Soil Salt Content Estimation in the Yellow River Delta with Satellite Hyperspectral Data," Canadian Journal of Remote Sensing, Vol. 34, No. 3, 2008, pp. 259-270.
19. S. Teggi, et al., "SPOT 5 Imagery for Soil Salinity Assessment in Iraq," Proceedings of SPIE—Earth Resources and Environmental Remote Sensing/GIS Applications III, Vol. 8538, 2012, pp. 85380V-85380V-12.
20. A. K. Koshal, "Spectral Characteristics of Soil Salinity Areas in Parts of South-West Punjab through Remote Sensing and GIS," International Journal of Remote Sensing and GIS, Vol. 1, No. 2, 2012, pp. 84-89.

21. A. Dehni and M. Lounis, "Remote Sensing Techniques for Salt Affected Soil Mapping: Application to the Oran Region of Algeria," Procedia Engineering, Vol. 33, 2012, pp. 188-198.
22. R. Setia, et al., "Severity of Salinity Accurately Detected and Classified on a Paddock Scale with High Resolution Multispectral Satellite Imagery," Land Degradation & Development, Vol. 24, No. 4, 2011, pp. 375-384.
23. R. Dwivedi, et al., "5 Generation of Farm-Level Information on Salt-Affected Soils Using IKONOS-II Multispectral Data," In: G. Metternicht and J. Zinck, Eds., Remote Sensing of Soil Salinization: Impact on Land Management, CRC Press, Boca Raton, 2008.
24. E. Weiss, et al., "Application of NOAA-AVHRR NDVI Time-Series Data to Assess Changes in Saudi Arabia's Rangelands," International Journal of Remote Sensing, Vol. 22, No. 6, 2001, pp. 1005-1027.
25. D. R. Tilley, et al., "Hyperspectral Reflectance Response of Freshwater Macrophytes to Salinity in a Brackish Subtropical Marsh," Journal of Environmental Quality, Vol. 36, No. 3, 2007, pp. 780-789.
26. H. R. Matinfar, et al., "Detection of Soil Salinity Changes and Mapping Land Cover Types Based upon Remotely Sensed Data," Arabian Journal of Geosciences, Vol. 6, No. 3, 2013, pp. 913-919.
27. M. S. Alhammadi and E. P. Glenn, "Detecting Date Palm Trees Health and Vegetation Greenness Change on the Eastern Coast of the United Arab Emirates Using SAVI," International Journal of Remote Sensing, Vol. 29, No. 6, 2008, pp. 1745-1765.
28. F. Iqbal, "Detection of Salt Affected Soil in Rice-Wheat Area Using Satellite Image," African Journal of Agricultural Research, Vol. 6, No. 21, 2011, pp. 4973-4982.
29. T. T. Zhang, et al., "Using Hyperspectral Vegetation Indices as a Proxy to Monitor Soil Salinity," Ecological Indicators, Vol. 11, No. 6, 2011, pp. 1552-1562.
30. Y. Y. Aldakheel, "Assessing NDVI Spatial Pattern as Related to Irrigation and Soil Salinity Management in Al-Hassa Oasis, Saudi Arabia," Journal of the Indian Society of Remote Sensing, Vol. 39, No. 2, 2011, pp. 171-180.
31. G. Metternicht and A. Zinck, "Remote Sensing of Soil Salinization: Impact on Land Management," CRC Press, 2008.
32. J. A. M. Demattê, et al., "Visible-NIR Reflectance: A New Approach on Soil Evaluation," Geoderma, Vol. 121, No. 1-2, 2004, pp. 95-112.

33. D. J. Brown, et al., "Global Soil Characterization with VNIR Diffuse Reflectance Spectroscopy," Geoderma, Vol. 132, No. 3, 2006, pp. 273-290.
34. M. Baumgardner, et al., "Reflectance Properties of Soils," Advances in Agronomy, Vol. 38, 1985, pp. 2-44.
35. S. De Jong, "The Analysis of Spectroscopical Data to Map Soil Types and Soil Crusts of Mediterranean Eroded Soils," Soil technology, Vol. 5, No. 3, 1992, pp. 199-211.
36. D. Shrestha, D. E. Margateb, F. van der Meera and H. V. Anhc, "Analysis and Classification of Hyperspectral Data for Mapping Land Degradation: An Application in Southern Spain," International Journal of Applied Earth Observation and Geoinformation, Vol. 7, No. 2, 2005, pp. 85-96.
37. T. Schmid, M. Koch and J. Gumuzzio, "Application of Hyperspectral Imagery to Soil Salinity Mapping," In: G. Metternicht and J. Zinck, Eds., Remote Sensing of Soil Salinization: Impact on Land Management, CRC Press, Boca Raton, 2008, pp. 113-137.
38. D. S. G. Thomas, "Arid Zone Geomorphology: Process, Form and Change in Drylands," John Wiley & Sons, Chichester, 2011.
39. R. P. Singh and A. Sirohi, "Spectral Reflectance Properties of Different Types of Soil Surfaces," ISPRS Journal of Photogrammetry and Remote Sensing, Vol. 49, No. 4, 1994, pp. 34-40.
40. B. R. M. Rao, R. C. Sharma, T. Ravi Sankar, S. N. Das, R. S. Dwivedi, S. S. Thammappa and L. Venkataratnam, "Spectral Behaviour of Salt-Affected Soils," International Journal of Remote Sensing, Vol. 16, No. 12, 1995, pp. 2125-2136.
41. G. Metternicht and J. A. Zinck, "Spatial Discrimination of Saltand Sodium-Affected Soil Surfaces," International Journal of Remote Sensing, Vol. 18, No. 12, 1997, pp. 2571-2586.
42. N. Fernandez-Buces, C. Siebe, S. Cram and J. L. Palacio, "Mapping Soil Salinity Using a Combined Spectral Response Index for Bare Soil and Vegetation: A Case Study in the Former Lake Texcoco, Mexico," Journal of Arid Environments, Vol. 65, No. 4, 2006, pp. 644-667.
43. E. P. Glenn, J. Jed Brown and E. Blumwald, "Salt Tolerance and Crop Potential of Halophytes," Critical Reviews in Plant Sciences, Vol. 18, No. 2, 1999, pp. 227-255.

44. G. I. Metternicht, "Analyzing the Relationship between Ground Based Reflectance and Environment Indicators of Salinity Processes in the Cochabamba Valley (Bolivia)," International Journal of Ecology and Environmental Sciences, Vol. 24, No. 4, 1998, pp. 359-370.
45. C. Wiegand, G. Anderson, S. Lingle and D. Escobar, "Soil Salinity Effects on Crop Growth and Yield-Illustration of an Analysis and Mapping Methodology for Sugarcane," Journal of Plant Physiology, Vol. 148, No. 3-4, 1996, pp. 418-424. http://dx.doi.org/10.1016/S0176-1617(96)80274-4
46. C. Wiegand, J. D. Rhoades, D. E. Escobar and J. H. Everitt, "Photographic and Videographic Observations for Determining and Mapping the Response of Cotton to Soil Salinity," Remote Sensing of Environment, Vol. 49, No. 3, 1994, pp. 212-223.
47. D. B. Lobell, J. I. Ortiz-Monasterio, F. C. Gurrola and L. Valenzuela, "Identification of Saline Soils with Multiyear Remote Sensing of Crop Yields," Soil Science Society of America Journal, Vol. 71, No. 3, 2007, pp. 777-783.
48. D. B. Lobell, S. M. Leschb, D. L. Corwinc, M. G. Ulmerd, K. A. Andersone, D. J. Pottsf, J. A. Doolittleg, M. R. Matosh and M. J. Baltes, "Regional-Scale Assessment of Soil Salinity in the Red River Valley Using Multi-Year MODIS EVI and NDVI," Journal of Environmental Quality, Vol. 39, No. 1, 2010, pp. 35-41.
49. R. S. Dwivedi, "Soil Resources Mapping: A Remote Sensing Perspective," Remote Sensing Reviews, Vol. 20, No. 2, 2001, pp. 89-122.
50. K. Verma, R. K. Saxena, A. K. Barthwal and S. N. Deshmukh, "Remote Sensing Technique for Mapping Salt Affected Soils," International Journal of Remote Sensing, Vol. 15, No. 9, 1994, pp. 1901-1914.
51. A. A. Elnaggar and J. S. Noller, "Application of RemoteSensing Data and Decision-Tree Analysis to Mapping SaltAffected Soils over Large Areas," Remote Sensing, Vol. 2, No. 1, 2009, pp. 151-165.
52. R. Katawatin and W. Kotrapat, "Use of LANDSAT-7 ETM+ with Ancillary Data for Soil Salinity Mapping in Northeast Thailand," Third International Conference on Experimental Mechanics and Third Conference of the Asian 2004, Sevilla, 3 June 2005, pp. 708-716.
53. R. T. Mehrjardi, Sh. Mahmoodi, M. Taze and E. Sahebjalal, "Accuracy Assessment of Soil Salinity Map in Yazd-Ardakan Plain, Central Iran, Based on Land sat ETM+ Imagery," American-Eurasian Journal of Agricultural & Environmental Sciences, Vol. 3, No. 5, 2008, pp. 708-712.

54. R. Yu, T. X. Liu, Y. P. Xu, C. Zhu, Q' Zhang, Z. Y. Qu, X. M. Liu and C. Y. Li, "Analysis of Salinization Dynamics by Remote Sensing in Hetao Irrigation District of North China," Agricultural Water Management, Vol. 97, No. 12, 2010, pp. 1952-1960.
55. A. E. K. Douaoui, H. Nicolasb and C. Walter, "Detecting Salinity Hazards within a Semiarid Context by Means of Combining Soil and Remote-Sensing Data," Geoderma, Vol. 134, No. 1-2, 2006, pp. 217-230.
56. J. Farifteh, A. Farshad and R. J. George, "Assessing SaltAffected Soils Using Remote Sensing, Solute Modelling, and Geophysics," Geoderma, Vol. 130, No. 3-4, 2006, pp. 191-206.
57. A. Eldeiry and L. A. Garcia, "Detecting Soil Salinity in Alfalfa Fields Using Spatial Modeling and Remote Sensing," Soil Science Society of America Journal, Vol. 72, No. 1, 2008, pp. 201-211.
58. R. Goossens, M. el Badawi, T. Ghabour and M. de Dapper, "A Simulation Model to Monitor the Soil Salinity in Irrigated Arable Land in Arid Areas Based upon Remote Sensing and GIS," EARSeL. Advances in Remote Sensing, Vol. 2, No. 3, 1993, pp. 165-171.
59. I. Ahmed and H. H. Andrianasolo, "Comparative Assessment of Multisensor Data for Suitability in Study of the Soil Salinity Using Remote Sensing and GIS in the Fordwah Irrigation Division," IEEE International Conference on Geoscience and Remote Sensing, Singapore, 3-8 August 1997, pp. 1627-1629.
60. S. K. Alavi Panah and R. Goossens, "Relationship between the Landsat TM, MSS Data and Soil Salinity," Journal of Agricultural Science and Technology (JAST), Vol. 3, No. 2, 2001, pp. 21-31.
61. R. Goossens, A. P. S. Kazem, M. De Dapper and O. Kissyar, "The Use of Thermal Band of Landsat TM for the Study of Soil Salinity in Iran (Ardakan Area) and Egypt (Ismailia Province)," International conference on Geoinformatics for Natural Resource Assessment, Monitoring and Management, Indian Institute of Remote Sensing, Dehradun, 3-11 March 1999, pp. 454-459.
62. S. Y. Huang, Q. H. Liu, X. W. Li and Q. H. Liu, "Spectral Model of Soil Salinity in Xinjiang of China," IEEE International Geoscience and Remote Sensing Symposium, Seoul, 25-29 July 2005, pp. 4458-4460.
63. K. Navulur, "Multispectral Image Analysis Using the Object-Oriented Paradigm," CRC Press, Boca Raton, 2006.

64. A. Elhaddad and L. Garcia, "Detecting Soil Salinity Levels in Agricultural Lands Using Satellite Imagery," Proceedings of the American Society for Photogrammetry and Remote Sensing Annual Conference, Reno, 2006.
65. B. Spies and P. Woodgate, "Salinity Mapping Methods in the Australian Context," Department of the Environment and Heritage, Canberra, 2005.
66. S. Furby, et al., "Detecting and Monitoring Salt-Affected Land: A Report from the LWRRDC Project Detecting and Monitoring Changes in Land Condition through Time Using Remotely Sensed Data," CSIRO, Division of Mathematics & Statistics, 1995.
67. F. Howari, "The Use of Remote Sensing Data to Extract Information from Agricultural Land with Emphasis on Soil Salinity," Australian Journal of Soil Research, Vol. 41, No. 7, 2003, pp. 1243-1253.
68. D. Fraser and S. Joseph, "Mapping Soil Salinity in the Murray Valley (NSW) Using Satellite Imagery," Proceeding of the 9th Australaian Remote Sensing and Photogrammetry Conference, Australia Causal Productions Sydney, Sydney, 1998.
69. P. Hick and W. Russell, "Remote Sensing of Agricultural Salinity," CSIRO, Division of Exploration Geoscience, Perth, 1987.
70. R. P. Gupta, "Remote Sensing Geology," Springer Verlag, New York, 2003.
71. A. Dutkiewicz, "Evaluating Hyperspectral Imagery for Mapping the Surface Symptoms of Dryland Salinity," The University of Adelaide, Adelaide, 2006.
72. G. R. Taylor, B. A. Bennett, A. H. Mah and R. D. Hewson, "Spectral Properties of Salinised Land and Implications for Interpretation of 24 Channel Imaging Spectrometry," Proceedings of the International Airborne Remote Sensing Conference and Exhibition, Taylor & Francis Strasbourg, France, 1994, pp. 504-513.
73. R. L. Dehaan and G. R. Taylor, "Field-Derived Spectra of Salinized Soils and Vegetation as Indicators of Irrigation-Induced Soil Salinization," Remote Sensing of Environment, Vol. 80, No. 3, 2002, pp. 406-417.
74. R. Dehaan and G. R. Taylor, "Image-Derived Spectral Endmembers as Indicators of Salinisation," International Journal of Remote Sensing, Vol. 24, No. 4, 2003, pp. 775- 794.

75. J. Farifteh, F. Van der Meera, C. Atzbergerb and E. J. M. Carranzaa, "Quantitative Analysis of Salt-Affected Soil Reflectance Spectra: A Comparison of Two Adaptive Methods (PLSR and ANN)," Remote Sensing of Environment, Vol. 110, No. 1, 2007, pp. 59-78.
76. G. Metternicht and A. Zinck, "Remote Sensing of Soil Salinization: Impact on Land Management," CRC Press, Taylor and Francis Publisher, Boca Raton, 2008.
77. O. Satir, S. Berberoglu, S. Kapur, T. Nagano, E. Akça, M. A. Erdogan, C. Donmez, N. Yonca Satir and K. Tanaka, "Soil Salinity Mapping Using Chris-Proba Hyperspectral Data," Proceedings of Hyperspectral Workshop 2010, Frascati, 17-19 March 2010, 8p.
78. S. Hamzeh, A. A. Naseri, S. K. Alavi Panah, B. Mojaradi, H. M. Bartholomeus and M. Herold, "Mapping Salinity Stress in Sugarcane Fields with Hyperspecteral Satellite Imagery," Proceedings SPIE 8531, Remote Sensing for Agriculture, Ecosystems, and Hydrology XIV, Edinburgh, 24 September 2012, pp. 85312B-1.
79. M. T. Jabbar and X. Chen, "Land Degradation Due to Salinization in Arid and Semi-Arid Regions with the Aid of Geo-Information Techniques," Geo-Spatial Information Science, Vol. 11, No. 2, 2008, pp. 112-120.
80. D. Wang, J. A. Poss, T. J. Donovan, M. C. Shannon and S. M. Lesch, "Biophysical Properties and Biomass Production of Elephant Grass under Saline Conditions," Journal of Arid Environments, Vol. 52, No. 4, 2002, pp. 447-456.
81. D. Lobell, S. M. Lesch, D. L. Corwin, M. G. Ulmer, K. A. Anderson, D. J. Potts, J. A. Doolittle, M. R. Matos and M. J. Baltes "Regional-Scale Assessment of Soil Salinity in the Red River Valley Using Multi-Year MODIS EVI and NDVI," Journal of Environmental Quality, Vol. 39, No. 1, 2010, pp. 35-41.
82. D. Deering and J. Rouse, "Measuring 'Forage Production' of Grazing Units from Landsat MSS Data," 10th International Symposium on Remote Sensing of Environment, ERIM, Ann Arbor, 1975, pp. 1169-1178.
83. M. E. Pérez González, M. del Pilar García Rodríguez, V. González-Quiñones and R. Jiménez Ballesta, "Spatial Variability of Soil Quality in the Surroundings of a Saline Lake Environment," Environmental Geology, Vol. 51, No. 1, 2006, pp. 143-149.

84. A. Bannari, A. M. Guedona, A. El-Hartib, F. Z. Cherkaouic and A. El-Ghmari, "Characterization of Slightly and Moderately Saline and Sodic Soils in Irrigated Agricultural Land using Simulated Data of Advanced Land Imaging (EO-1) Sensor," Communications in Soil Science and Plant Analysis, Vol. 39, No. 19-20, 2008, pp. 2795-2811.
85. H. Q. Liu and A. Huete, "A Feedback Based Modification of the NDVI to Minimize Canopy Background and Atmospheric Noise," IEEE Transactions on Geoscience and Remote Sensing, Vol. 33, No. 2, 1995, pp. 457-465.
86. D. Wang, C. Wilson and M. C. Shannon, "Interpretation of Salinity and Irrigation Effects on Soybean Canopy Reflectance in Visible and Near-Infrared Spectrum Domain," International Journal of Remote Sensing, Vol. 23, No. 5, 2002, pp. 811-824.
87. A. R. Huete, "A Soil-Adjusted Vegetation Index (SAVI)," Remote Sensing of Environment, Vol. 25, No. 3, 1988, pp. 295-309.
88. M. S. Alhammadi, "Using QuickBird Satellite Images to Study the Salinity Effect on Date Palm Field," International Congress Geotunis 2010, Tunis, 29 November-3 December 2010, pp. 1-6.
89. A. K. Koshal, "Indices Based Salinity Areas Detection through Remote Sensing and GIS In Parts of South Waste Punjab," 13th Annual International Conference and Exibition on Geospatial Information Technology and Applications, Gurgaon, 2010.
90. A. Masoud and K. Koike, "Arid Land Salinization Detected by Remotely-Sensed Landcover Changes: A Case Study in the Siwa Region, NW Egypt," Journal of Arid Environments, Vol. 66, No. 1, 2006, pp. 151-167.
91. D. Major, F. Baret and G. Guyot, "A Ratio Vegetation Index Adjusted for Soil Brightness," International Journal of Remote Sensing, Vol. 11, No. 5, 1990, pp. 727-740.
92. N. M. Khan, V. V. Rastoskuev, Y. Sato and S. Shiozawa, "Assessment of Hydrosaline Land Degradation by Using a Simple Approach of Remote Sensing Indicators," Agricultural Water Management, Vol. 77, No. 1, 2005, pp. 96-109.
93. A. Abbas and S. Khan, "Using Remote Sensing Techniques for Appraisal of Irrigated Soil Salinity," In: L. Oxley and D. Kulasiri, Eds., International Congress on Modelling and Simulation (MODSIM), Modelling and Simulation Society of Australia and New Zealand, Brighton, 2007, pp. 2632-2638.

94. A. Vidal, P. Maure, H. Durand and P. Strosser, "Remote Sensing Applied to Irrigation System Management: Example of Pakistan," EURISY Colloquium: Satellite Observation for Sustainable Development in the Mediterranean Area, Paris, 1996, pp. 132-142.
95. B. Vincent, A. Vidal, D. Tabbet, A. Baqri and M. Kuper, "Use of Satellite Remote Sensing for the Assessment of Water Logging or Salinity as an Indication of the Performance of Drained Systems," 16th Congress on Irrigation and Drainage, Cairo, 15-22 September 1996, pp. 203-216.
96. M. Bouaziz, J. Matschullat and R. Gloaguen, "Improved Remote Sensing Detection of Soil Salinity from a SemiArid Climate in Northeast Brazil," Comptes Rendus Geoscience, Vol. 343, No. 11-12, 2011, pp. 795-803.
97. X. Fan, B. Pedroli, G. Liu, Q. Liu, H. Liu and L. Shu, "Soil Salinity Development in the Yellow River Delta in Relation to Groundwater Dynamics," Land Degradation & Development, Vol. 23, No. 2, 2012, pp.175-189.
98. M. Sethi, D. S. Bundela, M. Sethi, K. Lal and S. K. Kamra, "Remote Sensing and Geographic Information System for Appraisal of Salt-Affected Soils in India," Journal of Environmental Quality, Vol. 39, No. 1, 2010, pp. 5-15.
99. T. J. Schmugge, W. P. Kustas, J. C. Ritchie, T. J. Jackson and A. Rango, "Remote Sensing in Hydrology," Advances in Water Resources, Vol. 25, No. 8-12, 2002, pp. 1367- 1385.
100. O. Dewitte, A. Jones, H. Elbelrhiti, S. Horion and L. Montanarella, "Satellite Remote Sensing for Soil Mapping in Africa An Overview," Progress in Physical Geography, Vol. 36, No. 4, 2012, pp. 514-538.
101. H. A. Ghrefat, P. C. Goodell, B. E. Hubbard, R. P. Langford and R. E. Aldouri, "Modeling Grain Size Variations of Aeolian Gypsum Deposits at White Sands, New Mexico, Using AVIRIS Imagery," Geomorphology, Vol. 88, No. 1-2, 2007, pp. 57-68.
102. A. L. Kaleita, L. F. Tian and M. C. Hirschi, "Relationship between Soil Moisture Content and Soil Surface Reflectance," Transactions of the ASAE, Vol. 48, No. 5, 2005, pp. 1979-1986.
103. J. Ding, M.-C. WU and T. Tiyip, "Study on Soil Salinization Information in Arid Region Using Remote Sensing Technique," Agricultural Sciences in China, Vol. 10, No. 3, 2011, pp. 404-411.

104. Y. Tashi, P. C. Chamard, M.-F. Courel, T. Tiyip, Y. Tuerxun and S. Drake, "The Recent Evolution of the Oasis Environment in the Taklimakan Desert, China," In: Water and Sustainability in Arid Regions, Springer, Netherlands, 2010, pp. 51-74.
105. V. Mulder, S. de Bruin, M. E. Schaepman and T. R. Mayr, "The Use of Remote Sensing in Soil and Terrain Mapping—A Review," Geoderma, Vol. 162, No. 1-2, 2011, pp. 1-19.
106. B. Mougenot, M. Pouget and G. F. Epema, "Remote Sensing of Salt Affected Soils," Remote Sensing Reviews, Vol. 7, No. 3-4, 1993, pp. 241-259.
107. E. Cloutis, "Review Article Hyperspectral Geological Remote Sensing: Evaluation of Analytical Techniques," International Journal of Remote Sensing, Vol. 17, No. 12, 1996, pp. 2215-2242.
108. M. Sethi, G. S. Dasog, A. Van Lieshoutc and S. B. Salimathd, "Salinity Appraisal Using IRS Images in Shorapur Taluka, Upper Krishna Irrigation Project, Phase I, Gulbarga District, Karnataka, India," International Journal of Remote Sensing, Vol. 27, No. 14, 2006, pp. 2917-2926.

9

CHAPTER

INVESTIGATING THE USE OF REMOTE SENSING AND GIS TECHNIQUES TO DETECT LAND USE AND LAND COVER CHANGE: A REVIEW

Abdullah F. Alqurashi[1,2], Lalit Kumar[1]

[1]School of Environmental and Rural Science, University of New England, Armidale, Australia

[2]Department of Geography, Umm Al-Qura University, Makkah, Saudi Arabia

ABSTRACT

The accuracy of change detection on the earth's surface is important for understanding the relationships and interactions between human and natural phenomena. Remote Sensing and Geographic Information Systems (GIS) have the potential to provide accurate information regarding land use and land cover changes. In this paper, we investigate the major techniques that are utilized to detect land use and land cover changes. Eleven change detection techniques are reviewed. An analysis of the related literature shows that the most used techniques are post-classification comparison and principle component analysis. Post-classification comparison can minimize the impacts of atmospheric and sensor differences between two dates. Image differencing and image ratioing are easy to implement, but at times they do not provide accurate results. Hybrid change detection is a useful technique that makes full use of the benefits of many techniques, but it is complex and depends on the characteristics of the other techniques such as supervised and unsupervised classifications. Change vector analysis is complicated to implement, but it is useful for providing the direction and magnitude of change. Recently, artificial neural networks, chi-square, decision tree and image fusion have been frequently used in change detection. Research on integrating remote sensing data and GIS into change detection has also increased.

Keywords: Change Detection Techniques; Remote Sensing; GIS; Land Use and Land Cover Change

1. INTRODUCTION

Change detection is the process of identifying differences in the state of an object or phenomenon by observing it at different times [1]. The time and accuracy of change detection on the earth's surface can provide a better understanding of the relationships and interactions between human and natural phenomena and can provide guidance in the management of the use of resources. In change detection applications, it is necessary to use multi-temporal datasets to analyze the temporal effects of the object or phenomenon [2]. Currently, with increased computer capability and data availability, Remote Sensing and Geographic Information Systems (GIS) have become effective tools for detecting objects and phenomena change.

Remote Sensing means the ability to detect change on the earth's surface through space-borne sensors [3]. The repetitive coverage of satellite images and the improvement of image quality can provide valuable assistance in the identification of changes [4]. Temporal and spatial resolutions allow scientists to monitor and detect changes over a broad scale and help planners to obtain or maintain information on various phenomena, such as shifting agricultural patterns, crop stress, disaster monitoring, land use and land cover changes [5].

Geographic Information System is a useful tool for measuring the change between two or more time periods. It has the ability to incorporate multi-sources of data into a change detection platform [2]. For example, the use of multiple layers, such as classified images, topographical maps, soil maps and hydrological maps, provides a greater ability to extract useful information about the changes over a particular area. Moreover, GIS can measure the trends in these changes by modeling the available data and using statistical and analytical functions. The benefit of GIS is the provision of different outputs in different formats (e.g. maps or tables), which allows users to select the appropriate output for extracting the desired information.

Remotely sensed data and GIS are widely used for detecting land use and land cover changes. Many studies have attempted to use remotely sensed data and GIS to address land use change detection e.g. [6-9]. A variety of procedures or methods of Remote Sensing technologies are used to detect land use and land cover changes. Some studies have actually utilized Remote Sensing techniques; others have integrated remotely sensed data with GIS data e.g. [10-12]. In addition, many studies have reviewed and summarized the various change detection techniques [1,2,5,13-15]. The articles reviewed here consider these change detection techniques in many applications, including land use and land cover changes, vegetation and forest changes, urban changes, environmental changes, crop monitoring, forest fires, deforestation and other applications. Table 1 shows examples of applications that are investigated using change detection techniques.

In this paper, we firstly review the major land use and land cover change detection techniques, including image differencing, image ratioing, change vector analysis (CVA), principal component analysis (PCA), chi-square, post-classification comparison, decision trees, image fusion, hybrid change detection, artificial neural networks (ANN) and Geographic Information Systems (GIS), by giving overview about the characteristics, strengths and weaknesses of each technique. Second, we discuss the importance of accuracy assessment in change detection and the use of different matrices in previous studies. Finally, we evaluate the change detection techniques according to the analysis of related literature.

2. LAND USE AND LAND COVER CHANGE TECHNIQUES

2.1. Image Differencing

Image differencing, also referred to as image delta [16], is a simple, easy-to-use technique for implementing and interpreting change detection [17]. It divides the image pixels into two results: change or no change. The process by which these results are obtained involves subtracting a pixel's digital number on the image for date one from the corresponding pixel's digital number on the image for date two. The general process for detecting the change in two dates in image differencing is extracting the change of the image of date 2 from the image of date 1 (e.g. image of date 1—image of date 2) [18]. In this technique, it is necessary to select thresholds for determining the changed area [16,19]. However, the image differencing technique cannot provide sufficient information about the change itself. Atmospheric and other non-surface radiance characteristics can affect the results of image differencing [20].

Image differencing is widely utilized for change detection in the geographical environment [16,21-23]. It has been used either as a single-band difference [24] or as a color composite of three bands [25]. Sohl [22] examined five change detection techniques, including univariate image differencing, "enhanced" image differencing, vegetation index differencing, post-classification differencing and change vector analysis, to detect the landscape change in the Abu Dhabi Emirate using Landsat Thematic Mapper (TM) data. He found that the "enhanced" image differencing technique provided the most accurate values of change when compared to other techniques, while change vector analysis was a useful technique for providing rich qualitative detail about the nature of the change. Ridd and Liu [26] compared four techniques for change detection in an urban environment; image differencing, image regression, tasseled cap transformation and chi-square were used. The results showed that the most accurate technique for detecting change was the regression of TM band 3, whereas the image differencing of TM 4 was found to be the least accurate.

In arid and semi-arid environments, red band image differencing is more effective than using the Normalized Difference Vegetation Index (NDVI) for vegetation change detection [27]. Also, Pilon, Howarth [28] concluded that visible red band data provided the most accurate identification of spectral change for their semi-arid study area in north-western Nigeria in Sub-Sahelian Africa.

TABLE 1. Examples of the applications that can be investigated using change detection techniques.

Application	Most Commonly Used Techniques	Examples
Land use/Land cover change	Image differencing, image ratioing, NDVI, CVA, PCA, chi-square, post-classification, hybrid change detection, ANN, decision tree, GIS	[21,22,26,31,33-41]
Urban change	Image differencing, post-classification, hybrid change detection, PCA, GIS, chi-square, image fusion	[26,42-44]
Environmental change	NDVI, ANN, CVA, post-classification, image differencing	[30,38,45]
Vegetation change	NDVI, CVA, image differencing, post-classification	[46-48]
Landscape change	Post-classification, GIS	[49-51]
Deforestation	Post-classification, NDVI, image differencing, PCA	[52-54]
Wetland change	Post-classification, GIS	[55-57]

2.2. Image Ratioing

Image ratioing is extracting the information between two or more different images by using the same bands of two or more images. For example, to compute changes between two times using band 2, the simple process of image ratioing could be: (Band 2 of t_1 divided by Band 2 of t_2). It is used to highlight subtle variations between the pixels of the various land covers. In image ratioing, the unchanged pixel takes the same number for both dates with a grey level. The changed pixel takes a different value and is displayed at a lighter or darker level. To display the different changes between two or more images, selecting the appropriate threshold value is important in image ratioing. The image ratioing technique is useful for extracting vegetation cover information. The advantage of this technique is that the effect of shadows, the radiation change, image noise and the sun angle can be reduced [29]. However, there are two major disadvantages: it is difficult to select the threshold value, and the types of changes cannot be analyzed [2].

Image ratioing was used by Howarth and Wickware [30] to generate three different color enhancements—an overlay of band 5, ratios of bands 5 and 7 and a vegetation index. The results showed that the major change was emphasized by using band ratioing and band 5 overlay, which gave additional information about the change, while the vegetation index enhancement results showed less information about the change. Nelson [31] tested three change detection techniques, including image differencing, image ratioing and vegetation index differencing, to determine the most appropriate technique for detecting changes due to gypsy moth defoliation. The results indicated that the vegetation index difference and the band 5 ratio were more accurate for delineation of forest canopy changes than other single bands or band combinations. Prakash and Gupta [32] used image ratioing and the normalized difference vegetation index image and they found that image ratioing provided very useful land use mapping information.

2.3. Change Vector Analysis (CVA)

Change vector analysis can represent both the direction and magnitude of a change. The total change magnitude per pixel (CM_{pixel}) can be computed by determining the Euclidean distance between end points through n-dimensional change space [58]:

$$CM_{\text{pixel}} = \sum_{i=1}^{n} (X_2 - X_1)^2 \quad (1)$$

where X_1 and X_2 are date one and date two pixel values in i band.

CVA is complicated to implement, but it is useful for providing information about the change. Change vector analysis was used by Schoppmann and Tyler [46] to monitor changes in and around the Chernobyl nuclear power plant. They concluded that CVA was able to isolate a number of phenomena, including construction and excavation, water level variations, changes in agricultural cover type and riparian vegetation. In addition, change vectors were consistent across both change images. Allen and Kupfer [36] used change vector analysis to examine spruce-fir ecosystems. The results showed the ability of change vectors in multiple biophysical dimensions to differentiate forest disturbance and regeneration trends as an alternative to nominal forest or land cover classifications. Berberoglu and Akin [59] used four change detection techniques, including image differencing, image ratioing, image regression and CVA to detect land use and land cover change in a Mediterranean environment. The results showed that CVA was the most accurate technique for handling the variability present in Mediterranean land use and land cover change.

Combining CVA with other techniques provides more information than either technique alone. For example, Silapaswan, Verbyla [37] used CVA, unsupervised classification and visual interpretation of aerial photographs to detect land cover changes on the Seward Peninsula. They found that, in a transitional region between tundra and boreal forest, the use of CVA and unsupervised classification together provided a more powerful interpretation of change than either method alone.

2.4. Principle Component Analysis (PCA)

Principle component analysis, also referred to as eigenvector transformation, Hotelling transformation and Karhunen Loeve transformation in remote sensing, is a multivariate technique [60] that is used to reduce dataset dimensionality. In this technique, the original remote sensing dataset, which is a correlated variable, is transformed into a simpler dataset for interpretation. This allows the dataset to be uncorrelated variables representing the most important information from the original [15]. The calculation of the variance-covariance matrix (C) of multiband images is expressed as:

$$C = \frac{\sum_{j=1}^{n}(X_j - M)(X_j - M)^T}{n-1} \quad (2)$$

where M and X are the multiband image mean and individual pixel value vectors respectively, and n is the number of pixels.

In change detection, there are two ways to apply PCA. The first method is adding two image dates to a single file, and the second methods is subtracting the second image date from the corresponding image of the first date after performing PCA separately. The disadvantages of PCA can be grouped into three issues: 1) it requires the selection of thresholds for identifying the change; 2) PCA results are difficult to interpret and label; and 3) it does not provide a complete matrix of change class [2].

Principal components analysis has been used in the case of land use and land cover change detection using standardized methods. For example, Baronti, Carla [33] applied PCA to investigate the changes occurring in multi-temporal polarimetric synthetic aperture radar (SAR) images. They used correlation instead of a covariance matrix in the transformation to reduce gain variations that are introduced by the imaging system and that give weight to each polarization. In another example, Liu, Nishiyama [43] compared four techniques, including image differencing, image ratioing, image regression and PCA, from a mathematical perspective. They recognized that standardized PCA achieved the best performance for change detection. Standardized PCA is better than unstandardized PCA for change detection because, if the images subjected to PCA are not measured in the same scale, the correlation matrix normalizes the data onto the same scale [61].

2.5. Chi-Square Transformation

Chi-square transformation uses multi-bands together to produce a single change image. The original chi-square test, also referred to as Pearson's chi-square, was presented by Karl Pearson in early 1990s [62]. Ridd and Liu [26] introduced the chi-square formula that is currently used in change detection:

$$Y = (X - M)^T \Sigma^{-1} (X - M) \quad (3)$$

where Y is the digital value of the changed image; X is the vector of the difference of the six digital values between the two dates; M is the vector of the mean residuals of each band; T is the transverse of the matrix; and Σ^{-1} is the inverse covariance matrix of the six bands [2]. Y is distributed as a chi-square random variable with P degrees of freedom, where P is the number of bands when $Y = 0$, meaning no change. Chi-square can produce a single change image from multiple bands. However, it is not true that $Y = 0$ represents an unchanged pixel when a large portion of the image is changed. In addition, a change related to a specific spectral direction is difficult to identify [2].

The chi-square transformation change detection technique was introduced by Ridd and Liu [26]. They also used other digital change detection techniques, including image differencing, image regression, and tasseled cap transformation. Chi-square was used with six reflective bands to create a single band change image.

2.6. Post-Classification Comparison

Post-classification comparison is a useful technique for extracting land use and land cover information. It is a common method in change detection. Post-classification requires the comparison of independently produced classified changes. It could be that the produced map shows the complete matrix of changes by coding the classification results for time 1 and time 2. Moreover, the subset of changes is observed by selective grouping of the classification results. In post-classification comparison, the images are classified separately, pixel by pixel, on two dates, minimizing the atmospheric and sensor differences between the two dates. In addition, the accurate registration problem of multi-date images is reduced in this technique [1]. However, a great deal of time and expertise is required for the performance of post-classification comparison and the quality of the classified image for each date affects the final accuracy.

Post-classification comparison includes two scheme classification and unsupervised classification algorithms. Supervised classification is a process when the analyst selects a number of areas for an image and then identifies the type of each phenomenon on the computer screen. In accordance with the phenomenon identified by the analyst, the computer identifies the characteristics of the data that comprise each type, and classifies the most similar remaining image pixels [14]. Supervised classification usually requires training data and prior knowledge of the objects that are selected for classification.

Unsupervised classification is a process by which the computer partitions the data without prior knowledge and then applies thematic labels [14].

Unsupervised classification does not normally require training data nor further initial input from the analyst [15].

Post-classification comparison is the most commonly used technique in change detection. [6,21,41,63-69]. The advantage of post-classification was recognized by Mas [70], who compared six change detection techniques to monitor land cover change. These techniques included image differencing, vegetative index differencing, selective PCA, direct multi-date unsupervised classification, post-classification differencing and a combination of image enhancement and post-classification comparison. He concluded that post-classification comparison was the most accurate procedure and presented the advantage of indicating the nature of the changes. Schulz, Cayuela [71] utilized post-classification for investigating land cover changes in dry-land forest landscapes in Central Chile. The maximum likelihood algorithm was used to provide a consistent classifier for multi-date classification. They indicated that the classification accuracy increased, especially after applying the post-classification procedure.

Combining classification with other techniques, such as classification with NDVI, or classification with PCA and CVA, has also been used in some studies to increase classification accuracy as well as to avoid the leak of training sample sets for image classification. For example, Li and Yeh [72] combined interactive supervised maximum likelihood classification with combined PCA to prevent over-estimation of the amount of change. Abd El-Kawy, Rød [73] also used map interpretation to improve classification accuracy and to identify areas with effective use of water for irrigation and areas of private land reclamation.

Unsupervised classification was performed to determine the change in eelgrass meadows with Landsat TM data by Macleod and Congalton [74]. They used three techniques including post-classification, image differencing and PCA. Classification was separately performed for each image using both supervised and unsupervised methods. Brink and Eva [75] also used an unsupervised classification algorithm to monitor 25 years of land cover change dynamics in Africa.

2.7. Artificial Neural Networks (ANNs)

A neural network is a computing process that provides a solution through an unstructured approach founded on the adjustment of the weights connecting the neurons in a network. The concept of a neuron was inspired by the study of biological neural processing [76].

There are three types of layers-an input layer, a hidden layer and an output layer-in which an artificial neural network (ANN) is typically contained. In the input layer, the neurons can be the multispectral reflectance values for each pixel with their texture, e.g. elevation, slope and aspect. In the hidden layers, the use of neurons enables the simulation of nonlinear patterns in the input data. And lastly, in the output layer, the neuron represents a single thematic map for each individual land cover class e.g. agricultural or urban. Each one of these layers consists of interconnected nodes. That interconnection allows the different information to flow in multiple directions [15]. The back-propagation algorithm is used to train the multi-layer perceptron neural model [2].

To extract useful information from the remotely sensed data and ancillary data, the ANN requires training and testing classifications [15]. However, an ANN is sensitive to the amount of training data that is used and requires a long training time. Moreover, ANNs are not common in remote sensing software [2].

Using ANNs in remote sensing has been widely recognized by many researchers [34,35,77-80]. They have also been used to detect changes in many applications. For example, Foody, McCULLOCH [34] utilized a feedforward ANN using a variant of the back-propagation algorithm to classify agricultural crops from synthetic aperture radar data and found that the ANN was able to characterize classes better than discriminant analysis. The ANN was used to classify image data into from-to classes for detecting urban change. Principle component analysis was also applied to extract the salient features and reduce the dimensionality of the input data prior to the ANN-based change detection. By using an ANN, the accuracy increased more than with post-classification comparison. Principle component analysis also improved change detection accuracy [38]. Artificial neural networks have been used to map land use and land cover change with a combination of satellite sensor data and GIS. The ANN approach was used to generate land use and land cover classes from ancillary and spectral data [81]. This combination increased the accuracy of classification compared with spectral classification alone and established relationships between land use and land cover and environmental variables specific to the mapped area.

2.8. Hybrid Change Detection

A hybrid technique uses two or more techniques (e.g. supervised and unsupervised classification, PCA and CVA) to detect a change. There are two types of hybrid techniques. The first type is procedure-based hybrid analysis, which involves the use of different detection techniques in different detection procedures. The second type is result-based hybrid analysis, which means using different change detection techniques successively and then analyzing the results. This technique makes full use of the benefits of many techniques to obtain significant change detection results. On the other hand, the hybrid change detection technique is complex, difficult to perform and not very effective because it depends on the characteristics of other techniques [82].

Hybrid change detection, combining the advantages of supervised and unsupervised classification, was used to derive a land cover map from three Landsat Thematic Mapper Plus (ETM+) images from the year 2000 in three Eastern Europe countries (Slovakia, Poland and the Ukraine). The results were more accurate than using individual techniques [7]. The same results were obtained by Zhang, Ma [83], who used hybrid change detection and a decision tree classifier based on a data mining algorithm. The results showed better accuracy than using individual change detection techniques.

2.9. Image Fusion

Image fusion is a technology that merges two or more images from the same area in different sensors and wavelengths [61,84]. In general, different spectral and spatial resolutions are helpful when high spectral resolution is used in discriminating land use and land cover type, whereas high spatial resolution holds advantages in identifying terrain features and the earth's structure. The purpose of using image fusion is to provide additional information to help users detect and identify targets on the earth's surface. Image fusion can be implemented along with PCA, the intensity hue-saturation (HIS) transform, Brovey's methods and wavelet transformation to provide more accurate information and at high resolution [61].

The advantage of using image fusion in change detection is that fusion techniques can provide high spatial and spectral resolution, which can easily and effectively allow users to extract land use and land cover information. However, it is obvious that resized and registered images, which are obtained from different sensors, may be difficult to implement. For example, in a fusion image from Landsat TM and a SPOT-5 image, the TM image is resampled at the SPOT size, and then the two images have the same size and are registered before image fusion [61]. Therefore, users should take this point into account before performing fusion technique activities.

In a change detection context, image fusion has been used to detect land use and land cover change over urban areas. For example, Zhou, Huang [40] compared three methods for land cover classification of shaded areas from high spatial resolution imagery in an urban environment, including combined spectral information, linear-correlation correction and multisource data fusion. The results indicated that data fusion achieved better accuracy when compared with combined spectral information and linear-correlation correction. Zeng, Zhang [85] presented the results of different temporal SAR and optical image fusion algorithms for land cover change detection. The data used in their research were from SPOT-5 imagery and RADARSAT-1. Peijun, Sicong [44] used feature and decision level fusion to combine simple change detectors and to build an automatic change detection procedure. This method was tested with multi-temporal CBERS and HJ-1 images. The results were satisfactory and more effective than other methods.

2.10. Decision Tree

A decision tree classifier is an algorithm for the labeling of an unknown pattern using a sequence of decisions [61]. The structure of a decision tree is organized with rules, conditions and hypotheses, where rules and conditions are evaluated in order to test the hypotheses. The concept of a hierarchical (or top-down) decision tree is referred to when each hypothesis appears as the trunk of the tree, each rule is a limb of the tree and each condition is a leaf [15]. A decision tree splits a complex decision into many simpler decisions to obtain a solution that is easier to interpret [86].

There are two approaches in the design of a decision tree. The first one is called the manual design approach; it is based on the user's knowledge and relies solely on user interaction. The second approach is based on an automatic procedure. The manual design approach is time-consuming and it does not provide satisfactory results, especially when the number of classes is large and there is spectral overlap between classes [61].

Hansen, Defries [87] used decision tree to classify land cover using data for 1992-1993 from the Advanced Very High Resolution Radiometer (AVHRR). Yang, Ren [88] used a comparison of post-classification and knowledgebased decision tree (K-DT) classification to quantify land use and land cover changes in the watershed of the Inner Mongolia Autonomic Region of China using Landsat TM and ETM+ images from 1987 to 2007. The results showed significant modification and conversion of land use and land cover of the watershed over the course of 20 years.

2.11. Geographic Information Systems (GIS)

All the techniques discussed above are based on pixel by pixel change detection analysis of satellite images as well as being based on images only. It is important to use the benefit of collateral information, such as digital elevation models, hydrology and soil maps [15], which can be provided with the extracted information from remotely sensed images into GIS platforms. Thus, the integration of remotely sensed data with GIS data has the potential to improve the accuracy of results. The main advantage of GIS is that changes can be detected more clearly than with other techniques using multi-source data. However, using different source data with different formats and accuracies may affect the change detection results [2].

In recent years, remote sensing and GIS have been commonly integrated for analyzing and mapping land use and land cover changes [9,11,12,89]. Driving land use and land cover change maps into GIS applications has been done using supervised classification algorithms through remotely sensed software [90,91]. Utilizing ancillary data with satellite imagery, (e.g. digital elevation models and soil maps) provides more accuracy in the detection of change [92]. The use of GIS has also been recognized for detecting and mapping land use and land cover changes. Spatial statistical analysis and advanced functions (e.g. hotspots) have been used for change detection [93]. In addition, screen digitizing of satellite images and previous land use and land cover maps have been used to detect land use and land cover change [94, 95].

3. CHANGE DETECTION ACCURACY ASSESSMENT

Accuracy assessment is very important for understanding the obtained results and making good decisions. Accuracy is determined by the quality of the thematic information and maps that are obtained from remotely sensed data. Accuracy assessment is required to minimize the common sources of error in remotely sensed data (e.g. sensor systems errors such as detectors and cameras, the movement of spacecraft platforms, ground control errors and classification errors). The statistical methods used to assess the accuracy of land cover maps include overall accuracy, user's accuracy, producer's accuracy and the Kappa coefficient.

The consideration of classification errors is common in previous studies and reviews [96-102]. Image classification is subject to a variety of errors that occur during the classification process. The difference between the class label assigned by the classification and objects in the real field is significant error. In this case, an error matrix provides an appropriate summary of the omission and commission errors [98]. An error matrix or a confusion matrix and its measures, such as user's accuracy and producer's accuracy, is the most common approach for determining the accuracy of remote sensing image classification [96,103].

Change detection accuracy assessment is particularly difficult and requires a large amount of work [104]. The matrices that are most commonly used are single date error matrices, binary change/no change error matrices and the transition error matrix. However, the transition error matrix is less commonly reported than single date error matrices and binary change/no change error matrices [105]. The single date classification error matrix needs to be adapted to a change detection error matrix. The characteristics of the new matrix are the same as the single date classification error matrix and assesses errors in changes between two time periods (time 1 and time 2) [104].

Many previous studies on change detection failed to provide the required information to produce a stratified estimator of change area and construct confidence intervals [106]. It is not possible to check whole mapped areas; sampling is the means from which accuracy from land cover maps can be derived [107]. Thematic map accuracy assessment begins with the generation of an error matrix. The factors that should be considered in the generation of an error matrix include ground data collection, classification and sample scheme, sample size and sample unit, and spatial autocorrelation [108]. It is sometimes possible to reduce or remove ground data errors or to derive accurate estimates of change detection without ground data [109].

4. EVALUATION OF CHANGE DETECTION TECHNIQUES

Providing details and determining the direction of change is important. Some change detection techniques, such as image differencing, image ratioing, and PCA, do not provide sufficient change trend information. These techniques only provide change or no change results, therefore, the trend and direction of the change is difficult to determine. Other techniques, such as post-classification comparison, provide more detail about the objects that are selected and provide more information about change trends (Table 2).

In addition, some change detection techniques, such as image differencing, image ratioing, PCA, chi-square, hybrid change detection and CVA, require the selection of a threshold (Table 2) [19]. In this case, selecting an appropriate threshold is important for obtaining good change detection results; however, selecting a suitable threshold is difficult and time-consuming. On the other hand, some techniques are affected by errors and require more time and effort to reduce such errors in order to produce high-quality thematic change detection maps. For example, post-classification comparison always contains omission and commission errors and needs the selection of a confusion matrix and its measures (e.g. user's and producer's accuracy) to minimize these errors.

TABLE 2. Characteristics of change detection techniques.

Technique	Providing Change Direction	Selecting Threshold	Selecting Training	Providing Change Matrix	Data Used to Preform
1. Image differencing	×	√	×	×	Satellite imagery
2. Image ratioing	×	√	×	×	Satellite imagery
3. Change vector analysis	√	√	×	×	Satellite imagery
4. Principle component analysis	×	√	×	×	Satellite imagery
5. Chi-square	×	√	×	×	Satellite imagery
6. Post-classification comparison	√	×	√	√	Satellite imagery
7. Hybrid change detection	×	√	×	×	Satellite imagery
8. Artificial neural networks	√	×	√	√	Satellite imagery & ancillary data
9. Decision tree	×	×	×	×	Satellite imagery
10. Image fusion	×	√	×	×	Satellite imagery
11. GIS	√	×	×	√	Satellite imagery & ancillary data

Furthermore, the strengths and limitations of individual change detection techniques should be taken into account. For example, post-classification comparison minimizes the impacts of atmospheric, sensor and environmental differences between multi-temporal images; image differencing is affected by these elements. Also, image ratioing can reduce the impact of sun angle, shadow and topography. Image differencing and image ratioing allow for easy interpretation of change detection results, whereas the results of principle component analysis are difficult to interpret. The use of additional data, such as digital elevation models, soil and hydrology maps, with satellite imagery may provide additional information about a change. However, the accuracy of different data may affect the quality of land use and land cover change results. It is also important to consider the level of difficulty in dealing with the various functions and software. For example, the limitation of ANNs could be in their functions, which are not common in image processing software [2].

5. SUMMARY

By analyzing the related literature, it seems that no one technique is suitable for all cases. Selection of an appropriate method for detecting change in an object or a phenomenon on the earth's surface depends on a number of elements, including the characteristics of the study area, the spatial resolution of the sensor, atmospheric effects and sun angle, which should be taken into account before applying a suitable technique for the detector. Furthermore, the characteristics objects themselves should be considered before selecting a method because some of these techniques showed more powerful in some cases than others. For instance, for detecting vegetation cover in arid environments, image differencing is more effective than Normalized Difference Vegetation Index. Also, post-classification comparison is useful in two ways. It provides details of changes, and avoids selecting appropriate thresholds. The accuracy of change detection results can be affected by these elements, as well as the resolution of spatial and spectral images.

Research using change detection techniques should consider the elements that may affect the results. Moreover, researchers should provide a quantitative analysis of the results with their conclusions. Many of the studies reviewed failed to do this; as these studies compare the performance of change detection techniques, there was no ground reference for accuracy assessment. It is necessary to link the results obtained using change detection techniques with the real changes occurring on the ground. Some of the previous studies failed to consider the real changes.

Most of the change detection studies considered the impact of urban expansion on tropical and temperate environments; however, unfortunately, few studies addressed changes in desert environments. Certain elements, including urban expansion and natural factors, may lead to changes in a desert environment. These changes are sometimes challenging to detect using satellite imagery. Vegetation cover, for example, can be easily determined from satellite images in a forested area, but not in a desert area. Therefore, further studies are warranted to apply change detection techniques to desert environments.

REFERENCES

1. A. Singh, "Review Article Digital Change Detection Techniques Using Remotely-Sensed Data," International Journal of Remote Sensing, Vol. 10, No. 6, 1989, pp. 989- 1003.
2. D. Lu, et al., "Change Detection Techniques," International Journal of Remote Sensing, Vol. 25, No. 12, 2004, pp. 2365-2401.
3. T. Ramachandra and U. Kumar, "Geographic Resources Decision Support System for Land Use, Land Cover Dynamics Analysis," Proceedings of the FOSS/GRASS Users Conference, Bangkok, September 2004.
4. J. R. E. Jensen, "Urban/Suburban Land Use Analysis," American Society of Photogrammetry, Falls Church, Virginia, Vol. 2, 1983, pp. 1571-1666.
5. J. Rogan and D. M. Chen, "Remote Sensing Technology for Mapping and Monitoring Land-Cover and Land-Use Change," Progress in Planning, Vol. 61, No. 4, 2004, pp. 301-325.
6. E. Brondizio, et al., "Land Use Change in the Amazon Estuary: Patterns of Caboclo Settlement and Landscape Management," Human Ecology, Vol. 22, No. 3, 1994, pp. 249-278.
7. T. Kuemmerle, et al., "Cross-Border Comparison of Land Cover and Landscape Pattern in Eastern Europe Using a Hybrid Classification Technique," Remote Sensing of Environment, Vol. 103, No. 4, 2006, pp. 449-464.
8. R. Pelorosso, et al., "Land Cover and Land Use Change in the Italian Central Apennines: A Comparison of Assessment Methods," Applied Geography, Vol. 29, No. 1, 2009, pp. 35-48.
9. R. B. Thapa and Y. Murayama, "Urban Mapping, Accuracy, & Image Classification: A Comparison of Multiple Approaches in Tsukuba City, Japan," Applied Geography, Vol. 29, No. 1, 2009, pp. 135-144.
10. W. Z. Michalak, "GIS in Land Use Change Analysis: Integration of Remotely Sensed Data into GIS," Applied Geography, Vol. 13, No. 1, 1993, pp. 28-44.
11. Q. Weng, "A Remote Sensing GIS Evaluation of Urban Expansion and Its Impact on Surface Temperature in the Zhujiang Delta, China," International Journal of Remote Sensing, Vol. 22, No. 10, 2001, pp. 1999-2014.

12. J. Rogan, et al., "Integrating GIS and Remotely Sensed Data for Mapping Forest Disturbance and Change," In: M. A. Wulder and S. E. Franklin, Eds., Understanding Forest Disturbance and Spatial Pattern: Remote Sensing and GIS Approaches, 2007, pp. 133-171.
13. D. A. Mouat, et al., "Remote Sensing Techniques in the Analysis of Change Detection," Geocarto International, Vol. 8, No. 2, 1993, pp. 39-50.
14. P. Deer, "Digital Change Detection Techniques in Remote Sensing," DTIC Document, 1995.
15. J. R. Jensen, "Introductory Digital Image Processing: A Remote Sensing Perspective," In: K. C. Clarke, Ed., 3rd Edition, Prentice Hall, The United States of America, 2005.
16. D. M. Muchoney and B. N. Haack, "Change Detection for Monitoring Forest Defoliation," Photogrammetric Engineering and Remote Sensing, Vol. 60, No. 10, 1994, pp. 1243-1251.
17. S. S. Podeh, et al., "Forest Change Detection in the North of Iran Using TM/ETM+ Imagery," Asian Journal of Applied Sciences, Vol. 2, No. 6, 2009, pp. 464-474.
18. M. Jahari, et al., "Change Detection Studies in Matang Mangrove Forest Area, Perak," Pertanika Journal of Science and Technology, Vol. 19, No. 2, 2011, pp. 307-327.
19. P. L. Rosin and T. Ellis, "Image Difference Threshold Strategies and Shadow Detection," Proceedings of the 6th British Machine Vision Conference, Citeseer, 1995.
20. P. A. Rogerson, "Change Detection Thresholds for Remotely Sensed Images," Journal of Geographical Systems, Vol. 4, No. 1, 2002, pp. 85-97.
21. R. Weismiller, et al., "Change Detection in Coastal Zone Environments," Photogrammetric Engineering and Remote Sensing, Vol. 43, No. 12, 1977, pp. 1533-1539.
22. T. L. Sohl, "Change Analysis in the United Arab Emirates: An Investigation of Techniques," Photogrammetric Engineering and Remote Sensing, Vol. 65, No. 4, 1999, pp. 475-484.
23. H. A. Afify, "Evaluation of Change Detection Techniques for Monitoring Land-Cover Changes: A Case Study in New Burg El-Arab Area," Alexandria Engineering Journal, Vol. 50, No. 2, 2011, pp. 187-195.
24. J. R. Jensen and D. Toll, "Detecting Residential Land-Use Development at the Urban Fringe," Photogrammetric Engineering & Remote Sensing, Vol. 48, No. 4, 1982, pp. 629-643.

25. D. Williams and M. Stauffer, "Monitoring Gypsy Moth Defoliation by Applying Change Detection Techniques to Landsat Imagery," Proceeding of the Sympsium on Vegetation Damage Assessment, American Society of Photogammetry, Falls Church, 1978, pp. 221-229.
26. M. K. Ridd and J. Liu, "A Comparison of Four Algorithms for Change Detection in an Urban Environment," Remote Sensing of Environment, Vol. 63, No. 2, 1998, pp. 95-100.
27. P. S. Chavez and D. J. MacKinnon, "Automatic Detection of Vegetation Changes in the Southwestern United States Using Remotely Sensed Images," Photogrammetric Engineering and Remote Sensing, Vol. 60, No. 5, 1994, pp. 571-583.
28. P. Pilon, et al., "An Enhanced Classification Approach to Change Detection in Semi-Arid Environments," Photogrammetric Engineering and Remote Sensing, Vol. 54, No. 12, 1988, pp. 1709-1716.
29. H. Alphan, "Comparing the Utility of Image Algebra Operations for Characterizing Landscape Changes: The Case of the Mediterranean Coast," Journal of Environmental Management, Vol. 92, No. 11, 2011, pp. 2961- 2971.
30. P. J. Howarth and G. M. Wickware, "Procedures for Change Detection Using Landsat Digital Data," International Journal of Remote Sensing, Vol. 2, No. 3, 1981, p. 277-291.
31. R. Nelson, "Detecting Forest Canopy Change Due to Insect Activity Using Landsat MSS," Photogrammetric Engineering and Remote Sensing, Vol. 49, 1983, pp. 1303- 1314.
32. A. Prakash and R. Gupta, "Land-Use Mapping and Change Detection in a Coal Mining Area—A Case Study in the Jharia Coalfield, India," International Journal of Remote Sensing, Vol. 19, No. 3, 1998, pp. 391-410.
33. S. Baronti, et al. "Principal Component Analysis for Change Detection on Polarimetric Multitemporal SAR Data," Geoscience and Remote Sensing Symposium, Surface and Atmospheric Remote Sensing: Technologies, Data Analysis and Interpretation, 1994.
34. G. M. Foody, et al., "Classification of Remotely Sensed Data by an Artificial Neural Network: Issues Related to Training Data Characteristics," Photogrammetric Engineering and Remote Sensing, Vol. 61, No. 4, 1995, pp. 391-401.
35. J. Jensen, et al., "Predictive Modelling of Coniferous Forest Age Using Statistical and Artificial Neural Network Approaches Applied to Remote

Sensor Data," International Journal of Remote Sensing, Vol. 20, No. 14, 1999, pp. 2805-2822.

36. T. R. Allen and J. A. Kupfer, "Application of Spherical Statistics to Change Vector Analysis of Landsat Data: Southern Appalachian Spruce-Fir Forests," Remote Sensing of Environment, Vol. 74, No. 3, 2000, pp. 482-493.
37. C. Silapaswan, et al., "Land Cover Change on the Seward Peninsula: The Use of Remote Sensing to Evaluate the Potential Influences of Climate Warming on Historical Vegetation Dynamics," Canadian Journal of Remote Sensing, Vol. 27, No. 5, 2001, pp. 542- 554.
38. X. Liu and R. G. Lathrop, "Urban Change Detection Based on an Artificial Neural Network," International Journal of Remote Sensing, Vol. 23, No. 12, 2002, pp. 2513-2518.
39. L. Castellana, A. d'Addabbo and G. Pasquariello, "A Composed Supervised/Unsupervised Approach to Improve Change Detection from Remote Sensing," Pattern Recognition Letters, Vol. 28, No. 4, 2007, pp. 405-413.
40. W. Zhou, et al., "Object-Based Land Cover Classification of Shaded Areas in High Spatial Resolution Imagery of Urban Areas: A Comparison Study," Remote Sensing of Environment, Vol. 113, No. 8, 2009, pp. 1769-1777.
41. J. Miettinen, et al., "2010 Land Cover Map of Insular Southeast Asia in 250-m Spatial Resolution," Remote Sensing Letters, Vol. 3, No. 1, 2011, pp. 11-20.
42. Q. Zhang, et al., "Urban Built-Up Land Change Detection with Road Density and Spectral Information from MultiTemporal Landsat TM Data," International Journal of Remote Sensing, Vol. 23, No. 15, 2002, pp. 3057-3078.
43. Y. Liu, et al., "Analysis of Four Change Detection Algorithms in Bi-Temporal Space with a Case Study," International Journal of Remote Sensing, Vol. 25, No. 11, 2004, pp. 2121-2139.
44. D. Peijun, et al., "Fusion of Difference Images for Change Detection over Urban Areas," IEEE Journal of Selected Topics in Applied Earth Observations and Remote Sensing, Vol. 5, No. 4, 2012, pp. 1076-1086.
45. A. J. Peters, et al., "Drought Monitoring with NDVIBased Standardized Vegetation Index," Photogrammetric Engineering and Remote Sensing, Vol. 68, No. 1, 2002, pp. 71-75.

46. M. W. Schoppmann and W. A. Tyler, "Chernobyl Revisited: Monitoring Change with Change Vector Analysis," Geocarto International, Vol. 11, No. 1, 1996, pp. 13-27.
47. S. E. Franklin and B. Wilson, "Vegetation Mapping and Change Detection Using SPOT MLA and Landsat Imagery in Kluane National Park," Canadian Journal of Remote Sensing, Vol. 17, No. 1, 1991, pp. 2-22.
48. J. B. Thayn, "Assessing Vegetation Cover on the Date of Satellite-Derived Start of Spring," Remote Sensing Letters, Vol. 3, No. 8, 2012, pp. 721-728.
49. D. Zheng, et al., "Rates and Patterns of Landscape Change between 1972 and 1988 in the Changbai Mountain Area of China and North Korea," Landscape Ecology, Vol. 12, No. 4, 1997, pp. 241-254.
50. W. G. Kepner, et al., "A Landscape Approach for Detecting and Evaluating Change in a Semi-Arid Environment," Environmental Monitoring and Assessment, Vol. 64, No. 1, 2000, pp. 179-195.
51. J. Taylor, et al., "Monitoring Landscape Change in the National Parks of England and Wales Using Aerial Photo Interpretation and GIS," International Journal of Remote Sensing, Vol. 21, No. 13-14, 2000, pp. 2737-2752.
52. R. M. Lucas, et al., "Characterizing Tropical Secondary Forests Using Multi-Temporal Landsat Sensor Imagery," International Journal of Remote Sensing, Vol. 14, No. 16, 1993, pp. 3061-3067.
53. D. J. Hayes and S. A. Sader, "Comparison of ChangeDetection Techniques for Monitoring Tropical Forest Clearing and Vegetation Regrowth in a Time Series," Photogrammetric Engineering and Remote Sensing, Vol. 67, No. 9, 2001, pp. 1067-1075.
54. E. H. Wilson and S. A. Sader, "Detection of Forest Harvest Type Using Multiple Dates of Landsat TM Imagery," Remote Sensing of Environment, Vol. 80, No. 3, 2002, pp. 385-396.
55. E. J. Christensen, et al., "Aircraft MSS Data Registration and Vegetation Classification for Wetland Change Detection," International Journal of Remote Sensing, Vol. 9, No. 1, 1988, pp. 23-38.
56. J. Jensen, et al., "An Evaluation of the Coast Watch Change Detection Protocol in South Carolina," Photogrammetric Engineering and Remote Sensing, Vol. 59, No. 6, 1993, pp. 1039-1044.
57. C. Munyati, "Wetland Change Detection on the Kafue Flats, Zambia, by Classification of a Multitemporal Remote Sensing Image Dataset,"

International Journal of Remote Sensing, Vol. 21, No. 9, 2000, pp. 1787-1806.
58. J. L. Michalek, et al., "Multispectral Change Vector Analysis for Monitoring Coastal Marine Environments," Photogrammetric Engineering and Remote Sensing, Vol. 59, No. 3, 1993, pp. 635-641.
59. S. Berberoglu and A. Akin, "Assessing Different Remote Sensing Techniques to Detect Land Use/Cover Changes in the Eastern Mediterranean," International Journal of Applied Earth Observation and Geoinformation, Vol. 11, No. 1, 2009, pp. 46-53.
60. A. Singh and A. Harrison, "Standardized Principal Components," International Journal of Remote Sensing, Vol. 6, No. 6, 1985, pp. 883-896.
61. B. Tso and P. P. Mather, "Classification Methods for Remotely Sensed Data," 2nd Edition, CRC Press, Boca Raton, 2009.
62. D. C. Howell, "Chi-Square Test-Analysis of Contingency Tables," Women, Vol. 35, No. 3, 2009, pp. 28-83.
63. S. I. Gordon, "Utilizing LANDSAT Imagery to Monitor Land-Use Change: A Case Study in Ohio," Remote Sensing of Environment, Vol. 9, No. 3, 1980, pp. 189-196.
64. J. Jensen, et al., "Inland Wetland Change Detection Using Aircraft MSS Data," Photogrammetric Engineering and Remote Sensing, Vol. 53, No. 5, 1987, pp. 521-529.
65. M. U. H. Dimyati, et al., "An Analysis of Land Use/Cover Change in Indonesia," International Journal of Remote Sensing, Vol. 17, No. 5, 1996, pp. 931-944.
66. C. Song, et al., "Classification and Change Detection Using Landsat TM Data: When and How to Correct Atmospheric Effects?" Remote Sensing of Environment, Vol. 75, No. 2, 2001, pp. 230-244.
67. D. L. Civco, et al., "A Comparison of Land Use and Land Cover Change Detection Methods," Proceedings of the 2002 ASPRS Annual Convention, Washington DC, 22-26 April 2002.
68. J. Aguirre-Gutiérrez, et al., "Optimizing Land Cover Classification Accuracy for Change Detection, a Combined Pixel-Based and Object-Based Approach in a Mountainous Area in Mexico," Applied Geography, Vol. 34, 2012, pp. 29-37.
69. J. O. Sexton, et al., "Long-Term Land Cover Dynamics by Multi-Temporal Classification across the Landsat-5 Record," Remote Sensing of Environment, Vol. 128, No. 21, 2013, pp. 246-258.

70. J. F. Mas, "Monitoring Land-Cover Changes: A Comparison of Change Detection Techniques," International Journal of Remote Sensing, Vol. 20, No. 1, 1999, pp. 139- 152.
71. J. J. Schulz, et al., "Monitoring Land Cover Change of the Dryland Forest Landscape of Central Chile (1975- 2008)," Applied Geography, Vol. 30, No. 3, 2010, pp. 436-447.
72. X. Li and A. Yeh, "Principal Component Analysis of Stacked Multi-Temporal Images for the Monitoring of Rapid Urban Expansion in the Pearl River Delta," International Journal of Remote Sensing, Vol. 19, No. 8, 1998, pp. 1501-1518.
73. O. R. Abd El-Kawy, et al., "Land Use and Land Cover Change Detection in the Western Nile Delta of Egypt Using Remote Sensing Data," Applied Geography, Vol. 31, No. 2, 2011, pp. 483-494.
74. R. D. Macleod and R. G. Congalton, "A Quantitative Comparison of Change-Detection Algorithms for Monitoring Eelgrass from Remotely Sensed Data," Photogrammetric Engineering and Remote Sensing, Vol. 64, No. 3, 1998, pp. 207-216.
75. A. B. Brink and H. D. Eva, "Monitoring 25 Years of Land Cover Change Dynamics in Africa: A Sample Based Remote Sensing Approach," Applied Geography, Vol. 29, No. 4, 2009, pp. 501-512.
76. V. B. Rao and H. V. Rao, "C++ Neural Networks and Fuzzy Logic," MIS-Press, New York, 1993.
77. G. Foody and M. Arora, "An Evaluation of Some Factors Affecting the Accuracy of Classification by an Artificial Neural Network," International Journal of Remote Sensing, Vol. 18, No. 4, 1997, pp. 799-810.
78. J. Benediktsson and J. Sveinsson, "Feature Extraction for Multisource Data Classification with Artificial Neural Networks," International Journal of Remote Sensing, Vol. 18, No. 4, 1997, pp. 727-740.
79. J. R. Jensen, F. Qiu and K. Patterson, "A Neural Network Image Interpretation System to Extract Rural and Urban Land Use and Land Cover Information from Remote Sensor Data," Geocarto International, Vol. 16, No. 1, 2001, pp. 21-30.
80. F. Qiu and J. Jensen, "Opening the Black Box of Neural Networks for Remote Sensing Image Classification," International Journal of Remote Sensing, Vol. 25, No. 9, 2004, pp. 1749-1768.

81. J. F. Mas, "Mapping Land Use/Cover in a Tropical Coastal Area Using Satellite Sensor Data, GIS and Artificial Neural Networks," Estuarine, Coastal and Shelf Science, Vol. 59, No. 2, 2004, pp. 219-230.
82. J. Y. G., et al., "A Review of Multi-Temporal Remote Sensing Data Change Detection Algorithms," The International Archives of the Photogrammetry, Remote Sensing and Spatial Information Sciences, Vol. 37, No. B7, 2008, pp. 757-762.
83. Y. Zhang, et al., "Hybrid Change Detection for Watershed Impervious Surface Using Multi-Time Remotely Sensed Data," IEEE International of Geoscience and Remote Sensing Symposium, Barcelona, 23-28 July 2007, pp. 1939-1942.
84. L. Wald, "Some Terms of Reference in Data Fusion," IEEE Transactions on Geoscience and Remote Sensing, Vol. 37, No. 3, 1999, pp. 1190-1193.
85. Y. Zeng, et al., "Image Fusion for Land Cover Change Detection," International Journal of Image and Data Fusion, Vol. 1, No. 2, 2010, pp. 193-215.
86. M. Xu, et al., "Decision Tree Regression for Soft Classification of Remote Sensing Data," Remote Sensing of Environment, Vol. 97, No. 3, 2005, pp. 322-336.
87. M. C. Hansen, et al., "Global Land Cover Classification at 1 km Spatial Resolution Using a Classification Tree Approach," International Journal of Remote Sensing, Vol. 21, No. 6-7, 2000, pp. 1331-1364.
88. X. Yang, et al., "Impacts of Land Use and Land Cover Changes on Evapotranspiration and Runoff at Shalamulun River Watershed, China," Hydrology Research, Vol. 43, No. 1-2, 2012, pp. 23-37.
89. R. Efe, et al., "Land Use and Land Cover Change Detection in Karinca River Catchment (NW Turkey) Using GIS and RS Techniques," Journal of Environmental Biology, Vol. 33, No. 2, 2012, pp. 439-447.
90. D. Tripathi and M. Kumar, "Remote Sensing Based Analysis of Land Use/Land Cover Dynamics in Takula Block, Almora District(Uttarakhand)," Journal of Human Ecology, Vol. 38, No. 3, 2012, pp. 207-212.
91. S. Gajbhiye and S. K. Sharma, "Land Use and Land Cover Change Detection of Indra River Watershed through Remote Sensing Using Multi-Temporal satellite Data," International Journal of Geomatics and Geosciences, Vol. 3, No. 1, 2012, pp. 89-96.

92. S. Reis, "Analyzing Land Use/Land Cover Changes Using Remote Sensing and GIS in Rize, North-East Turkey," Sensors, Vol. 8, No. 10, 2008, pp. 6188-6202.
93. P. P. Serra, et al., "Land-Cover and Land-Use Change in a Mediterranean Landscape: A Spatial Analysis of Driving Forces Integrating Biophysical and Human Factors," Applied Geography, Vol. 28, No. 3, 2008, pp. 189-209.
94. I. Abbas, et al., "Mapping Land Use-Land Cover and Change Detection in Kafur Local Government, Katsina, Nigeria (1995-2008) Using Remote Sensing and GIS," Research Journal of Environmental and Earth Sciences, Vol. 2, No. 1, 2010, pp. 6-12.
95. N. Nagarajan and S. Poongothai, "Effect of Land Use/ Land Cover Change Detection of Ungauged Watershed," World Applied Sciences Journal, Vol. 17, No. 6, 2012, pp. 718-723.
96. R. G. Congalton, "A Review of Assessing the Accuracy of Classifications of Remotely Sensed Data," Remote Sensing of Environment, Vol. 37, No. 1, 1991, pp. 35-46.
97. C. Conese and F. Maselli, "Use of Error Matrices to Improve Area Estimates with Maximum Likelihood Classification Procedures," Remote Sensing of Environment, Vol. 40, No. 2, 1992, pp. 113-124.
98. G. M. Foody, "Status of Land Cover Classification Accuracy Assessment," Remote Sensing of Environment, Vol. 80, No. 1, 2002, pp. 185-201.
99. R. Latifovic and I. Olthof, "Accuracy Assessment Using Sub-Pixel Fractional Error Matrices of Global Land Cover Products Derived from Satellite Data," Remote Sensing of Environment, Vol. 90, No. 2, 2004, pp. 153-165.
100. H. Liu and Q. Zhou, "Accuracy Analysis of Remote Sensing Change Detection by Rule-Based Rationality Evaluation with Post-Classification Comparison," International Journal of Remote Sensing, Vol. 25, No. 5, 2004, pp. 1037-1050.
101. C. Liu, et al., "Comparative Assessment of the Measures of Thematic Classification Accuracy," Remote Sensing of Environment, Vol. 107, No. 4, 2007, pp. 606-616.
102. P. P. L. Zimmerman, et al., "An Accuracy Assessment of Forest Disturbance Mapping in the Western Great Lakes," Remote Sensing of Environment, Vol. 128, No. 21, 2013, pp. 176-185.

103. A. Comber, et al., "Spatial Analysis of Remote Sensing Image Classification Accuracy," Remote Sensing of Environment, Vo. 127, 2012, pp. 237-246.
104. R. G. Congalton and K. Green, "Assessing the Accuracy of Remotely Sensed Data: Principles and Practices," 2th Edition, CRC Press, Boca Raton, 2008.
105. P. A. J. van Oort, "Interpreting the Change Detection Error Matrix," Remote Sensing of Environment, Vol. 108, No. 1, 2007, pp. 1-8.
106. P. P. Olofsson, et al., "Making Better Use of Accuracy Data in Land Change Studies: Estimating Accuracy and Area and Quantifying Uncertainty Using Stratified Estimation," Remote Sensing of Environment, Vol. 129, No. 15, 2013, pp. 122-131.
107. G. Banko, "A Review of Assessing the Accuracy of Classifications of Remotely Sensed Data and of Methods Including Remote Sensing Data in Forest Inventory," International Institute for Applied Systems Analysis, Laxenburg, 1998.
108. R. G. Congalton, et al., "Quality Assurance and Accuracy Assessment of Information Derived from Remotely Sensed Data," In: J. D. Bossler, Ed., Manual of Geospatial Science and Technology, CRC Press, London, 2001, pp. 349-361.
109. G. M. Foody, "Assessing the Accuracy of Land Cover Change with Imperfect Ground Reference Data," Remote Sensing of Environment, Vol. 114, No. 10, 2010, pp. 2271-2285.

10

CHAPTER

APPLICATION OF REMOTE SENSING AND GEOGRAPHICAL INFORMATION SYSTEMS IN DETERMINING THE GROUNDWATER POTENTIAL IN THE CRYSTALLINE BASEMENT OF BULAWAYO METROPOLITAN AREA, ZIMBABWE

Constant Chuma[1,2*], Oluwagbenga O. I. Orimoogunje[3], Dumisani J. Hlatywayo[2], Joseph O. Akinyede[1]

[1]*African Regional Centre for Space Science and Technology Education in English, Obafemi Awolowo University, Ile-Ife, Nigeria*

[2]*Department of Applied Physics, Faculty of Applied Sciences, National University of Science and Technology, Bulawayo, Zimbabwe*

[3]*Department of Geography, Faculty of Social Sciences, Obafemi Awolowo University, Ile-Ife, Nigeria*

ABSTRACT

This study is concerned with the identification and delineation of aquiferous zones for potential groundwater development across Bulawayo Metropolitan from remotely sensed data and geological inference. Attempts have been made to review literatures on groundwater exploitation in the study area and the constraints to effective and sustainable management of underground water in the study area. Remote Sensing and Geographical Information Systems (GIS) is useful in the recognition and delineation of aquiferous zones for potential groundwater in crystalline basement aquifers. LANDSAT ETM+ image, SRTM data, aeromagnetic data and other ancillary data sets were utilized to extract information on the groundwater storing controlling features of this study area. Six thematic maps were produced from remote sensing data and other ancillary data—Land use/landcover, drainage density, slope map, contact density, lithology and lineament density. GIS modeling technique of the index overlay method was used to produce the groundwater potential map. The study revealed that the regional lineaments correlate with faults, fracture zones, and lithological contrasts along fold belts in the crystalline basement rocks, while the main direction of faulting and jointing is north-north-west to north with several faults oriented, to the north-north-east, parallel to the Great Dyke. Proximity to lineaments is the highest zone of increased porosity and permeability which in turn have a greater chance of accumulating groundwater. The results have shown massive spatial variability of ground water potential ranging from very good to poor. The variability closely followed variations in the structures, geology, topography/slope, drainage density and land use/land cover in the project area. This work is a reconnaissance which needs to be validated by the use of high-resolution terrain data and satellite imagery and the quantitative analysis should be done using geophysical and hydrogeological surveys.

Keywords: Groundwater Potential; Lineaments; Remote Sensing; GIS; Crystalline Basement; Aquifer; Bulawayo Metropolitan

1. INTRODUCTION

Groundwater plays a crucial role in arid and semi-arid regions in most parts of the world especially in Africa. However, in Africa most arid and semi-arid regions are underlain by crystalline basement rocks and groundwater development is complicated by the high variability of geological and hydrogeological settings of the basement aquifers [1-3]. The situation is not good for Bulawayo Metropolitan which is covered by crystalline complex rock and frequent water shortages are being faced annually due to unreliable surface water supplies [4].

There has been comparatively little research conducted on the occurrence of groundwater in the whole crystalline basement aquifer of the Bulawayo Metropolitan, despite its significance in mitigating problems of water shortages in semi-arid regions. Most of the research conducted on part of the area has focused on groundwater quality monitoring, groundwater resource evaluation and partial characterisation of the subsurface e.g. [4-11]. Consequently, there is insufficient information to explain the occurrence and distribution of groundwater in the Bulawayo basement complex aquifer. This has slowed down the exploitation and development of groundwater in the Bulawayo City as a complimentary water source to the surface water. In [2], it was also reported that the lack of understanding of the occurrence, movement and recharge of groundwater in basement terrain has frequently contributed to the unsustainable use of this resource.

Through the use remote sensing and GIS, suitable areas for recharging basement aquifers can be delineated which have porous lithologies, maximum fractures, highly weathered region and regions of null slope. The exploration of surface features by integrating the remote sensing data and GIS techniques proved to be an essential in identifying aquiferous zones for potential groundwater exploitation in Bulawayo crystalline basement rock. This study provided preliminary information which will enhance the success of the quantitative subsurface geophysical investigation that will subsequently follow the study. The regional lineament map produced in this study is not only essential to groundwater but will be used as a reconnaissance study in mineral exploration, as well as in identifying earthquake potential zones. Zimbabwe has been facing tectonic movements in most parts of the country including the western part of Bulawayo recent that are being investigated [12,13].

The objectives of this study, therefore were to produce a regional structural lineament map of Bulawayo Metropolitan from remotely sensed data, topographic, geological inference and structural data; prepare thematic maps of the area such as lithology, lineaments, drainage density, slope and land use/land cover from remotely sensed data and other data sources; assess the hydrogeological implication of the lineaments by integrating them with the available ancillary data especially digital elevation models (DEM), and finally identify and delineate groundwater potential zones through integration of various thematic maps within a GIS environment.

2. DESCRIPTION OF STUDY AREA

The study area lies in the water-distressed, western region of Zimbabwe. It is located between longitude 28°22'05"S and 28°46'12"S and between latitude 20°18'04"E and 20°00'00"E. The gross command area is approximately 400 km^2 in areal extent. Bulawayo is the second largest city in Zimbabwe with a population of over 1 million people, and is regarded as the industrial and business capital of Zimbabwe and is home to the National Railways of Zimbabwe because of its strategic position near Botswana and South Africa. It is the nearest large city to most tourist resort areas in Zimbabwe such as Hwange National Park, Matopo National Park and Victoria Falls. Figure 1 shows the study area and inserts of the map of Africa and the climate is semi-arid with dry winters and wet summers. The mean annual rainfall is 550 mm [14]. Rainfall is not evenly distributed throughout the year and droughts are a frequent phenomenon. Mean annual maximum and minimum temperatures for the study area is 25.8°C and 12.7°C, respectively.

3. MATERIALS AND METHODS

3.1. Geology of the Study Area

The geology of Bulawayo is mainly constituted by the crystalline basement rock. The crystalline basement encompasses the Bulawayo greenstone belt and the surrounding granite rock. According to the geological report by [15] the greenstone belt is flanked by a belt of phyroblastic tonalitic to granodioritic gneisses, penetrated by later granitic bodies. The greenstones comprise a succession of tightly-folded metavolcanic and metasedimentary rocks with generally steep dips away from the margins of the granitic rocks. A second phase of deformation is associated with thrusting directed from the eastern part of the area over a western hinterland consisting of granitic gneiss and lower greenstone rocks. Superimposed in this phase is an episode in which a strong east-west cleavage was developed. Faulting and shearing later occurred mainly along north to north-north-west trends with some off-setting boundaries. The general lithology underlay the study area shown in Figure 2.

3.2. Preparing the Thematic Maps

This study was mainly carried out using Remote Sensing (RS) and Geographical Information Systems (GIS). In addition to that, an extensive desk study was done in collection, collation and analyzing aeromagnetic data. Remote sensing have the advantages of spatial, spectral and temporal availability of data covering large and inaccessible areas within a short time and has become a very rapid and cost effective tool in assessing, monitoring and conserving groundwater resources [16]. On the other hand, GIS is powerful environment for real time database development, especially in studies such as delineating groundwater pollution potential zones [17-19] and recharge sites [20], groundwater modeling studies [21,22] among others. The data for this study were obtained mainly from geologic maps, topographic maps, LANDSAT 7 enhanced thematic mapper plus (ETM+) image of October 2005, ASTER DEM, aeromagnetic data, geologic reports and climatological records coupled with intensive fieldwork.

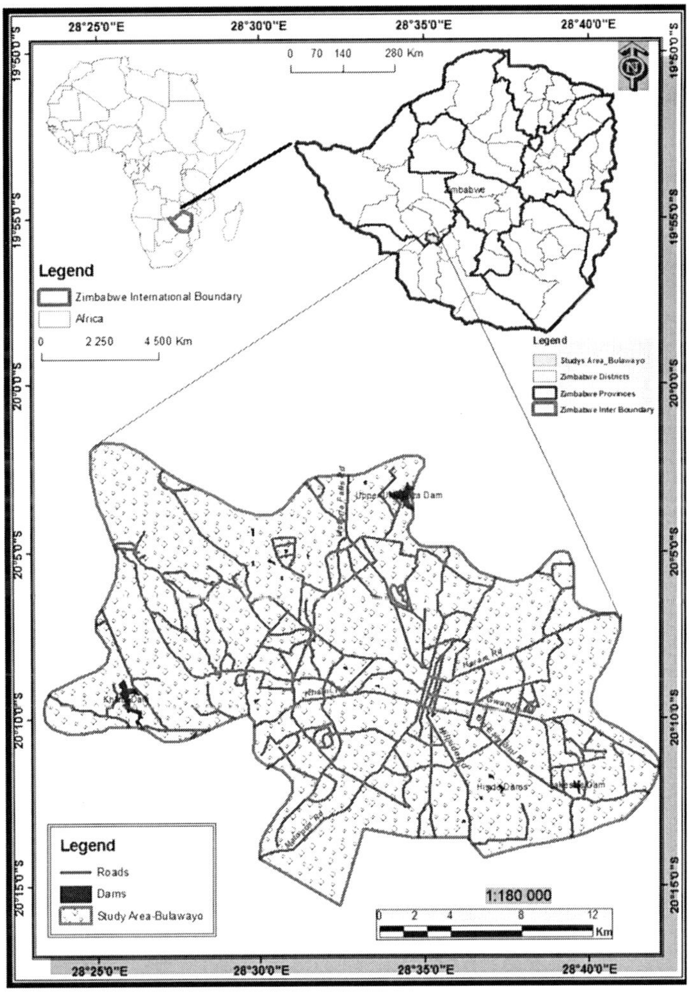

FIGURE 1. Study area showing the roads, dams and inserts of Africa and Zimbabwe.

The methodology adopted in this study is simplified by the schematic flow chart in Figure 3. There were four correlated processing pathways followed: the LANDSAT ETM+ data, ASTER DEM data, aeromagnetic data and geological data. Remote sensing data were processed separately from the other data from which land-cover classes were determined through image classification. Additionally, lineaments in the study area were identified visually. Ancillary data were further processed to provide the layers, and classes for each layer needed for the ArcGIS Model.

The lineaments were identified by visual interpretation of both LANDSAT ETM+ and ASTER DEM processed images (Figure 4). In order to clearly identify the lineaments from LANDSAT ETM+ data, a false color composite band combination of 754 (RGB) was used which was preceded with principal component analysis and directional filtering using Environment for Visualizing Image software (ENVI 4.7). The lineaments observed on the imagery are shadowed with radiation from the sun, and were traced onto the base map. Since the area was built up there, very few lineaments were identified from

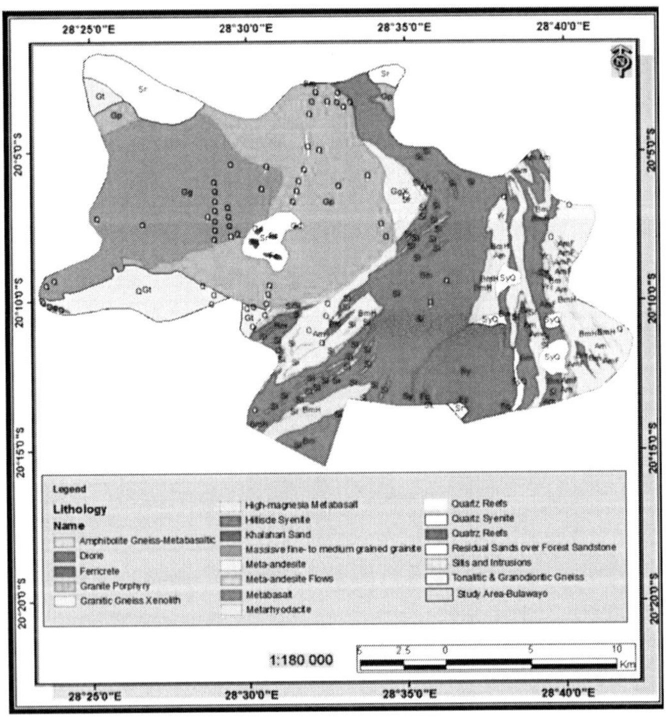

FIGURE 2. Map showing the geological pattern in the study area.

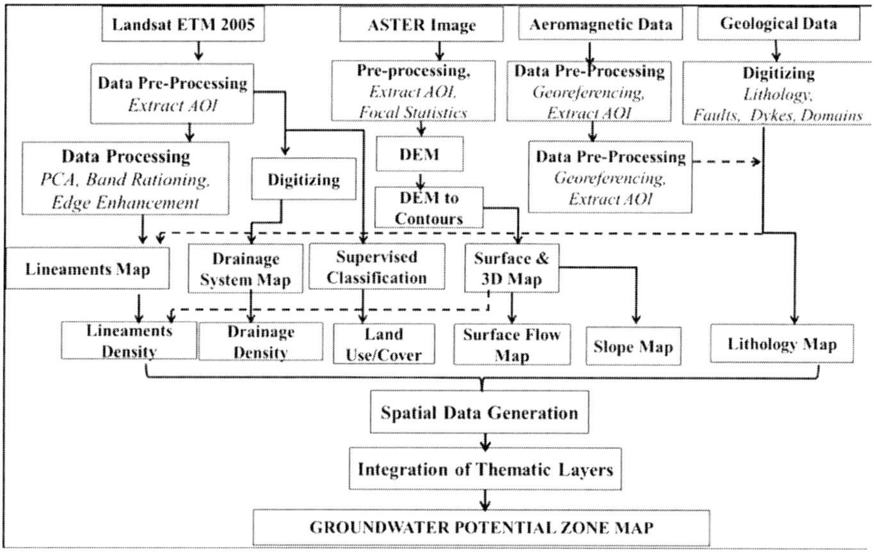

FIGURE 3. The methodology used to determine the groundwater potential zones in the study area.

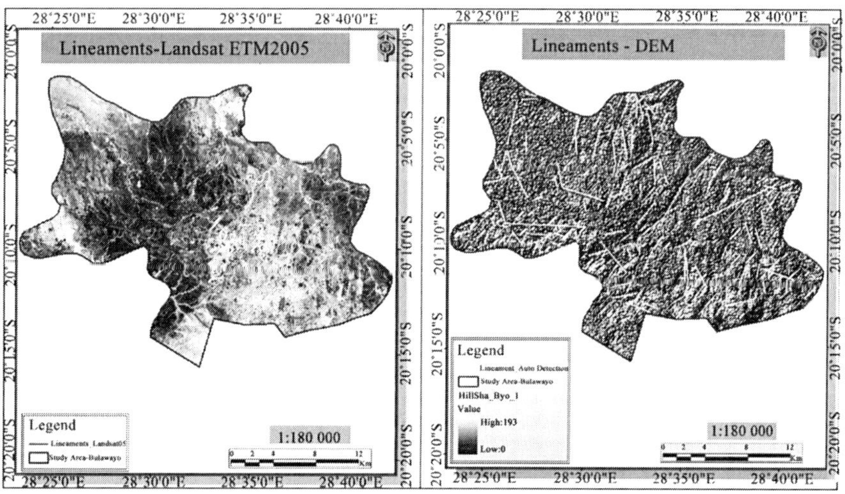

FIGURE 4. The lineaments observed from the LANDSAT EMT (right) and ASTER DEM data (left).

the LANDSAT ETM+ image. The ASTER DEM data were converted to a hillside terrain map using ArcGIS (ArcMap). From the hillshade surface lineaments were observed by a drop in elevation for a short distance.

Drainage pattern is one of the most important indicators of hydrogeological features, because drainage pattern and density are controlled in a fundamental way by the underlying lithology [23]. The drainage map was prepared by hand digitizing the drainage lines from processed LANDSAT ETM+ and geological map. Drainage density is the length of streams per unit area (Equation (1)). It is usually measured in km/km^2.

$$D_d = L_s / A_s \tag{1}$$

where D_d is the drainage density, L_s is the total length of streams, and A_s is the total area of the whole basin.

The slope amount map was prepared using contours produced from ASTER DEM 90m data and topographic map. Surfer 8 was then used to produce a three dimension surface map.

Land use and land cover (LULC) influences the occurrence of sub-surface groundwater occurrence, as a result supervised classification using the maximum likelihood classification was performed on the LANDSAT EMT+ image using ERDAS Imagine version 9.2. A false color combination of 432 (RGB) was used to identify the different themes in the image.

3.3. Groundwater Potential Modelling

In order to delineate potential groundwater sites in the project area, all the data sets were integrated using the model constructed in ArcGIS Model builder engine [24]. The final map produced by Weighted Linear Combination (WLC) where each class individual's weight was multiplied by the map scores and then adds to the results as in Equation (2).

$$S = \sum W_i X_i \tag{2}$$

where S is the suitability, W_i is the weight for each map score and X_i is the individual map.

3.4. Weighting of Thematic Maps

Finally the six thematic layers were converted into a grid with related item weight generated from the Analytic Hierarchy Process (AHP) by [25,26].

The final six thematic maps were reclassified in the ArcMap according to the importance of each factor in each raster image as follows:

- Reclassification of land use/cover for analysis was done based on their character to infiltrate water into the ground and to hold water on the ground.
- Drainage density raster was reclassified based on the recharge potential.
- Pairwise comparison done and the weight calculated for slope angle was based on the fact that the flatter the topography (low slope angle) is the better are the chances for groundwater accumulation.
- Reclassified map of lineament was produced based on the weight calculated after a pair-wise comparison done based on the fact that areas closer to lineaments are the highest zone of increased porosity and permeability which in turn have greater chance of accumulating groundwater.
- Reclassification of the lithologies weight was based on the susceptibility of the rock to weathering.
- Contacts density map was based on the nature of the rocks in contact and the density of the contacts. Contacts normally act as conduit of groundwater movement in basement rocks.

In the AHP method the relative importance of each individual class within the same map and vector maps were compared to each other. As a result important matrices were produced using calculated weights from the IDRSIS module (Table 1). MatLab software was used to solve the matrices and calculate the consistence value. The grids in the integrated layer were grouped into different groundwater potential zones by a suitable logical reasoning and conditioning with emphasis on the hydrogeological condition of each grid cell. The layers were then overlaid using a weighted sum in the ArcMap module. Figure 5 illustrates the overlaying of the six factors that affects the occurrence of groundwater in the ArcSene environment.

The different polygons in the thematic layers were labeled separately and then registered. In the final thematic layer initially each of the polygons was designated on the basis of interpretation of groundwater occurrence and weights in terms of interpreting the probability level of occurrence that was assigned to each polygon.

4. RESULTS AND DISCUSSION

Geospatial analysis of the aeromagnetic data was useful in delineating the extent of the Bulawayo crystalline basement rock as shown in Figure 6. The high magnetic susceptibility values observed indicate the presence of igneous rocks which are mainly composed of basalt and granite. These two rock formations gave what is known as the Bulawayo Crystalline Basement mentioned in [27]. Therefore, the study area is mainly underlaid by a basement aquifer. Groundwater is mainly controlled by secondary porosity, which is due to mechanical and chemical weathering.

The high magnetic values are associated with serpentinised volcanic activity whilst the sills and other intrusions that developed over the years. Crystalline basement rocks (i.e. usually of Precambrian age) have generally been subjected to multiple tectonic events under varying stress conditions that resulted in complex patterns of ductile folding and brittle fracturing in the near-surface regions of the earth's crust [28]. According to [1] basement aquifers are essentially phreatic in character, but may respond to localized abstraction in a semi-confined fashion, if the rest water level occurs in a low permeability horizon (e.g., in a clayey regolith). The greenstone belt in Bulawayo forms the crystalline rock formation which favors the occurrence of the basement aquifer in the study area.

4.1. Regional Structural Lineaments Analysis

Lineaments being a surface manifestation of structurally controlled features such as faults, fractures and rock contacts their high density may present incipient highly connected fractures that are favorable for the accumulation of groundwater. Therefore, lineament analysis of the study area extracted from both the remotely sensed data and geological images gave important information on subsurface features that controls the movement and or storage of groundwater (Figure 7). In addition to that, lineaments like joints, fractures etc., developing generally due to tectonic stress and strain, provide important clue on surface features and are responsible for the infiltration of surface runoff into the subsurface and also for the movement and storage of groundwater [29]. Most topographic lineaments in the study area correlate with faults, fracture zones, and lithological contrasts along fold belts and in crystalline basement rocks.

A detailed structure of the Bulawayo Greenstone Belt evidently is elucidated within the overall setting and framework of deformation of the Zimbabwean craton (Figure 8). Dolerite dykes; quartz and pegmatite veins traverse the area both in concordant and discordant fashions, which represent a later phase of basic igneous activity. All the rock types except the younger intrusive have suffered several phases of deformation and have been metamorphosed. These deep-seated structures and tectonic features include, for example, buried uplifts, buried igneous plugs and salt diapirs, subsurface folds and faults (both thrust and normal faults), strata affected by subtle regional warping, and stratigraphic pinch-outs [30].

TABLE 1. Weight applied in ArcMap weight sum scheme.

FACTORS	Lineaments	Lithology	Contacts	Slope	Drainage	LULC	Weight	Weight [%]
Lineaments	1	2	2	3	8	9	0.3455	35
Lithology	1/2	1	2	4	7	8	0.2905	29
Contacts	1/2	1/2	1	3	5	7	0.1995	20
Slope	1/7	1/4	1/3	1	3	6	0.0946	9
Drainage	1/8	1/7	1/5	1/3	1	5	0.0483	5
Land Use	1/9	1/8	1/7	1/6	1/5	1	0.0216	2

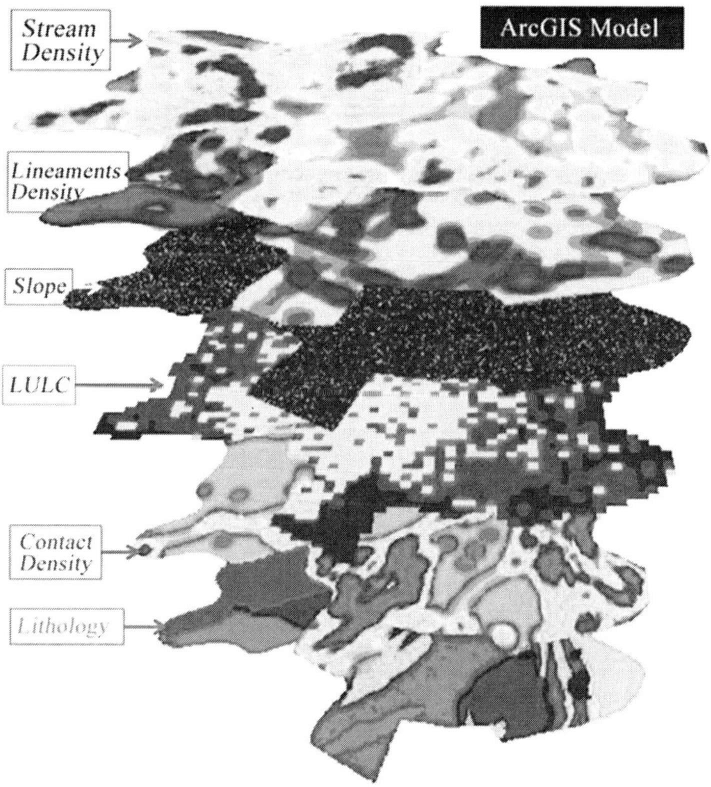

FIGURE 5. Integration of the thematic maps which gives the potential zones.

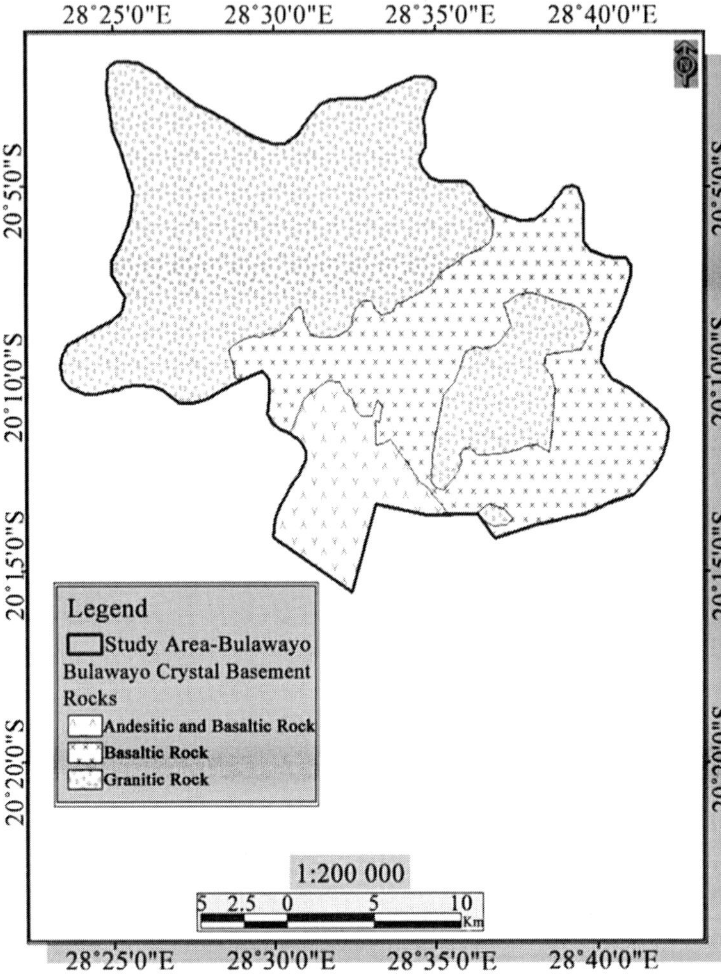

FIGURE 6. The geospatial analysis of aeromagnetic data indicating Bulawayo crystalline basement rocks.

From the regional structural lineament map a rose diagram was produced using Integrated Land and Water Information System 3.3 (ILWIS) software which shows the direction and magnitude of lineaments (Figure 8). Fracture zones usually occur along lineaments and often correspond to the surface drainage patterns. Fractures, dykes and zones of weathering and/or overburden deposits are targets for the exploitation of groundwater.

The main direction of the lineaments is north-westerly with more west-north-west trends in the west of the study area. The ENE-WSW to NE-SW lineaments are largely due normal faults with possibly of a small amount of transcurrent movement. As a result most of the rivers flow in the western direction of the study area as shown by the drainage map overlaid on the DEM in Figure 9. Bulawayo has four major rivers; Umatseumhlope River which runs from south to north, Khami River which runs east to west, Mpopoma River and Umguza River which run from south-east to north-west. All the small river and large rivers, which are found in the project area drain from central part to west and north direction.

4.2. Analysis of Factors Affecting the Occurrence of Groundwater

All surface and subsurface factors that affect the occurrence of groundwater were acquired, processed and analyzed. The reclassified thematic maps of all the six parameters which were incooperated in the APH system are shown in Figure 10.

Comparison of the drainage system of the area and structure has shown that the drainage system of the area is structurally controlled following lineaments directions. Dendritic and parallel drainage pattern are recognized, which are indicative of the presence of structures that act as conduits or storage for sub-surface water. Structurally controlled drainage patterns are observed in western and eastern part of the project area. Dendritic are formed on impervious rock whereas parallel drainage system form on uniform gentle slope. Overally, the drainage pattern in Bulawayo shows that the area very high drainage density is found in the west and north parts of the study area. Moderate and low drainage density concentrates in the southern and central part of the project area. The distribution is mainly attributed to the weak zones that are created due faulting, shearing and geological contacts.

Evaluation of LANDSAT ETM+ image and topographic map of year 2005 has shown that there is remarkable expansion of settlements since the independence of Zimbabwe in 1980, which negatively affect the groundwater recharge of the area. The original vegetation is more preserved in area far from the City Centre. This is due to the fact that people had made the natural forest to disappear on the flat lands near to the main road and cleared the land for construction. Generally settlements are found to be the least suitable for infiltration and after pair-wise comparison of each class weight for each class was calculated.

Based on the susceptibility of the rock to weathering, about nine classes were observed which shows have an intensive analysis of the crystalline basement rock. Results have shown that basalt rocks are easily weathered compared to the surrounding granite. Basalt rocks are associated with presence of structures such fractures, fissures, veins, joint and other structural deformations that occur in basement complex which controls the flow of ground water and also influence the rate of recharge and discharge of the main aquiferous units. This agrees with the aeromagnetic imaging of the area which shows a contact between the granite rocks and the metabasalt rocks which are penetrated intrusions in the south west part of the study area.

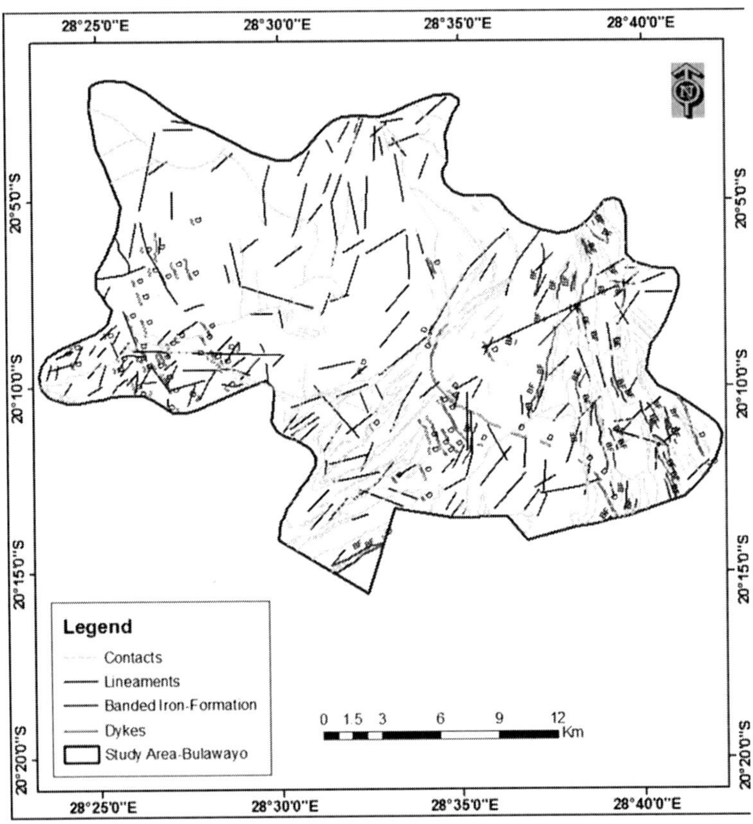

FIGURE 7. Geological structures, extracted lineaments and contacts.

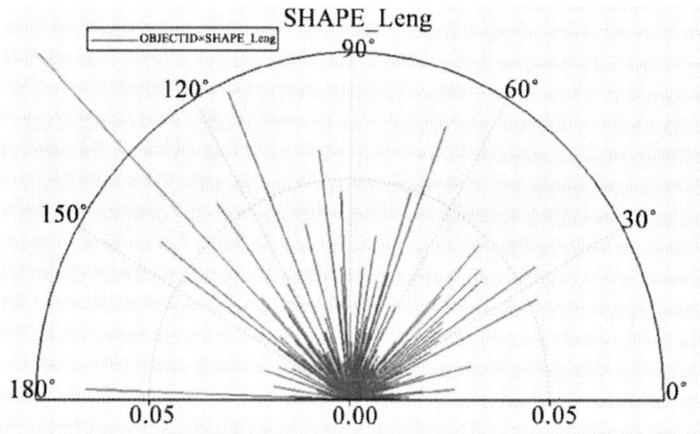

FIGURE 8. Rose diagram showing directions and lengths of lineaments in Bulawayo.

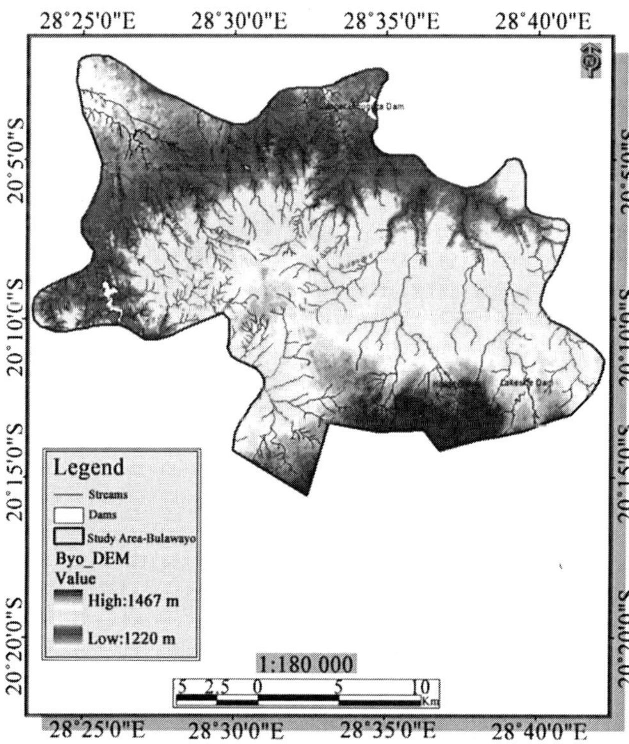

FIGURE 9. Drainage pattern overlaid on the DEM.

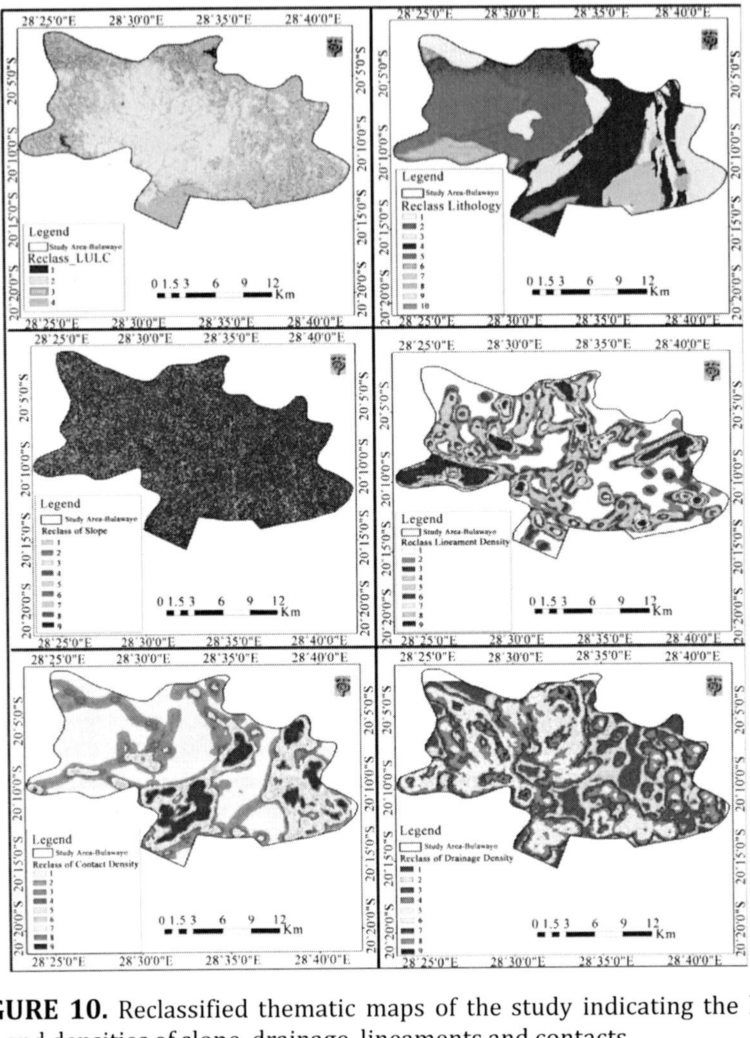

FIGURE 10. Reclassified thematic maps of the study indicating the land use and densities of slope, drainage, lineaments and contacts.

Figure 11 portrays a three dimensional surface map of the study area that was produced from the ASTER DEM data which was processed and analyzed in the Surfer 8 software. The surface map is showing an abrupt change in terrain which explains the variability of the slope in all directions. This is also an indication of the possible direction of geological structures in the study which control the movement and storage of groundwater. The slope was generated from the (Triangulated Irregular Network) TIN in the

ArcMap environment. Slope map layer indicates the relative time available for runoff via land to infiltrate. The region is mainly hilly with an extensive plateau to the south and small plateaus to the north and east. The valley itself is quite narrow and elongates, with steep ridge slopes that grade into a portion of terraces that, in turn, gently slope towards the main river channel. Topography of the land surface determines the general direction of groundwater flow, and it influences groundwater recharge and discharge.

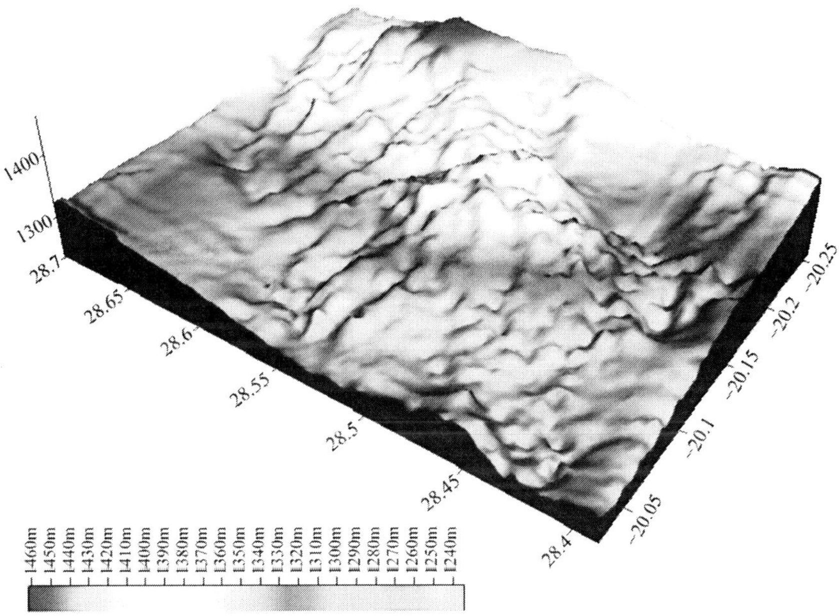

FIGURE 11. The three dimensional analysis of the study area.

4.3. Occurrence of Groundwater in Bulawayo

The map produced from the integration of the six thematic maps was reclassified and five categories of groundwater potential zones were produced as shown in Figure 12. The amount of groundwater is mainly determined by the structural geology environment and the topographic effect. The spatial distributions of the various groundwater potential zones obtained from the model generally show regional patterns of lineaments, drainage, landform and lithology.

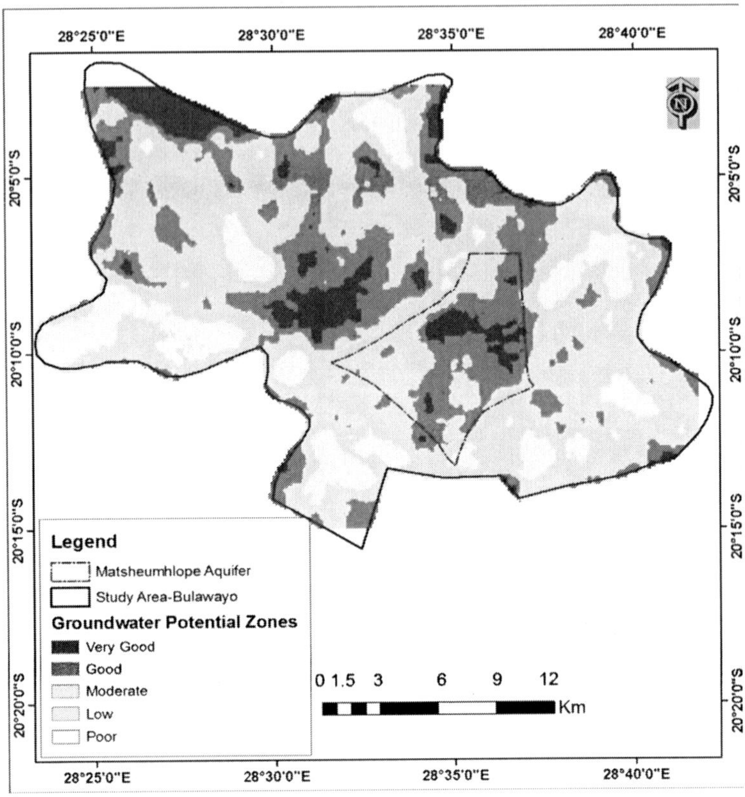

FIGURE 12. Groundwater potential zones identified from the weighted factors.

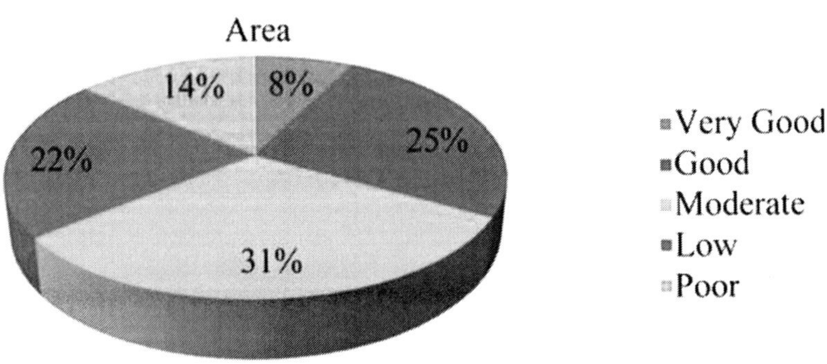

FIGURE 13. Pie chart of the classified categories obtained from groundwater potential model.

Spatially, very good and good categories are distributed along areas near to lineaments and less drainage density and where the lithology is affected by secondary structure and having interconnected pore spaces. This highlights importance of lineaments, geology and hydro-geomorphological parameters in the project area. Areas with moderate groundwater prospects are attributed to contributions from combinations of the land use/cover, lithology, slope and landform. The low to poor categories of groundwater potential zones are spatially distributed mainly along ridges where slope class is very high, the lithology is compact/massive and far from lineaments.

The greenstone belt in Bulawayo forms the crystalline rock formation which favors the occurrence of the basement aquifer in the study area. Groundwater is mainly controlled by secondary porosity, which is due to mechanical and chemical weathering. The fractures and faults that are produced during the course of these processes act as conduit for groundwater movement.

Figure 13 shows the pie chart of the classified categories obtained from groundwater potential model. About 36% of the study area has high potential of groundwater and is mainly concentrated in the southern and southerneastern part of the study area. There are also high groundwater potential zone in the northern direction, however the sources area likely to be high vulnerable to sewage system contamination. A great correlation can be noticed between areas assigned as high potential areas and the location of the Matsheumhlope aquifer [4].

Areas with moderate groundwater prospects are attributed to contributions from combinations of the land use/cover, lithology, slope and landform. The low to poor categories of groundwater potential zones are spatially distributed mainly along ridges where slope class is very high, the lithology is compact/massive and far from lineaments.

The variability in flow direction maybe attributed to variation and localization of the fractures within the basement rock. This is one of the characteristics of the basement rock which makes it difficult to exploit the groundwater available especially if the information is inadequate. Overally, the direction of flow is north and north-west as shown in Figure 14. This is mainly attributed to the drainage pattern in the study area.

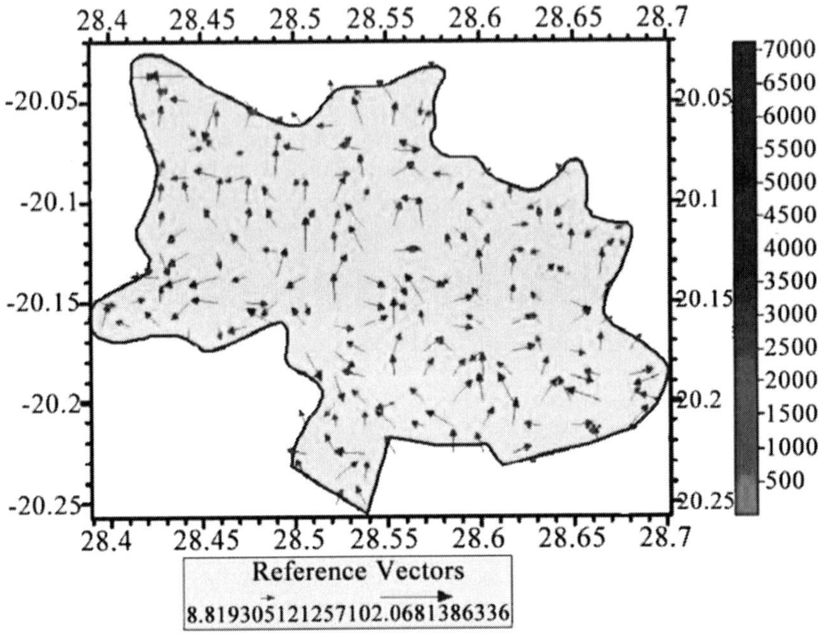

FIGURE 14. Possible flow direction of groundwater in the study area.

5. CONCLUSIONS

The picture that emerges from this study has shown that

there is large spatial variability of ground water potential while the most promising potential zone in the area is related to volcanic rock of which is affected, by secondary structure and having interconnected pore spaces, with gentle slope and less drainage density. It can also be inferred that most of the zones with low to poor groundwater potential lie in the massive basements unit which is far from lineaments. Areas close the sewer drainage system should be avoided or buffered to prevent contaminated water.

It can also be concluded that parts of the area that are characterized with surface expression of lineament are considered to be hydrogeological insignificant due to high slope and hence constitute the poor groundwater potential zone for the area. The low drainage density areas cause more infiltration and result in good groundwater potential zones as compared to high drainage density areas. High drainage density indicates unfavourable site for groundwater occurrence especially in area the northwestern part of Bulawayo Urban.

It must, however, be added that the results of this study is qualitative, hence, it is recommended that more quantitative investigations involving geophysical technique(s), borehole drilling and pump test must be carried out to fully understand the prospect of the groundwater resources of the area and consequently design a model for the effective groundwater use and management in the area.

6. ACKNOWLEDGEMENTS

I want to extend my profound gratitude to African Regional Centre for Space Science and Technology in English (ARCSSTE-E) for according me an opportunity to study and the financial support you gave throughout this course, National University of Science and Technology (NUST) and Zimbabwe Government for the support they rendered to the success of this study. Special thanks go to Dr. O. O. Orimoogunje, the supervisor for his expert advice in remote sensing and GIS and guidance at every stage of the study. Dr. D. J. Hlatywayo, my co-supervisor, requires special mention for his untiring supervision and guidance during field investigations. All the lecturers and ARCSSTE-E staff members including the Director, Dr. J.O Akenyede H.O.D and Administrative Course Coordinator I salute you all for imparting precious knowledge in my life.

REFERENCES

1. E. P. Wright, "The Hydrogeology of Crystalline Basement Aquifers in Africa," In: E. P. Wright and W. G. Burgess, Eds., The Hydrogeology of Crystalline Basement Aquifers in Africa, Geological Society Special Publication, London, 1992, pp. 1-27.
2. P. J. Chilton and S. S. D. Foster, "Hydrogeological Characterisation and Water-Supply Potential of Basement Aquifers in Tropical Africa," Hydrologeology Journal, Vol. 3, No. 1, 1995, pp. 36-49.
3. D. Banks and N. S. Robins, "An Introduction to Groundwater in Crystalline Bedrock," Geological Survey of Norway, Trondheim, 2002.
4. F. Rusinga and A. E. Taigbenu, "Groundwater Resource Evaluation of Urban Bulawayo Aquifer," Water SA, Vol. 31, No. 1, 2004, pp. 23-34.

5. E. Martinelli and G. L. Hubert, "Hydrogeology. A Report to the Ministry of Energy and Water Resources and Development, National Master Plan for Rural Water Supply and Sanitation," Interconsult/Norad, Vol. 2, 1985, pp. 109-114.
6. J. M. C. Weaver, J. Conrad and J. F. P. Engelbrecht, "The Matsheumhlope Wellfield Project, Wellfield Evaluation after Phase 1 Drilling for Bulawayo City Water Supply," Matabeleland Chamber of Industries, Bulawayo, 1992.
7. E. Mangore and A. E. Taigbenu, "Land-Use Impacts on the Quality of Groundwater in Bulawayo," Water SA, Vol. 30, No. 4, 2004, pp. 453-464
8. E. Mangore, "Monitoring and Evaluation of Bulawayo Aquifers," Unpublished M.Phil. Thesis, Library of National University of Science and Technology, Bulawayo, 2002.
9. I. Muchingami, J. D. Hlatywayo and J. M. Nel, "Electrical Resistivity Survey for Groundwater Investigations and Shallow Subsurface Evaluation of the Basaltic-Greenstone Formation of the Urban Bulawayo Aquifer," 11th WaterNet/WARFSA/GWP-SA Symposium, Victoria Falls, 11 November 2010, pp. 279-291.
10. C. Chuma, "Investigating the Occurrence of Groundwater in the Low Yield Aquifer within Matsheumhlope Wellfield," Unpublished M.Sc. Thesis, National University of Science and Technology Library, Bulawayo, 2011.
11. I. Muchingami, J. D. Hlatywayo, J. M. Nel and C. Chuma, "Electrical Resistivity Survey for Groundwater Investigations and Shallow Subsurface Evaluation of the BasalticGreenstone Formation of the Urban Bulawayo Aquifer," Journal for Physics and Chemistry of the Earth, Vol. 50- 52, 2012, pp. 44-51.
12. B. T. Shumba, D. J. Hlatywayo and V. Midzi, "Focal Mechanism Solution of The 15th March 2008, Nyamandlovu Earthquake," South African Journal of Geology, Vol. 112, No. 3, 2009, pp. 381-386.
13. D. J. Hlatywayo and V. Midzi, "Report on the Investigations into the Location, Causes and the Effects of the 25th June 2004 Bulawayo-Nyamandlovu Earthquake," Goverment of Zimbabwe, Harare, 2005, 13p.
14. Department of Meteorological Services, "Climate Data Records of Bulawayo," Goetz Observatory Weather Station, Bulawayo, 2003.

15. M. S. Garson and N. A. Mtsvangwa, "The Geology of Bulawayo Greenstone Belt and the Surrounding Granitic Terrain," Zimbabwe Geological Survey, Harare, 1995.
16. K. Sankar, "Evaluation of Groundwater Potential Zones Using Remote Sensing Data in Upper Vaigai River Basin, Tamil Nadu, India," Journal of the Indian Society of Remote Sensing, Vol. 30, No. 3, 2002, pp. 119-129.
17. B. M. Evans and W. L. Myers, "A GIS-based approach to evaluating regional groundwater pollution potential with DRASTIC," Journal of Soil and Water Conservation, Vol. 45, No. 2, 1990 pp. 242-245.
18. J. W. Merchant, "GIS-Based Groundwater Pollution Hazard Assessment: A Critical Review of the DRASTIC Model," Photogrammetric Engineering and Remote Sensing, Vol. 60, No. 9, 1994, pp. 1117-1128.
19. G. P. Panagopoulos, A. K. Antonakos and N. J. Lambrakis, "Optimization of the DRASTIC Method for Groundwater Vulnerability Assessment via the Use of Simple Statistical Methods and GIS," Hydrogeology Journal, Vol. 14, No. 6, 2006, pp. 894-911.
20. T. O. Tweed, M. Leblanc, J. A. Webb and M. W. Lubczynski, "Remote Sensing and GIS for Mapping Groundwater Recharge and Discharge Areas in Salinity Prone Catchments, Southeastern Australia," Hydrogeology Journal, Vol. 15, No. 1, 2007, pp. 75-96.
21. E. Sener, A. Davraz and M. Ozcelik, "An Integration of GIS and Remote Sensing in Groundwater Investigations: A Case Study in Burdur, Turkey," Hydrogeology Journal, Vol. 13, No. 5-6, 2005, pp. 826-834.
22. S. Shahid, S. K. Nath and J. Roy, "Groundwater Potential Modelling in a Soft Rock Area Using a GIS," International Journal of Remote Sensing, Vol. 21, No. 9, 2000, pp. 1919-1924.
23. J. E. Charon, "Hydrogeological Applications of ERTS Satellite Imagery," Proceedings of UN/FAO Regional Seminar on Remote Sensing of Earth Resources and Environment, Commonwealth Science Council, Cairo, 1974, pp. 439-456.
24. G. F. Bonham-Carter, "Geographic Information Systems for geoscientists: Modelling with GIS," Elseviers Science Ltd., Kidlington, 1994, 398p.
25. T. L. Saaty, "A Scaling Method for Priorities in Hierarchical Structures," Journal of Mathematical Psychology, Vol. 15, No. 3, 1997, pp. 234-281.
26. T. L. Saaty, "Fundamentals of Decision Making and Priority Theory with the AHP," RWS Publications, Pittsburgh, 1994.

27. J. E. Danielsen, T. Dahlin, R. Owen, P. Mangeya and E. Auken, "Geophysical and Hydrogeologic Investigation of Groundwater in the Karoo Stratigraphic Sequence at Sawmills in Northern Matabeleland, Zimbabwe: A Case History," Hydrogeology Journal, Vol. 15, No. 5, 2007, pp. 945-960.
28. J. W. Lloyd, "Water Resources in Hard Rock Aquifers in Arid and Semi-Arid Areas," UNESCO Studies and Reports in Hydrogeology No. 58, United Nations Educational, Scientific and Cultural Organization (UNESCO), Paris, 1999
29. N. S. Rao, G. K. J. Chakradhar and V. Srinivas, "Identification of Groundwater Potential Zones Using Remote Sensing Techniques in and around Gunur Town, Andhra Pradesh, India," Journal of the Indian Society of Remote Sensing, Vol. 29, No. 1-2, 2001, pp. 69-78.
30. S. C. Caran, C. M. Woodruff and E. J. Thompson, "Lineament Analysis and Inference of Geologic Structure Examples from the Balcones/Ouachita Trend of Texas," Gulf Coast Association of Geological Societies Transactions, Vol. 31, 1981, pp. 59-69.

11

CHAPTER

WIDE-AREA LANDSLIDE DEFORMATION MAPPING WITH MULTI-PATH ALOS PALSAR DATA STACKS: A CASE STUDY OF THREE GORGES AREA, CHINA

Xuguo Shi [1], Mingsheng Liao [1,2], Menghua Li [1], Lu Zhang [1,2,*] and Cory Cunningham [3]

[1] *State Key Laboratory of Information Engineering in Surveying, Mapping and Remote Sensing, Wuhan University, 129 Luoyu Road, Wuhan 430079, China*

[2] *Collaborative Innovation Center for Geospatial Technology, 129 Luoyu Road, Wuhan 430079, China*

[3] *SAIT Polytechnic, 1301-16 Avenue NW Calgary, Calgary, AB, T2M 0L4, Canada*

ABSTRACT

In recent years, satellite synthetic aperture radar interferometry (InSAR) has been adopted as a spaceborne geodetic tool to successfully measure surface deformation of a few well-known landslides in the Three Gorges area. In consideration of the fact that most events of slope failure happened at places other than those famous landslides since the reservoir impoundment in 2003, focusing on a limited number of slopes is insufficient to meet the requirements of regional-scale landslide disaster prevention and early warning. As a result, it has become a vital task to evaluate the overall stability of slopes across the vast area of Three Gorges using wide-coverage InSAR datasets. In this study, we explored the approach of carrying out joint analysis of multi-path InSAR data stacks for wide-area landslide deformation mapping. As an example, three ALOS (Advanced Land Observing Satellite) PALSAR (Phased Array type L-band Synthetic Aperture Radar) data stacks of neighboring ascending paths covering the area along the Yangtze River from Fengjie to Zigui were analyzed. A key problem to be solved is the separation of the tropospheric signal from the interferometric phase, for which we employed a hybrid description model of the atmospheric phase screen (APS) to improve APS estimation from time series interferograms. The estimated atmospheric phase was largely correlated with the seasonal rainfall in the temporal dimension. The experimental results show that about 30 slopes covering total areas of 48 km^2 were identified to be landslides in active deformation and should be kept under routine surveillance. Analyses of time series displacement measurements revealed that most landslides in the mountainous area far away from Yangtze River suffered from linear deformation, whereas landslides located on the river bank were destabilized predominantly by the influences of reservoir water level fluctuation and rainfall.

Keywords: slow moving landslide; InSAR; Three Gorges; displacement; water level; rainfall

1. INTRODUCTION

Landslides including rock falls and debris flows are one of the most destructive geological hazards, accounting for around 80% of geological disasters [1]. In the Three Gorges area landslides are a major threat to navigation, people's lives and properties. Wide areas of the Three Gorges are recognized as being at high risk, particularly areas near the bank of the Yangtze River [2]. The significant increase of the water level causes overland flooding in many areas in the Three Gorges which significantly affects the stability of the slopes. Additionally, periodical fluctuation of the water level and seasonal rainfall can make the situation even worse. Therefore, the detection and mapping of active slopes is a critical task for landslide disaster prevention and developing early warnings.

The Chinese government has spent many resources to find effective methods for landslide monitoring and reinforcement in the Three Gorges area [3]. Among the tools employed in landslide deformation monitoring, Synthetic Aperture Radar Interferometry (InSAR) as an important complement to traditional techniques has been applied to landslide investigation in the Three Gorges area. Serious decorrelation caused by dense vegetation coverage makes the application very difficult [4,5]. Advanced InSAR methods such as Persistent Scatterers InSAR (PSI) [6] and small baselines subset InSAR (SBAS) [7,8], proposed to overcome the limitations of decorrelation and atmospheric influence, were also used in landslide detection in Three Gorges area. Sparse point-like targets identified on landslides caused by decorrelation significantly affect the interpretation with TerraSAR-X and ENVISAT ASAR data [9]. Successful applications with advanced InSAR methods such as PSI, Quasi Permanent Scatterers (QPS) and SBAS were achieved on single landslides such as the Huangtupo landslide [10,11,12,13] and Fanjiaping landslide [14].

Until now, successful InSAR-based studies have focused only on small areas such as the Huangtupo and Fanjiaping landslides. This may be a result of two important reasons: (1) the dense vegetation coverage within the Three Gorges area causes serious decorrelation, leaving a limited number of sparse point-like targets and therefore resulting in unreliable measurements over large regions; (2) dynamic variation of the heterogeneous atmospheric condition across the vast area cannot be simply estimated using spatial and temporal filters, which can further reduce the accuracy of landslide deformation measurements. However, it remains insufficient to focus on just a few known landslides that have been closely monitored with traditional techniques such as GPS and extensometer. Identification of unknown active slopes is more important for risk management and government action, thus it is a main focus of this study. Furthermore, time series analysis will also be extremely helpful in evaluating the potential driving factors on slope instability.

In order to perform large-area landslide deformation mapping, the problems of decorrelation and atmospheric disturbances must be overcome. To mitigate the decorrelation, L-band PALSAR data of longer wavelength are used rather than C-band or X-band SAR data to ensure better coherence in mountainous areas covered by dense vegetation [15]. To deal with the problem of atmospheric disturbances, the Stanford Method for Persistent Scatterers (StaMPS) SBAS, developed by Hooper [8] for natural terrain displacement monitoring, was employed for time series analysis. In particular, since the unfavorable atmospheric phase screen (APS) induced by tropospheric path delay in mountainous areas usually consists of two components, namely stratified APS which is correlated with topography and turbulent APS which is independent of topography, a hybrid description model of APS was used to estimate APS from time series interferograms in the StaMPS SBAS method.

In this study, we investigated an approach for carrying out wide-area landslide deformation mapping with multi-path satellite SAR data stacks. Three ALOS PALSAR datasets acquired from adjacent paths were analyzed to detect landslides in the Three Gorges area along the Yangtze River. Deformation patterns of these landslides detected were characterized and the impacts of driving factors were evaluated qualitatively. Reservoir water level fluctuation and rainfall have been considered two major driving factors for slope instability [16], and correlations between water level/rainfall and displacements are analyzed and discussed.

2. STUDY AREA AND TEST DATASETS

2.1. Three Gorges Area

The Three Gorges Reservoir refers to the upstream region of the Yangtze River involved in the water impoundment of the Three Gorges Project within the Chongqing municipality and Hubei Province. The Three Gorges is a smaller region spanning from Fengjie County in Chongqing municipality to Yichang city in Hubei Province, as shown in Figure 1. This region lies in the transition zone from the second to the third topographic step of China, with more than 70% being mountainous areas. The maximum elevation in our study area is greater than 2000 m.

Construction of the Three Gorges Dam was finished in 2006, and the reservoir water level rose to 175 m for the first time in 2009. According to the operational scheme of the Three Gorges Reservoir, large seasonal water level fluctuations occur annually in three periods: a rise from 145 m to 175 m between early October and early November, a decrease from 175 m to 156 m between early January and April, and a further decrease from 156 m to 145 m between May and mid-June [17]. During the flooding season (June to September) when heavy rainfall occurs, the water level is maintained at 145 m [17]. The water level of the upstream Yangtze River and the weekly cumulative rainfall in Fengjie during the period from 2007 to 2011 are plotted in Figure 2.

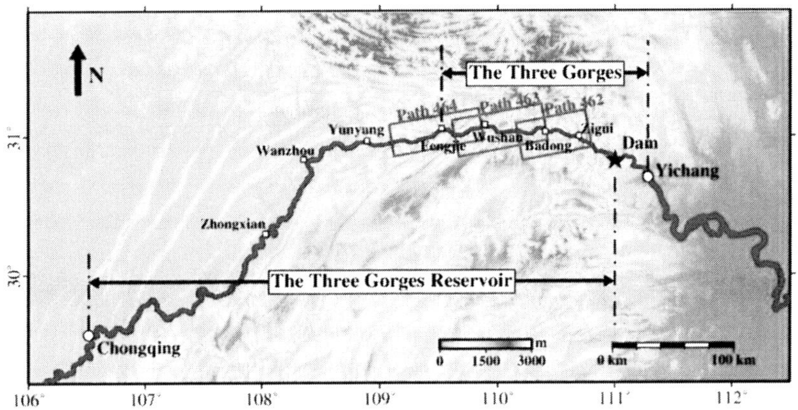

FIGURE 1. Geographic coverage of the Three Gorges area, with red rectangles indicating our test areas.

Many landslide disasters have happened in this area, such as the Xintan landslide in 1985 [2] and the Qianjiangping landslide in 2003 [18]. The wide distribution of landslides can be attributed in large part to a combination of lithological composition, planar geological structures, and climate conditions [2]. The huge water level fluctuations since the initial impoundment in 2003 make the conditions in this area more complex. The Qianjiangping landslide occurred a few days after the impoundment [18].

The climate in this area is categorized as subtropical monsoon with a hot and rainy summer. As we can see from Figure 2, the rainfall is mainly concentrated in summer with weekly cumulative rainfall at Fengjie reaching more than 150 mm. Intuitively, seasonal rainfall could introduce seasonal atmospheric signals into InSAR measurements and should be removed accordingly.

2.2. Datasets

Three ascending paths, path 462, path 463 and path 464, of the ALOS PALSAR datasets were collected over the Three Gorges area, spanning from the end of 2006 to early 2011. There are overlaps between adjacent paths (Figure 1). A similar look angle of approximately 34 degrees was applied in these three paths and they are all in the same flight direction. The images acquired by fine beam dual-polarization (FBD) mode are oversampled to match the pixel spacing of the fine beam single polarization (FBS) mode for cross-mode interferometry. The SRTM (Shuttle Radar Topography Mission)

DEM (Digital Elevation Model) of 1-arc-second resolution covering our study area was employed in the generation of single look differential interferograms and geocoding. Thresholds of 1000 days and 1000 m are used for temporal and perpendicular baselines separately during interferogram generation. As a result, 40 interferograms, 37 interferograms and 36 interferograms were produced from path 462, 463 and 464 datasets, respectively, for the final time series analysis (Figure 3).

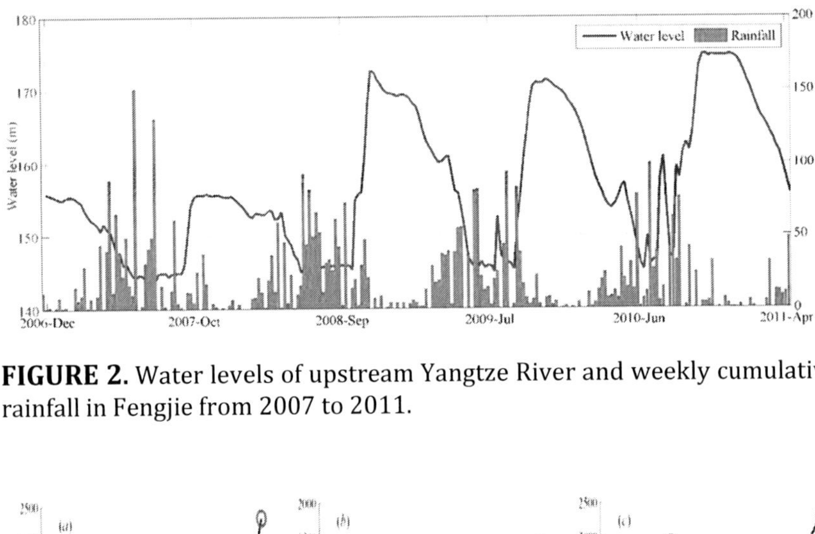

FIGURE 2. Water levels of upstream Yangtze River and weekly cumulative rainfall in Fengjie from 2007 to 2011.

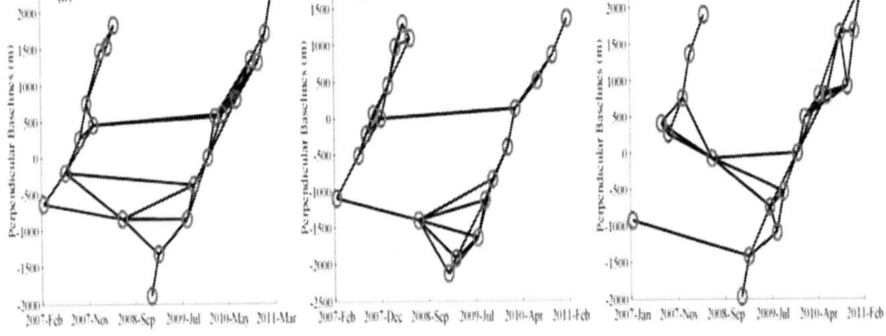

FIGURE 3. Interferograms used for small baselines analysis (**a**) path 462, (**b**) path 463 and (**c**) path 464. The circles represent the images and solid lines represent the interferograms.

3. METHODOLOGY

3.1. Small Baselines Subset InSAR Analysis

The small baseline subset method was initially proposed to overcome the problem of decorrelation by making full use of the interferograms with both small temporal baselines and short perpendicular baselines [7]. Since then it has been frequently applied in displacement monitoring in rural areas [11,19,20,21,22]. Compared with the original method which can operate with separated subsets of SAR acquisitions, isolated interferograms are not allowed in StaMPS SBAS analysis. Interferograms with a temporal baseline and perpendicular baseline below given thresholds were first generated to maximize the coherence and then grouped by common master or slave images to form subsets. Separated subsets were connected by interforgrams with large baselines to ensure the temporal continuity.

SAR amplitude information was initially exploited to exclude bad-quality pixels based on amplitude dispersion values [8]. The pixels with a slow decorrelating filtered phase (SDFP) that are widely distributed in natural terrain and have maintained good coherence during a short time interval are identified and analyzed to monitor surface displacements [8]. Phase characteristics are used to select the SDFP pixels. The spatially-correlated phase component and spatially-uncorrelated look angle error are estimated from the wrapped phase. The spatially-correlated contribution to the interferometric phase of a pixel, due to spatially-correlated signals including height error, orbit error and ground displacement, etc., is estimated by the bandpass filtering of surrounding pixels in the frequency domain. A spatially-uncorrelated look angle error, which is mainly composed of spatially-uncorrelated height error, is then estimated through its correlation with the perpendicular baseline. The residual, after removal of these two terms from the wrapped phase, gives an estimate of the decorrelation noise γ [8].

$$\gamma = \frac{1}{N}\left|\sum_{i=1}^{N} \exp\left\{\sqrt{-1}(\psi - \Delta\hat{\psi}^{u} - \Delta\widetilde{\psi})\right\}\right| \tag{1}$$

where N is the interferograms, ψ is the wrapped phase, $\Delta\psi$^u is the spatially-uncorrelated look angle error and $\Delta\psi$˜ is the spatially-correlated term.

Once the SDFP pixels were selected according to γ, three-dimensional phase unwrapping was carried out on all the pixels [23]. The unwrapped phase on a given pixel can be expressed as:

$$\varphi = \varphi_d + \varphi_T + \varphi_a + \varphi_o + \varphi_n \tag{2}$$

where φ is the unwrapped phase, φ_d, φ_T, φ_a, φ_o and φ_n are the phase components due, respectively, to surface displacements, topographic error, atmospheric disturbance, inaccurate orbit information and noise. In the StaMPS SBAS analysis, these different components were estimated by iterative filtering with consideration of their identifying characteristics in spatial-temporal domains.

Phase ramps due to orbital errors are spatially correlated and should be removed preferentially before the estimation of other terms. Orbital phase ramps caused by inaccurate ephemeris records in interferograms can be corrected using a bilinear or biquadratic model [19]. Usually, ALOS PALSAR data with large perpendicular baselines suffer considerably from orbital phase ramps. In most cases, a full second-order polynomial phase ramp correction method should be sufficient [24]. Therefore, a biquadratic model was used to correct the phase ramps in our study.

$$\varphi_o = a_0 + a_1 \cdot x + a_2 \cdot y + a_3 \cdot xy + a_4 \cdot x^2 + a_5 \cdot y^2 \tag{3}$$

where a_i (i= 0, 1,..., 5) are the coefficients used in orbital error phase estimation; x and y are the coordinates in azimuth and range direction on a SDFP pixel. Then phase ramps were estimated over all interferograms using the SDFP pixels. Typically, the deformation signal will leak into the orbital errors when there are displacements distributed in large areas, as in the case of earthquakes. If this occurs, these areas with displacements should be masked out before performing orbital error estimation.

The 1-arc-second SRTM DEM was used to remove the topographic phase in our study. The topographic changes since 2000 cannot be neglected, especially in developing countries such as China [25]. There is a linear relationship between topographic error and baseline.

$$\varphi_T = \frac{4\pi}{\lambda} \frac{B}{R\sin\theta} \Delta h \tag{4}$$

B, Δh, λ, R, θ stand for perpendicular baseline, topographic error, radar wavelength, slant range and incidence angle, respectively. Thus, the topographic error can be accurately estimated by using a least squares approach with multi-baseline interferograms. Interferograms with long temporal baselines sensitive to displacements were excluded from topographic error estimation.

According to previous studies [19,21,26], atmospheric delays in mountainous areas are normally composed of stratified APS and turbulent APS. Usually there is a linear relationship between phase delays introduced by stratified APS and terrain elevation.

$$\varphi_a = \varphi_s + \varphi_{a\_r} \tag{5}$$

$$\varphi_s = b \cdot h \tag{6}$$

where φ_s, φ_{a_r} represent the phase delays introduced by the stratified APS and residual, respectively; b is the coefficient used to represent the relationship between elevation and phase delays caused by stratified APS; and h is the elevation. A hybrid description model of the atmospheric phase screen (APS) was incorporated into StaMPS SBAS analysis to improve APS estimation from time series interferograms. The phase delays introduced by stratified APS were estimated by a least squares procedure using all the unwrapped SDFP pixels. Then the atmospheric phase residuals were estimated with temporally high-pass and spatially low-pass filters. Afterwards, time series deformation can be inverted using singular value decomposition (SVD) [7,8].

An iterative procedure was employed to ensure the estimation accuracy of each term in Equation (2). Since unwrapping errors are more prevalent in interferograms with longer perpendicular baselines, the estimated DEM error using Equation (4) was reduced before phase unwrapping to improve the unwrapping accuracy. Re-estimation of each term in Equation (2) was carried out with the newly unwrapped phase. This procedure can be repeated until the change of the residual is less than a target threshold.

3.2. Integration of Multi-Path Displacement Measurements

Since our objective is to detect potential landslides across the wide area of the Three Gorges Reservoir, the resultant displacement maps derived from the three paths of ALOS PALSAR data separately should be further integrated into one map. Usually, an assumed stable point is chosen as a reference point by which all the other parameters are estimated [7,27]. In this study, different reference points were used in our process for each path. Since no in-situ data was available, reference points located at Badong and the Fengjie city council building were used. The Badong reference point was used for both path 462 and path 463. The Fengjie reference point was used to calibrate the SBAS results from path 464. Biases between path 463 and the other two paths in the line-of-sight (LOS) direction are estimated and corrected as [27] did based on the average displacement rate differences in the overlapping area with respect to the Badong reference point.

It is worth noting that besides a displacement rate map in the LOS direction, generally a displacement map along the slope direction can also be obtained by a simple projection. However, here the displacement rate map in the along-slope direction was not available for two reasons: (1) the LOS deformation could be projected to the slope direction if the landslide is a translational landslide based on both geomorphological and DEM models [28], but we have no idea of the specific landslide category of each active slope detected; (2) the slopes along the Yangtze River are mostly south-north oriented while the LOS deformation is primarily sensitive to the deformation in the east-west and vertical directions, hence the projection between the LOS and slope direction would be inaccurate.

4. EXPERIMENTAL RESULTS AND ANALYSES

4.1. Mean Displacement Rate Map

The obtained displacement rate map in the LOS direction over the Three Gorges area from Fengjie County to Zigui County is shown in Figure 4. A total of 1,775,238 points were identified in our analysis, covering an area of 4800 km^2 from Fengjie County to Zigui County along the Yangtze River. Approximately 96% of the points show displacements rates between −10 mm/year and 10 mm/year, which indicated the overall stability of the study area. About 30 slopes with total areas of 48 km^2 were identified as being in active deformation.

The distribution of landslides in Fengjie is strongly connected with the orientation of synclinal stratigraphic structures. The stratum which initiates the instability is located at the transitional zone which is steeply dipping in the elevated regions and gently dipping in the lower regions [29]. As a famous karst region, stratigraphic formations in the west of Fengjie mainly consist of Jurassic red beds, while formations in the east mostly consist of Triassic carbonate rock [1]. Approximately half of the detected active landslides in our study area are distributed in Fengjie County (Figure 5). The SDFP pixels with positive values in Figure 5 are located on west-facing slopes. This is reasonable since most movements were in the downslope direction. According to [29], there are many landslides distributed along the banks of the Yangtze River, and it seems that most of them are stable and only six active landslides were detected along the river judging from our results. The slopes along the Yangtze River are generally facing north-south, which is insensitive for InSAR measurements. Nevertheless, local variations in slope orientation allow InSAR to detect some movements, such as the landslide where P4 is located. According to a previous study in Wanzhou, located in the west of the Three Gorges area, slopes of active landslides concentrate within the interval from 16 to 26 degrees [30]. Here the detected active landslides in Fengjie are also distributed on gentle slopes with angles varying from 15 to 30 degrees.

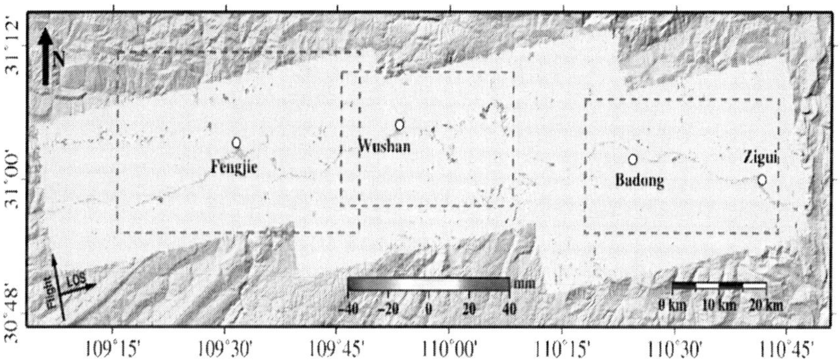

FIGURE 4. Mean displacement rate map of Three Gorges area generated from three ALOS PALSAR data stacks. Three selected areas are highlighted by dashed rectangles.

Figure 6 shows the displacement rate over the area of Wushan and three active landslides were detected. The landslide susceptibility map given by [31] shows that most of the landslide-prone areas were distributed around the south of the town of Wushan. All three landslides are located within the areas prone to landslides according to the susceptibility map [31]. However, those areas with steep slopes of more than 50 degrees seem generally stable, according to our results. Most of the areas along the bank of the Yangtze River are stable according to our InSAR results, which is consistent with the susceptibility map.

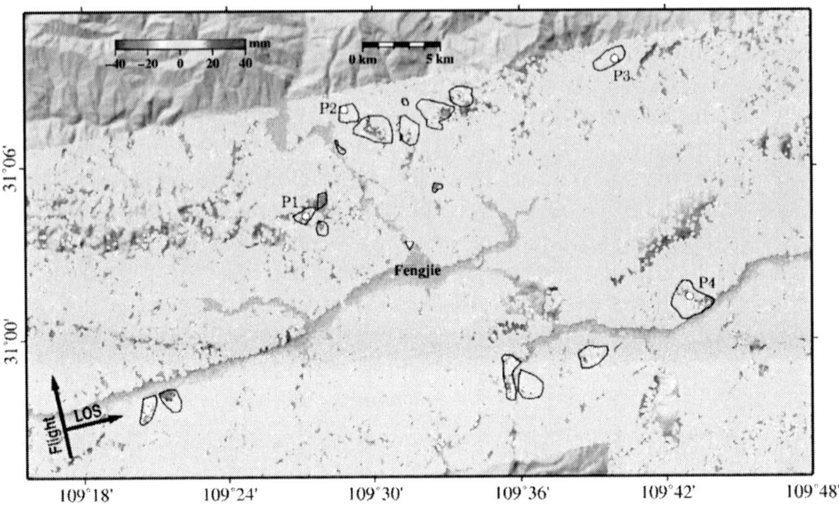

FIGURE 5. Mean displacement rate map measured at Fengjie. P1, P2, P3, P4 are points used for time series analysis shown in later sections.

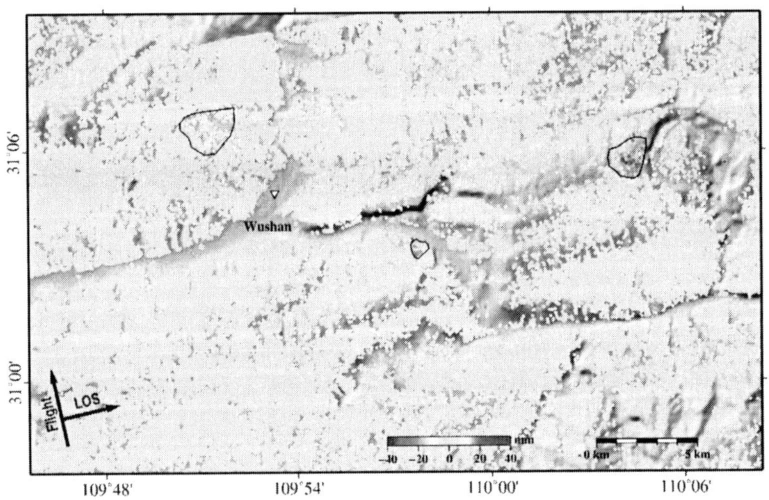

FIGURE 6. Mean displacement rate map of Wushan.

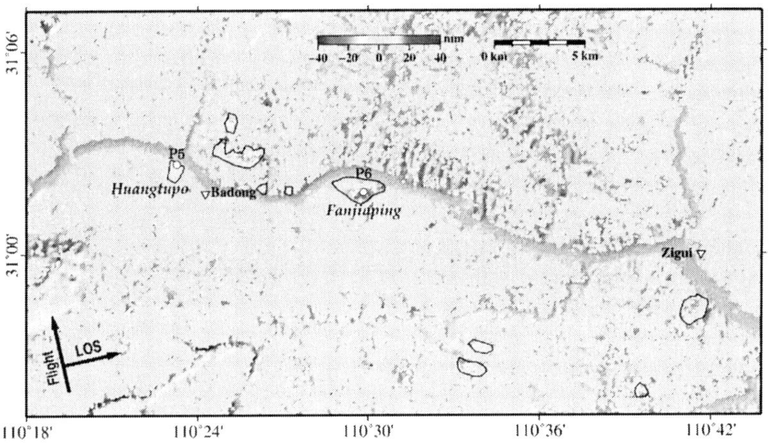

FIGURE 7. Mean displacement rate map covering areas around Badong and Zigui. P5 and P6 are points used in time series analysis shown in later sections.

Figure 7 shows the mean displacement rate map covering areas around Badong and Zigui. According to the landslide susceptibility map [31,32], the Badong blocks consisting of the Badong formation are prone to landslides. About five active landslides were detected within Badong County, among which the Huangtupo landslide was the most famous that has been studied using the InSAR method [10,11,12,13,14]. The mean displacement velocity of the Huangtupo landslide is about 10~15 mm/year. The active Huangtupo landslide has forced the government to change the relocation plan of the town of Badong. The Fanjiaping landslide that is located at the south bank of the Yangtze River and approximately 8 km east of the town of Badong was another one in active deformation [9,14,33]. The displacement rate of the Fanjiaping landslide can reach nearly 40 mm/year and is possibly correlated with the water level fluctuations and rainfall, which will be analyzed in the following sections.

4.2. Measurement Consistency Evaluation

In order to evaluate the consistency of the results, a quantitative analysis of the differences within the overlapping areas from different paths was performed. Figure 8 displays the difference in displacement rate values between adjacent paths. The standard deviations of the displacement rate differences on the common pixels between path 464 and path 463, and path

463 and path 462 are 4.7 mm/year and 4.9 mm/year, respectively. Possible factors which caused the disparity between different paths over the overlapping areas have been analyzed in detail in [27]. In brief, the main causes were identified as: (1) A slight difference in incidence angle and satellite heading angle caused by the different observation geometries at corresponding points on different paths. In our case, this difference was found to reach a maximum of more than 2 degrees. (2) The observation time series of three paths are different, resulting in different temporal samplings of the landslide progression. Since the progression of landslides is usually nonlinear, sampling time differences may cause inconsistencies in mean velocity estimation. (3) The SRTM DEM acquired in 2000 was used in our process and ground surface changes could introduce DEM uncertainties. This could cause the estimation of the displacement velocity to be inaccurate, in addition to causing a mismatch of points between adjacent paths. (4) There are some active landslides in the overlapping areas. In our case, the number of landslides in the overlapping areas is no more than three, so their contributions to this disparity may be limited.

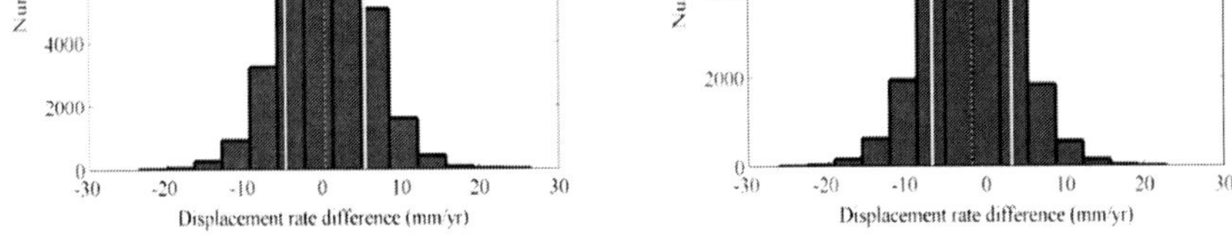

FIGURE 8. The distribution of displacement rate difference for the overlapping areas between (**a**) path 464 and path 463 datasets; (**b**) path 463 and path 462 datasets.

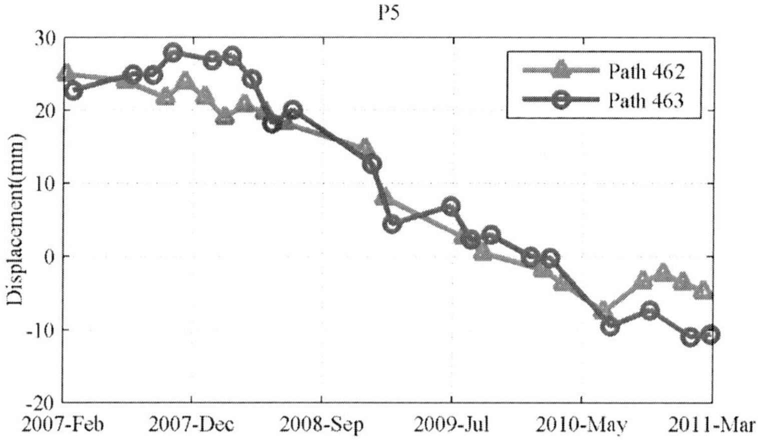

FIGURE 9. Time series displacement of point P5 along the slope direction located at the Huangtupo landslide.

Unfortunately, we have no in-situ measurement data in the Three Gorgers area, so it is impossible to evaluate the absolute accuracy of our results. To evaluate the consistency of time series measurements, time series displacements from path 462 and path 463 at point P5 (refer to Figure 7) located on the Huangtupo landslide were illustrated in Figure 9. The Huangtupo landslide is a translational landslide [34]. The LOS displacements were transformed onto the downslope direction. Slight discrepancies can be observed between the measurements from path 462 and path 463, for which the reasons might be the same as discussed above. On the other hand, the measurements of path 462 were referenced with respect to 4 January 2010 while the measurements of path 463 were referenced with respect to 21 January 2010. This could also be an important cause for the discrepancies in displacement differences between the two paths. Additionally, phase unwrapping artifacts induced by different datasets might also be considered.

4.3. Effect of Seasonal Rainfall on APS Variation

Preceding the analysis of the time series results of SDFT pixels, the influence of atmospheric turbulence to our displacement estimation was discussed. In a mountainous area such as the Three Gorges region along the Yangtze River, the phase distortions induced by atmospheric turbulence are very significant. Figure 10a gives the phase signal induced by atmospheric turbulence on point P2 for each slave image with respect to the master image. Strong correlation can be observed between atmospheric signal and seasonal rainfall. More specifically, we can see that the seasonal rainfall caused large atmospheric artifacts in the rainy seasons while the effect is relatively small in dry seasons. On the other hand, water level fluctuation might also contribute to the amount of water vapor in the Three Gorges area, whereby the large water level fluctuation in the Three Gorges area might result in a change in water vapor concentration. However, compared to the effect of rainfall on atmospheric turbulence, it seems that the effect of water level fluctuation is relatively minor.

The differences among estimations of APS with or without stratified APS corrections are very small for pixels with heights close to the reference point. For pixels with large elevation differences with respect to the reference point, if the atmospheric signal is not properly estimated and removed, the seasonal effects will remain in the displacement signal and significantly hinder the interpretation of the results. Figure 10b shows the time series displacements signal after removing the spatially-correlated atmospheric signal with and without stratified APS correction. We can see that without stratified APS correction, part of the APS signal still remains in the time series displacements, while the results appear to be more reasonable after correction.

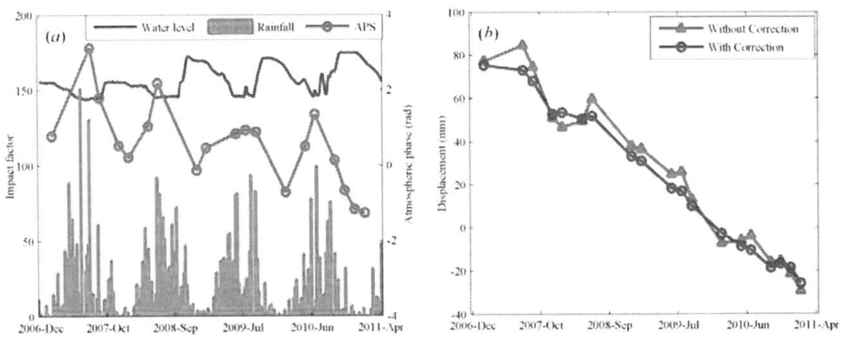

FIGURE 10. (**a**) Estimated atmospheric signal on P1 and impact factors: water level (unit: m) and rainfall (unit: mm), the dashed line represents the temporal location of the master image. (**b**) Time series displacements on P1 with and without stratified APS correction.

4.4. Impact Factors for Landslide Activities

According to our results, most of the landslides have a nearly linear displacement trend as shown in Figure 11a, particularly landslides located farther away from the Yangtze River. While small variations in the displacement trend might be occurring throughout the entire time span, the influence of water level variation is not obvious and can be neglected on these landslides.

Considering Figure 11b, the displacements of P3 can be divided into two periods. The displacement of P3 is very small with almost a linear rate before August 2008. Acceleration of displacement occurred during the second period, but the displacement trend is still linear. The change in displacement rate between these two periods might be triggered by the consecutive rainfall in the summer of 2008.

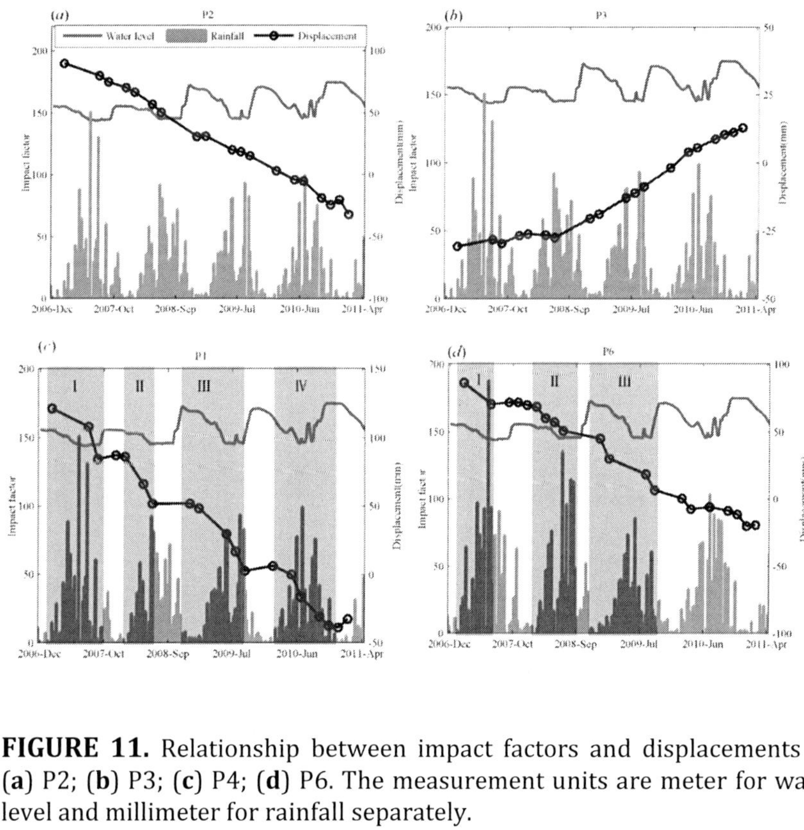

FIGURE 11. Relationship between impact factors and displacements on (**a**) P2; (**b**) P3; (**c**) P4; (**d**) P6. The measurement units are meter for water level and millimeter for rainfall separately.

Water level fluctuations and rainfall are the two most important impact factors of slope stabilities in the Three Gorges region for slopes near the river [16]. The upstream water level of Three Gorges and rainfall records from Badong and Fengjie were used to investigate the relationship between displacements and water level fluctuation/rainfall. Since the temporal sampling rate of ALOS PALSAR is very low, the relationship between displacements and water level/rainfall was analyzed by visual inspection. As we can see from Figure 11c, the landslides located at P4 experienced stepwise displacements. The landslides in the Three Gorges areas with stepwise displacement progression mainly undergo two main processes: the sliding process and the dormant process [16]. During the sliding process, the landslide moves at inconsistent rates, whereas during the dormant stage, shear strength has been recovered and stability is achieved for some duration of time.

There are generally four periods of movement at P4. Among the four periods, period II could be caused by the decline of the water level when the rainfall is relatively small. Periods I, III and IV could be caused by the joint effect of water level decline and rainfall. During these three periods, the rainfall was relatively low at first while the fast decline of the reservoir water level triggered the movement by changing the pressure difference between the groundwater inside the slope and the reservoir waterbody outside [35]. The following heavy rainfall caused the displacement to accelerate, especially during period I when the weekly rainfall reached more than 150 mm. It is also worth noting that a lag effect of water level may be present. For periods III and IV, although heavy rainfall followed when the decline of the water level ceased, the lag effect of the water level decline might also have contributed to the movement at the end of periods III and IV.

The movement of P6 located at the Fangjiaping landslide was fairly complicated, as shown in Figure 11d. There may be a constant displacement trend that exists during the entire progression of the Fangjiaping landslide; nevertheless, we can find three periods with where accelerations occur. Periods I and II may be mainly caused by the reservoir water level decline. Although the weekly rainfall at the end of period I reached more than 180 mm, no acceleration was observed during the following process. The movement of periods II and III might be accelerated by the water level decline and the following movements at the end of these two periods might be caused by the rainfall and lag effect of the water level decline.

5. CONCLUSIONS

In this paper, we investigated the stability of slopes from Fengjie to Zigui within the Three Gorges area by time series InSAR analysis. InSAR data with its large coverage and millimeter accuracy is suitable for displacement monitoring in the vast Three Gorges area. However, we have to overcome decorrelation effects and atmospheric disturbances to achieve good results. To avoid decorrelation as much as possible, L-band SAR data with longer wavelengths are preferred. To overcome atmospheric disturbances, a hybrid description model of atmospheric phase screen (APS) is necessary to improve APS estimation from time series interferograms, particularly in mountainous areas such as the Three Gorges region. The estimated atmospheric phase was mainly correlated with the seasonal rainfall in the temporal dimension.

About 30 slopes covering areas of 48 km^2 in total were identified to be active with slopes ranging from 15 to 30 degrees. Through time series analysis, active landslides far away from the Yangtze River exhibit a linear displacement trend. Rainfall plays a key role in the change of the displacement trend. For those active landslides located near the river, rapid water level decline and consecutive rainfall are potentially two key factors to increase slope instability.

It is worth noting that the limitations of the InSAR method can make it blind to several active phenomena. InSAR methods applied to polar-orbit satellite SAR datasets are only sensitive to displacement in the LOS direction, while they hardly detect movements parallel to the flight direction, such as the north-south direction in our study. Geometric distortions such as shadow and obscuring landscape features can also make many slopes invisible in SAR images [22]. Descending and ascending datasets can be combined to increase the effective coverage of InSAR measurements. Additionally, for fast-moving landslides, a large displacement gradient might go beyond the upper detectability of InSAR methods, which would make interferometric phase observation completely fail to measure these displacements. An effective tool for fast-moving landslide detection is the pixel offset tracking method with high-resolution SAR data. Successful applications have also been achieved on the Shuping landslide in the Three Gorges area with high-resolution TerraSAR-X data [22,35,36,37]. A combination of the InSAR method and pixel offset tracking methods can be carried out with high-resolution SAR datasets in our future work.

ACKNOWLEDGMENTS

This work was financially supported by the National Key Basic Research Program of China (Grant No. 2013CB733205 and No. 2013CB733204) and the National Natural Science Foundation of China (Grant No. 61331016 and 41271457). The PALSAR datasets were provided by Japan Aerospace Exploration Agency (JAXA) through the ALOS-RA4 project (PI1247, PI1440). Ziyun Wang from Wuhan University was appreciated for helping to collect the water level data of the upstream of Yangtze River.

REFERENCES

1. Liu, C.; Liu, Y.; Wen, M.; Li, T.; Lian, J.; Qin, S. Geo-hazard initiation and assessment in the Three Gorges Reservoir. In Landslide Disaster Mitigation in Three Gorges Reservoir, China; Wang, F., Li, T., Eds.; Springer Berlin: Heidelberg, Germany, 2009; pp. 3–40.
2. Peng, L.; Niu, R.; Huang, B.; Wu, X.; Zhao, Y.; Ye, R. Landslide susceptibility mapping based on rough set theory and support vector machines: A case of the Three Gorges area, China. Geomorphology **2014**, 204, 287–301.
3. Plateau, T. Three Gorges Dam: Into the unknown. Science **2006**, 25, 1034.
4. Xia, Y.; Kaufmann, H.; Guo, X. Differential SAR interferometry using corner reflectors. IEEE Int. Geosci. Remote Sens. Symp. **2002**, 2, 1243–1246.
5. Xia, Y.; Kaufmann, H.; Guo, X. Landslide monitoring in the Three Gorges area using D-InSAR and corner reflectors. Photogramm. Eng. Remote Sens. **2004**, 70, 1167–1172.
6. Ferretti, A.; Prati, C.; Rocca, F. Permanent scatterers in SAR interferometry. IEEE Trans. Geosci. Remote Sens. **2001**, 39, 8–20.
7. Berardino, P.; Fornaro, G.; Lanari, R.; Sansosti, E. A new algorithm for surface deformation monitoring based on small baseline differential SAR interferograms. IEEE Trans. Geosci. Remote Sens. **2002**, 40, 2375–2383.
8. Hooper, A. A multi-temporal InSAR method incorporating both persistent scatterer and small baseline approaches. Geophys. Res. Lett. **2008**, 35, L16302.

9. Liao, M.S.; Tang, J.; Wang, T.; Balz, T.; Zhang, L. Landslide monitoring with high-resolution SAR data in the Three Gorges region. Sci. China Earth Sci. **2012**, 55, 1–12.
10. Perissin, D.; Wang, T. Repeat-Pass SAR Interferometry With Partially Coherent Targets. IEEE Trans. Geosci. Remote Sens. **2012**, 50, 271–280.
11. Liu, P.; Li, Z.; Hoey, T.; Kincal, C.; Zhang, J.; Zeng, Q.; Muller, J.-P. Using advanced InSAR time series techniques to monitor landslide movements in Badong of the Three Gorges region, China. Int. J. Appl. Earth Observ. Geoinf. **2013**, 21, 253–264.
12. Tantianuparp, P.; Shi, X.; Zhang, L.; Balz, T.; Liao, M. Characterization of Landslide Deformations in Three Gorges Area Using Multiple InSAR Data Stacks. Remote Sens. **2013**, 5, 2704–2719.
13. Shi, X.; Zhang, L.; Liao, M.; Balz, T. Deformation monitoring of slow-moving landslide with L- and C-band SAR interferometry. Remote Sens. Lett. **2014**, 5, 951–960.
14. Zhang, L.; Liao, M.; Balz, T.; Shi, X.; Jiang, Y. Monitoring landslide activities in the Three Gorges Area with multi-frequency satellite SAR data sets. In Modern Technologies for Landslide Monitoring and Prediction; Scaioni, M., Ed.; Springer Berlin Heidelberg: Berlin, Germany, 2015; pp. 181–208.
15. Ng, A.H.-M.; Chang, H.-C.; Ge, L.; Rizos, C.; Omura, M. Assessment of radar interferometry performance for ground subsidence monitoring due to underground mining. Earth Planets Space (EPS) **2009**, 61, 733.
16. Miao, H.; Wang, G.; Yin, K.; Kama, T.I.; Li, Y. Mechanism of the slow-moving landslides in Jurassic red-strata in the Three Gorges Reservoir, China. Eng. Geol. **2014**, 171, 59–69.
17. Cojean, R.; Caï, Y.J. Analysis and modeling of slope stability in the Three-Gorges Dam reservoir (China)—The case of Huangtupo landslide. J. Mountain Sci. **2011**, 8, 166–175.
18. Wang, F.; Zhang, Y.; Huo, Z.; Peng, X.; Wang, S.; Yamasaki, S. Mechanism for the rapid motion of the Qianjiangping landslide during reactivation by the first impoundment of the Three Gorges Dam reservoir, China. Landslides **2008**, 5, 379–386.
19. Sun, Q.; Zhang, L.; Ding, X.L.; Hu, J.; Li, Z.W.; Zhu, J.J. Slope deformation prior to Zhouqu, China landslide from InSAR time series analysis. Remote Sens. Environ. **2015**, 156, 45–57.
20. Teshebaeva, K.; Roessner, S.; Echtler, H.; Motagh, M.; Wetzel, H.-U.; Molodbekov, B. ALOS/PALSAR InSAR Time series Analysis for

Detecting Very Slow-Moving Landslides in Southern Kyrgyzstan. Remote Sens. **2015**, 7, 8973–8994.

21. Tang, P.; Chen, F.; Guo, H.; Tian, B.; Wang, X.; Ishwaran, N. Large-Area Landslides Monitoring Using Advanced Multi-Temporal InSAR Technique over the Giant Panda Habitat, Sichuan, China. Remote Sens. **2015**, 7, 8925–8949.
22. Wasowski, J.; Bovenga, F. Investigating landslides and unstable slopes with satellite Multi Temporal Interferometry: Current issues and future perspectives. Eng. Geol. **2014**, 174, 103–138.
23. Hooper, A.; Zebker, H.A. Phase unwrapping in three dimensions with application to InSAR time series. J. Opt. Soc. Am. A **2007**, 24, 2737–2747.
24. Zhang, L.; Ding, X.; Lu, Z.; Jung, H.-S.; Hu, J.; Feng, G. A novel multitemporal InSAR model for joint estimation of deformation rates and orbital errors. IEEE Trans. Geosci. Remote Sens. **2014**, 52, 3529–3540.
25. Perissin, D.; Wang, T. Time series InSAR Applications Over Urban Areas in China. IEEE J. Sel. Top. Appl. Earth Observ. Remote Sens. **2011**, 4, 92–100.
26. Liao, M.; Jiang, H.; Wang, Y.; Wang, T.; Zhang, L. Improved topographic mapping through high-resolution SAR interferometry with atmospheric effect removal. ISPRS J. Photogramm. Remote Sens. **2013**, 80, 72–79.
27. Ng, A.H.-M.; Ge, L.; Li, X. Assessments of land subsidence in the Gippsland Basin of Australia using ALOS PALSAR data. Remote Sens. Environ. **2015**, 159, 86–101.
28. Cascini, L.; Fornaro, G.; Peduto, D. Advanced low- and full-resolution DInSAR map generation for slow-moving landslide analysis at different scales. Eng. Geol. **2010**, 112, 29–42.
29. Qiao, J.; Wang, M. Distribution Features of Landslides in Three Gorges Area and the Contribution of Basic Factors. In Landslide Disaster Mitigation in Three Gorges Reservoir, China; Springer: Berlin, Germany, 2009; pp. 173–192.
30. Gui, L. Research on Landslide Development Regularities and Risk in Wan Zhou District, Three Gorges Reservoir. PhD Thesis, China University of Geosciences, Wuhan, China, 31 May 2014.
31. Fourniadis, I.G.; Liu, J.G.; Mason, P.J. Landslide hazard assessment in the Three Gorges area, China, using ASTER imagery: Wushan–Badong. Geomorphology **2007**, 84, 126–144.

32. Liu, J.G.; Mason, P.J.; Cleric, N.; Chen, S.; Davis, A.; Miao, F.; Deng, H.; Liang, L. Landslide hazard assessment in the Three Gorges area of the Yangtze river using ASTER imagery: Zigui–Badong. Geomorphology **2004**, 61, 171–187.
33. Xia, Y. Synthetic Aperture Radar Interferometry. In Sciences of Geodesy—I; Xu, G., Ed.; Springer Berlin Heidelberg: Berlin, Germany, 2010; pp. 415–474.
34. Chai, B.; Yin, K.; Du, J.; Xiao, L. Correlation between incompetent beds and slope deformation at Badong town in the Three Gorges reservoir, China. Environ. Earth Sci. **2013**, 69, 209–223.
35. Shi, X.; Zhang, L.; Balz, T.; Liao, M. Landslide deformation monitoring using point-like target offset tracking with multi-mode high-resolution TerraSAR-X data. ISPRS J. Photogramm. Remote Sens. **2015**, 105, 128–140.
36. Singleton, A.G.; Li, Z.; Hoey, T.; Muller, J.P. Evaluating sub-pixel offset techniques as an alternative to D-InSAR for monitoring episodic landslide movements in vegetated terrain. Remote Sens. Environ. **2014**, 147, 133–144.
37. Singleton, A.G. Analysing Landslides in the Three Gorges Region (China) Using frequently Acquired SAR Images. PhD Thesis, University of Glasgow, Glasgow, UK, 2014.

12

CHAPTER

A REVIEW ON EXTRACTION OF LAKES FROM REMOTELY SENSED OPTICAL SATELLITE DATA WITH A SPECIAL FOCUS ON CRYOSPHERIC LAKES

Shridhar D. Jawak[1], Kamana Kulkarni[2], Alvarinho J. Luis[1]

[1]Polar Remote Sensing Division, Earth System Science Organization (ESSO), National Centre for Antarctic & Ocean Research (NCAOR), Ministry of Earth Sciences, Vasco da Gama, India

[2]School of Electronics, Devi Ahilya Vishwavidyalaya, Indore, India

ABSTRACT

Water on the Earth's surface is an essential part of the hydrological cycle. Water resources include surface waters, groundwater, lakes, inland waters, rivers, coastal waters, and aquifers. Monitoring lake dynamics is critical to favor sustainable management of water resources on Earth. In cryos- phere, lake ice cover is a robust indicator of local climate variability and change. Therefore, it is necessary to review recent methods, technologies, and satellite sensors employed for the extrac- tion of lakes from satellite imagery. The present review focuses on the comprehensive evaluation of existing methods for extraction of lake or water body features from remotely sensed optical data. We summarize pixel-based, object-based, hybrid, spectral index based, target and spectral matching methods employed in extracting lake features in urban and cryospheric environments. To our knowledge, almost all of the published research studies on the extraction of surface lakes in cryospheric environments have essentially used satellite remote sensing data and geospatial methods. Satellite sensors of varying spatial, temporal and spectral resolutions have been used to extract and analyze the information regarding surface water. Multispectral remote sensing has been widely utilized in cryospheric studies and has employed a variety of electro-optical satellite sensor systems for characterization and extraction of various cryospheric features, such as glaci- ers, sea ice, lakes and rivers, the extent of snow and ice, and icebergs. It is apparent that the most common methods for extracting water bodies use single band-based threshold methods, spectral index ratio (SIR)-based multiband methods, image segmentation methods, spectral-matching methods, and target detection methods (unsupervised, supervised and hybrid). A Synergetic fusion of various remote sensing methods is also proposed to improve water information extraction accuracies. The methods developed so far are not generic rather they are specific to either the location or satellite imagery or to the type of the feature to be extracted. Lots of factors are responsible for leading to inaccurate results of lake-feature extraction in cryospheric regions, e.g. the mountain shadow which also appears as a dark pixel is often misclassified as an open lake. The methods which are working well in the cryospheric environment for feature extraction or landcover classification does not really guarantee that they will be working in the same manner for the urban environment. Thus, in coming years, it is expected that much of the work will be done on object-based approach or hybrid approach involving both pixel as well as object-based technology. A more accurate, versatile and robust method is necessary to be developed that would work independent of geographical location (for both urban and cryosphere) and type of optical sensor.

Keywords: Cryospehere, Remote Sensing, Semi-Automatic Extraction, Lakes, Spectral Index Ratio

1. INTRODUCTION

Water on the Earth's surface is an essential part of the hydrological cycle. Water resources include surface water, groundwater, inland water, rivers, lakes, transitional waters, coastal waters and aquifers [1] . In this study, we are essentially looking at the extraction of information on lakes because water resources have been degraded and exhausted in last few years. It is necessary to understand the changes in spatial extents of water resources. Lakes are inland bodies of standing water. Lakes are an essential component of the hydrological cycle and hence key tools for the management of water resources [2] - [4] . Although millions of lakes are scattered over the earth's surface, most are located at higher latitudes and mountainous areas. Lakes may be classified according to their manner of formation or characteristics (Table 1). Lakes can be formed by glaciers, tectonic plate movements, river and wind currents, and volcanic or meteorite activity. Some lakes are only seasonal, drying up during parts of the year. All lakes are either open or closed. If water leaves a lake by a river or other outlet, it is said to be open. If water leaves a lake only by evaporation, the lake is closed. Closed lakes usually become saline. Lakes usually show circular, semi-circular or elongated shape. Understanding and evaluating lake dynamics is nece- ssary to conduct the sustainable management of water resources [5]. In addition, lake surface areas (especially closed lakes) are sensitive to natural changes and thus may serve as significant proxies for variations in regional environmental and fluctuations in global climate [3] [6] . Changes in the areal extent of lake surface water may occur due to various factors, including the progressive unveiling of the lake basin by sediments, climate change, tectonic activity causing uplift or subsidence, and the development of drainage faults [7] [8] . Being able to access the spatial distribution and geographical extent information on lakes in real time has great significance in limnology and for understanding interactions between regional hydrology and climate change [9] . Satellite remote sensing (RS) has advantages because it can track land surface information in real-time macroscopically, multitemporally, multispectrally, dynamically, and repetitively; hence, it is appropriate for surveying and mapping surface water bodies [10] . The present review especially focuses on the broader category of cryo- spheric lakes and their extraction using optical RS methods. The cryosphere refers to those parts of the Earth containing water in its frozen state: snow, glaciers, permafrost, seasonally frozen ground, lake and river ice, sea ice, ice sheets, and shelves. The cryosphere holds a significant amount of the Earth's total supply of freshwater. About 77% of Earth's freshwater is frozen, 91% of which is contained in the Antarctic ice sheet, 8% in the Greenland ice sheet, and the remaining 1% is contained in glaciers [11] . Many studies and research works have been carried out on cryospheric lakes in order to track environmental changes and behavior. Cryospheric lakes are classified into 5 types as depicted on Table 2. Inclement weather in the polar regions (Arctic and Antarctic), few numbers of fine-weather days in summer, and high logistic cost restricts research trips to polar Regions. Therefore, satellite RS data and aerial photography are important sources of information for monitoring the short-term and long-term changes that occur at a specific location in cryospheric regions over time. Although RS data can never replace aerial photographs, which provide images at a resolution as high as 0.2 - 0.3 m, it is suitable for lake-feature extraction in the cryospheric regions, where frequent aerial photography is difficult because of the extremely harsh environment and the high logistical costs. Hence, development of automated or semi-automated feature extraction methods using RS data is much needed for continuous monitoring of the geographical features in a cryospheric environment.

TABLE 1. Types of lakes (Source: [12]).

Type of lake	Description	Examples
Artificial lake	They may be constructed for various purposes, such as hydroelectric power generation, recreation, industrial use, agricultural use, or domestic water supply.	Lake Mead and Lake Powell, USA
Crater lake	A lake that is formed in a volcanic crater after the volcano has been inactive for some time. Lake water may be fresh or highly acidic and may contain various dissolved minerals.	Mount Aso crater lake, Japan Taal Volcano, Phillippines
Endorheic lake	A lake that has no significant outflow, either through rivers or underground diffusion.	Lake Eyre, central Australia, Aral Sea in central Asia
Fjord lake	A lake in a glacially eroded valley that has been eroded below sea level.	Geirangerfjord, Norway Tracy Arm fjord, Alaska
Former lake	Prehistoric lakes and those that have permanently dried up through evaporation or human intervention.	Owens Lake in California, USA
Underground lake	A lake that is formed beneath the surface of the Earth's crust. Such a lake may be associated with caves, aquifers, or springs.	Reed Flute cave, China Lake Vostok, Antarctica
Seasonal lake	A lake that exists as a body of water during only part of the year.	Badhkal lake and Sambhar lake, Rajasthan, India
Oxbow lake	Characterized by a distinctive curved shape, it is formed when a wide meander from a stream or a river is cut off.	Gambi lake on River Tana, Kenya
Lava lake	This term refers to a pool of molten lava in a volcanic crater or other depression. The term lava lake may also be used after the lava has partly or completely solidified.	Erta Ale, Ethiopia Nyiragongo, Democratic Republic of the Congo

2. BRIEF REVIEW ON RS METHODS USED FOR LAKE FEATURE EXTRACTION

A brief review of the most commonly used methods employed in mapping of water feature from urban and cryospheric regions is depicted in Table 3 and Table 4. Optical satellite systems have most frequently been applied to lake or water body extraction research. The parts of the electromagnetic (EM) spectrum covered by these sensors include the visible and near-infrared (NIR) ranging from 0.4 to 1.3 μm, the short wave infrared (SWIR) between 1.3 and 3.0 μm, the thermal infrared (TIR) from 3.0 to 15.0 μm and the long wavelength infrared (LWIR) from (7 - 14 μm). The decision tree and programming method are used for extracting water body information from the flood affected region [14] . The semi-automated change detection approach is used for extracting water feature form satellite image [15] . An automatic extraction method is used for extracting water body from IKONOS and other high resolution satellite image [16] . Thresholding and multivariate regression method [17] , a conceptual clustering technique and dynamic thresholding [18] , an original entropy-based me thod [19] , are also used for extraction of water bodies. The water body can be extracted by classification; unsupervised classification [20] ; the support vector machine (SVM) with one-against-one (1A1) and one-against-all (1AA) techniques are used for land cover mapping [21] . A supervised classification algorithm [22] of RS satellite image that uses the average fuzzy intra-cluster distance within the Bayesian algorithm [23] ; sometimes combinations of supervised and unsupervised classification [24] are used for water information extraction. A new index normalized optical water index (NOWI) was proposed to accurately discriminate between land and water regions in multi-spectral satellite imagery data from DubaiSat-1 [25] .

TABLE 2. Types of cryospheric lakes (Source [13]).

Type of lake	Description	Examples
Supraglacial/Epiglacial lakes (SGLs)/	The lakes formed over the glacial surface due to the processes active on the glacial surface. Shifting, merging, and draining are characteristics of SGLs. SGLs are more dynamic and vary in time and space.	SGL on Milam glacier
Open lake (Landlocked lakes)	A body of water that is surrounded completely by land. A landlocked lake has no water source such as a river. It is fed by water seepage in the ground and water runoff with the surrounding land.	Lake Vanda in Antarctica
Epishelf lake	When ice shelves completely block the mouth of a fjord, an epishelf lake is created. This is caused by melt water that flows into the fjord every summer, but is impounded behind the ice shelf.	The largest epishelf lake, Disraeli Fiord, Arctic Ocean.
Pro-glacial lake	Pro-glacial lakes are ice-contact lakes occurring adjacent to the frontal margin/snout of a glacier. Many such types of lake are ice-core moraine-dammed or ice-dammed and show ephemeral or perpetual nature.	Vasundhara Tal at Raikana glacier, Himalayas
Moraine-dammed lake	A lake formed as a glacier recedes from its terminal moraine, the moraine acting as an unstable dam. Most of these lakes are formed when valley and cirque glaciers retreated from advanced positions achieved during the Little Ice Age.	Laguna Paron, Peru

TABLE 3. Methods used for extraction of urban lakes.

Satellite Used	Study area	Methods	Advantages
Landsat 5 (TM) Landsat 7 (ETM+) Landsat 8 (OLI) Resolution: 30 m [35]	Lake Urmia located in the northwest of Iran	Normalized Difference Water Index (NDWI) and Principal Component Analysis (PCA)	NDWI performed slightly better than the NDWI-PCs. NDWI-PCs have an advantage over the NDWI, that it detects the surface water changes of two and three different times simultaneously by applying a single threshold to the selected PC.
Landsat TM ETM+ Resolution: 30 m [36]	Rift Valley lakes in Kenya	Water Index (WI) using Tasseled Cap Wetness (TCW) index and NDWI	WI detected the shorelines with an accuracy of 98.4%, which was 22.3% higher than the TCW, and 43.2% more accurate than the NDWI.
Landsat 4 - 5 Resolution: 30 m [37]	Huanghe river delta, China	B2 + B3 > B4 + B5 B4 < 60 B4 > B5 and B5 < B2	The decision tree algorithm failed to extract small water bodies at scales below the sensor resolution.
Landsat TM and ETM+ Resolution: 30 m [38]	Hebei, Jiangxi, Ningxia, China	A watershed segmentation method is adopted to detect mixed water pixels at the edges of lakes or rivers	Automatic method for extracting rivers and lakes (AMERL) successfully extracted most of the narrow rivers and lakes.
SPOT 4 Resolution: 10 m [39]	Jiangning county of Jiangsu, China	Decision Tree (DT) model based on both spectral and auxiliary information of Digital Elevation Model (DEM) and Slope (DTDS).	It is difficult to extract water bodies effectively by applying a single technique due to effects of shadows. Unsupervised classification yields result with low accuracy.
Landsat 5 TM Resolution: 30 m [40]	Murrumbidgee, Wagga Wagga, Australia	Single band density slicing and Maximum Likelihood (MXL).	MXL proved to be more accurate than density slicing to detect water bodies. Density slicing yielded a less speckled output image as compared to MXL.
Landsat 5 TM Resolution: 30 m [41]	Denmark, Switzerland, Ethiopia, South Africa, New Zealand.	$AWEI_{sh} = Blue + 0.25*Green - 1.5*(NIR + SWIR1) - 0.25*SWIR2$ $AWEIN_{sh} = 4*(Green - SWIR1) - (0.25*NIR + 2.75*SWIR2)$	Automatic Water Extraction Index (AWEI) successfully extracted surface water with high accuracy, particularly in mountainous regions where hills cast shadows on background surfaces and in urban areas with complex land cover. It is a simple technique to extract water in different environmental conditions.
MODIS (250 - 500 m) and ASTER (15 - 90 m) [42]	Bihor, Romania	Threshold method and supervised classification	This approach is useful in providing information about water classification from different resolution data.
MODIS (250 - 500 m) and ASTER (15 - 90 m) [42]	Bihor, Romania	Threshold method and supervised classification	This approach is useful in providing information about water classification from different resolution data.
ASTER (15 m) and MODIS [43]	Koros basin, Roma-nian-Humgarian border	$NDVI = \frac{R0.87 - R0.66}{R0.87 + R0.66}$ Band 1 (0.66 μm) - 250 m Band 2 (0.87 μm) - 250 m	Cloud shadow and water pixels are not completely separated out.
ASTER Resolution: 15 m [44]	Beijing, China	4 segmentation levels were created for differentiation between water, vegetation and non-vegetation	Accuracy of object-oriented classification is higher than the accuracy of MXL classification.

TABLE 4. Methods used for extraction of cryospheric lakes.

Satellite Data	Study area	Specifications	Remarks
Landsat (30 m) and ASTER (15 m) [71]	Gangotri glacier, Himalayas	$NDWI = \frac{NIR1 - blue}{NIR1 + blue}$ $ASTER\ ratio = \frac{Green}{NIR}$	Results led to the accurate identification of glacial lakes using the NDWI. Lake identification based on ASTER dataset is slightly more accurate than Landsat dataset.
World View 2 (WV2) (2 m) [72]	Greenland Ice Sheet (GrIS)	$NDWI = \frac{blue - red}{blue + red}$	The slush elimination process was not accurate, but it can be made accurately by fusing morphological procedures using edge detection with multi-threshold procedures (accuracy = 85.2%).
WV2 (2 m) [73] [74]	Larsemann Hills, Antarctica	$NDWI1 = \frac{Coastal - NIR2}{Coastal + NIR2}$ $NDWI2 = \frac{Coastal - NIR1}{Coastal + NIR1}$ $NDWI3 = \frac{Blue - NIR1}{Blue + NIR1}$ $NDWI4 = \frac{Blue - NIR2}{Blue + NIR2}$	NDWI-based indices approach offers several advantages as it is consistent in extracting water bodies against noise, flexible as the threshold values can be adjusted to remove background noise by not affecting the target extraction, time efficient in processing and producing the output and comparatively easier to implement.
Landsat ETM+ (15 m and 30 m) [75]	Lake Merzbacher, border of China	$NDWI = \frac{NIR - Blue}{NIR + Blue}$ Threshold 0.35 to 0.55 To accurately extract area of floating ice, ratio is used: $\frac{NIR}{Red}$	Results show that the index can predict the outburst process of the lake and determine the outburst period of the lake after its drainage had occurred. The automatic program is not so perfect due to the Interferential factors such as snow or spissatus in images (accuracy= 92.0%).
Landsat TM and SPOT (Fused 10 m) [76]	Swiss Alps	$NDWI = \frac{NIR - Blue}{NIR + Blue}$ $NDWI = \frac{TM4 - TM1}{TM4 + TM1}$ Simple channel ratio: $\frac{TM4}{TM1}$	TM4 and TM1 bands can be replaced by TM5 and TM7, but TM4 and TM1 are capable of discriminating water from snow and ice more accurately. The simple channel ratio works well, but NDWI showed an enhanced contrast between water and surrounding environment.

3. IMPORTANCE OF CRYOSPHERIC LAKES

Lake ice cover has been established as a robust indicator of local climate variability and fluctuations. Lake ice forms an essential component of the cryosphere, especially at high mountainous latitudes where a large number of lakes exist. Long-term records of lake ice have been significantly used as a proxy indicator of winter climate conditions. Previous studies have identified lake ice as a highly sensitive cryospheric component to climatic conditions [26] . Spatio-temporal changes in lake ice cover have an imperative feedback on energy exchanges between the lake surface and the atmosphere. Persistent warmer air temperatures [27] and raised snowfall had been observed in the Arctic over the last decades [28] , and found to be associated with an amplified reduction of sea-ice concentrations, thickness and extent [29] , which had been accelerated during recent years [30] . These spatio-temporal changes in the Arctic climate system have likely had an impact on ice phenology of lakes in coastal regions adjoining to the Arctic Ocean. SGLs play an important role in establishing hydrological connections that allow lubricating seasonal meltwater to reach the base of the ice sheet [33] [34] .

Among all components of glacier system, SGLs are the most straightforward to be recognized. SGLs have already been researched on the Greenland ice sheet, Svalbard, and Himalaya. SGLs on Greenland characteristically form as a response to summer melting. SGLs typically form below 1500 m altitude, in topographic low regimes in the ablation zone and can expand to numerous kilometers in size on the surface of the Greenland ice sheet. SGLs can also form in the lower ablation area of debris-covered valley glaciers. The life span of SGLs is unpredictable and in situ monitoring is less practical. Hence, RS can be effectively used for studying SGLs and their seasonal variations to address the status of glacier or ice sheet. SGLs and firn aquifers store a substantial amount of meltwater, providing a buffer between melting and mass loss to the ocean [31] [32].

4. RS FOR EXTRACTING CRYOSPHERIC LAKE FEATURES

To our knowledge, almost all of the published works on an extraction of surface lakes in cryospheric environments have used the satellite RS data. Multispectral RS has been widely utilized in cryospheric studies and have employed a variety of electro-optical satellite sensor systems for characterization and extraction of various cryospheric features, such as glaciers, sea ice, lakes and rivers, the extent of snow and ice, and icebergs. Aerial photography of ice-covered terrain began during the early 20th century by expeditions to the high altitudes and was used primarily to document the progress of the expedition. Wilkins documented ice cover in the Antarctic Peninsula during the first successful flight in Antarctica by using a handheld, folding Kodak 3A camera [45] [46] . The quality of the photographs is often exceptional, but the reason behind opting for satellite RS instead of aerial photography is the harsh environment of cryosphere where frequent monitoring is difficult by the use of aerial photography which adds up to high logistics costs. Cryospheric RS applications initiated as early as 1962 with the launch of the Argon satellite. Thereafter, in the 1970s Landsat 1, 2, and 3 Multi-Spectral Scanner (MSS) images constituted an important glaciological resource [47] . Initial successes in large-scale mapping were achieved through the use of the moderate spatial resolution (1 - 2.5 km) and wide swath (2400 km) advanced very high resolution radiometer (AVHRR) images, [48] which helped reveal details about ice stream flow in West Antarctica [49] . After the original AVHRR mosaic of Antarctica, the United State Geological Survey (USGS) made subsequent improvements to the mosaic by eliminating more clouds, separating the thermal band information to illustrate surface features more clearly, and correcting the coastline of the mosaic to include the grounded ice while excluding thin, floating fast ice [50] . Another large-scale mapping has been completed with MODIS for both the Arctic and the Antarctic [51] . Most recently, Landsat imagery of Antarctica has been compiled into a single, easily accessible map-quality data set [52] and SPOT stereo imagery have been used to derive DEM of ice sheets, ice caps, and glaciers [53] . In the snow and glaciated terrain of the Himalayas, satellite RS was established as the best tool because many of the glacial lakes are located at very high altitude, cold weather, and rugged terrain conditions, making it a tedious, hazardous and time-consuming task to monitor by conventional field methods. Satellite RS technology facilitates the study of initial and qualitative hazard assessment of glacial lakes of the Himalayas systematically with a cost-to-time benefit ratio [13] .

Cryospheric lake features have been researched primarily by means of multispectral satellite images from the ASTER [54] [55] , Landsat-7 [56] , and MODIS imagery [57] instead of aerial photography despite its very high spatial resolution. The relatively high spectral reflectance response from a water body feature in the ASTER, MODIS, and Landsat multispectral bands is the foundation for employing these sensors in water body mapping and surveying applications. An accurate manual delineation of lake extent is used [55] [58] [59] when lakes are easily identifiable on images. Applications of methods that discriminate water from surrounding ice and snow are possible on optical images using semi-automatic methods that employ different spectral bands of the satellite sensors [60] - [62] . Many research studies have surveyed methods for semi-automatic or automatic lake feature extraction using medium and coarse resolution satellite RS data (e.g., [57] [63] [64]). Automated or semi-auto- mated methods have the advantage of rapid extraction of lakes from multi-temporal images with large areal coverage [57] [62] . Although manual delineation is highly accurate [65] , it is time-consuming and thus unsuitable for wide geographical areas. Therefore, accurate manual delineation is preferable for studies over smaller areas, but studies that encompass larger areas would benefit from automatic or semi-automatic methods [63] . The most common method for deriving surface water bodies from satellite images is the density slice method, which uses single or multiple spectral bands, and multi-spectral classification [66] [67] . Frazier and Page reviewed numerous methods employed by various authors to extract water bodies from Landsat TM and MSS image classifications [40] . Yu et al. [68] investigated and discussed a few methods for deriving water information using SPOT-4 images. The thresholding and multivariate regression method, a conceptual clustering technique, the dynamic thresholding method, and entropy-based method, have been successfully implemented in surface water extraction studies [17] [18] . The water body features can be extracted by unsupervised classification [20] , supervised classification (e.g., a Bayesian algorithm) [23] , and a combination of both supervised and unsupervised classification [24] . Waldemark et al. [69] proposed a neural network (NN) approach for extracting water bodies from satellite images. In addition to the aforementioned water classification methods, there are several other original methods, such as the Decision Tree (DT) method and the step iterative method [39] [70] . It is evident that the most common methods for extracting lakes are, single band?based threshold methods, spectral index ratio (SIR)-based multiband methods, image segmentation methods, spectral-matching methods, and supervised target detection methods.

SIRs are utilized to extract a specific target or feature, and they are computed from the difference in reflectance values of the bands used to formulate the ratio [77] . Conventionally, water and vegetation have been the primary focus of normalized difference SIRs because they are simple to classify based on the difference in reflectance values, ranging from 450 nm to 750 nm. Presently, the methods for extracting lakes are based on a spectral index or multiband techniques, which are spectrum property?based methods [78] , such as the NDVI [79] and NDWI [80] - [83] . Since a single spectral index could not demarcate lakes effectively in different environments, many improved indices have been proposed to yield better results in specific environments [84] . Ouma and Tateishi [36] proposed a novel water extraction index for shoreline delineation by combining the TCW index (TCWI) and the NDWI.

A comprehensive water body information extraction technique was proposed by Wu et al. [85] through the fusion of the spectral relationships between various bands with supervised classification methods. Rogers and Kearney [86] proposed the NDWI for medium spatial resolution and high temporal resolution with MODIS multispectral satellite images (MSI). Furthermore, Xiao et al. [87] proposed a land surface water index (LSWI), while Mo et al. [88] proposed a mixed water index (MWI) by combining the NDVI and NIR data to identify water bodies in MODIS images. Lu et al. [89] recommended an integrated water body extraction technique with HJ-1A/B satellite imagery by utilizing differences between NDVI and NDWI. These modified indices have been commonly used to map surface water bodies using Landsat and MODIS images [90] - [97] . These indices are normalized, ranging from −1 to 1, in which zero acts as a threshold to discriminate water from vegetation and land surface. However, because of the complexity of cryospheric environments, various ground targets may have the same spectrum characteristics. Therefore, only one type of spectral index method cannot extract water bodies under all environmental conditions [98] .

5. RS METHODOLOGIES FOR EXTRACTING LAKE FEATURES

Methodologies of a lake or water body extraction can be summarized into three groups: feature extraction using pixel-based and object-based classification, and SIRs. Nath and Deb [66] provided a comprehensive overview of methods for water extraction from high resolution satellite images. June et al. [37] developed an automatic extraction of water bodies from a Landsat TM image using DT algorithm. The proposed algorithm was based on spectral characteristics of the water body in TM images. Wang et al. [99] developed water extraction method based on texture analysis. Luo et al. [100] developed an algorithm for water extraction using Landsat TM which combines water index computation, whole-scale segmentation, whole-scale classification and local scale segmentation and classification to achieve highly-precise water extraction result [101] . The traditional pixel based digital image classification has been and is still being used for characterization and mapping the spatial extent of forests, urban, water bodies, coastal, and wetland areas [102] . In principle, three types of classification methods exist, namely unsupervised, supervised and hybrid [103] . Unsupervised classification clusters pixels in a dataset based on statistics only, without any user-defined training classes. The most commonly used unsupervised land cover/land use (LC/LU) classifier is the Iterative Self-Organizing Data Analysis (ISODATA) classification algorithm. On the other hand, several types of statistics-based supervised classification algorithms have been developed. Examples of the more popular classifiers (in increasing complexity) are parallelepiped, minimum distance, MXL, and Mahalanobis distance [104] . Hybrid methods can combine the advantages of manual, parametric and non-parametric methods in various combinations to optimize the classification process [105] . Object- based Image Analysis (OBIA), a recent image analysis approach appears to be more popularly used for the classification of LC/LU of urban areas [106] . Object-based method considers image classification based on objects such as topologic (neighborhood, context) and geometric (form, size) information [107] . The object-oriented approach analyzes objects within images as a processing unit instead of using pixels. Geographic object-based image analysis (GEOBIA), as opposed to pixel-based image processing, is also emerging as a popular classification method [108] . Studies that monitored extreme cold areas using new satellite sensors were initiated by using medium and low spatial resolution images [109] . Sundal et al. [62] proposed an automatic method based on a set of fuzzy logic membership functions to identify and map lakes. The method used a single threshold developed by Box and Ski [60] to differentiate between meltwater and ice, exploiting the different sensitivity to water of MODIS bands 1 and 3. Certain types of lakes, such as deep lakes, were particularly hard to identify. A semi-automatic method to track lakes was developed by Selmes et al. [57] . Liang et al. [64] developed an automatic method for lake identification, mapping, and tracking. Exploiting the characteristics of the changing nature of the lakes and their surroundings, Johannson and Brown [63] developed a method known as Adaptive Lake Classification (ALC); ALC, specifically targeting the identified problem lakes [110] . Methods used for water feature extraction in an urban environment and cryospheric environment are summarized in Table 5 and Table 6.

TABLE 5. Technologies used for water body extraction in cryospheric environment.

Satellite data	Method name	Remarks
WV2 (0.5 m) [72]	Automated Spectral-Shape Procedure	Average positional offset between delineated and manually digitized streams is about 1 pixel. $NDWI_{ice}$ was discovered in order to increase the contrast between ice and water. (accuracy = 92.1%).
MODIS (250 m) [63]	ALC approach based on object-oriented classification.	ALC mainly focuses on the changing nature of lakes. Difficulty in detecting small lakes (<0.1 km^2) cannot be resolved. Lakes with partial or total ice cover posed a challenge for ALC as both lake ice and glacier ice have similar reflectance.
ASTER (15 m) [111]	Non-parametric classification. Spectral indices NDWI using green and NIR band, and square pixel metric (SqP) method	99% accuracy attained by applying NDWI, elevation and NIR/Red band ratio to separate water features from ice debris
Landsat (30 m) RADARSAT (7 m) [112]	Supervised MXL Classification	Difficulty in mapping small ponds which can be overcome by using high resolution imagery or a high resolution aerial photograph. User's accuracy for water class = 95%.
MODIS (250 m) ASTER (10 m) Landsat ETM+ (10 m) [113]	Lakes were delineated automatically using object oriented segmentation and classification methods [57] [62] [63] [110].	Sundal [110] had difficulty in resolving ice-covered lakes. Johannson and Brown [63] reported that as many as 18% of reported SGLs are likely to be false positive. In Selmes *et al.* [57] any lake <0.125 km^2 does not feature in the dataset. It is found to report lake area most accurately.
WV2 (0.5 m) [77]	Spectral indices using customized NDWIs having Coastal and Blue band against NIR1 and NIR2	The coastal band produced less false positive results in comparison to the blue band during the detection of lakes. The PAN-sharpening process does affect the accuracy of feature classification.
WV2 (0.5 m) [98]	Support vector machine (SVM), Spectral angle mapper (SAM), MXL, NN classifier, Winner takes all (WTA)	The study concluded that WTA (accuracy = 97.23%) was better for mapping water and land and SVM and NN classifier for mapping snow/ice.
IKONOS (4 m) Hyperion [109]	For improving the accuracy of water classification. IKONOS tasseled cap transformation is applied to wetness	Small lakes were detected due to the high spatial resolution of IKONOS image. The water class was defined by an IKONOS NDVI of greater than −0.1 (accuracy = 86.70%).

TABLE 6. Technologies used for water body extraction in an urban environment.

Satellite data	Method name	Remarks
Landsat TM (30 m) [114]	Object-oriented classification, Pixel-based supervised MXL.	MXL classification produced salt and pepper image, whereas the classified image derived from polygon-based classification is closer to human visual interpretation.
Landsat TM (30 m) [115]	Supervised MXL classification, Unsupervised ISODATA classification.	Accuracy assessment showed that the ISODATA could multispectrally classify the urban water successfully. Even smaller ponds or rivers (<30 m) can be extracted if the high resolution imagery is used.
Quickbird (0.61 m) [116]	Statistical Regional Merging (SRM) for image segmentation, NDWI and Normalized Saturation-value Difference Index (NSVDI) for water extraction.	The results prove that the accuracy of the extracted water features can be significantly improved and shadows can be effectively eliminated.
Zi-Yuan 3 (ZY-3) (2.1 m sharpened) [117]	Object oriented multi-resolution segmentation, Edge detection using Canny-edge detector.	The extraction results on the high resolution remotely sensed image are significant by taking into account the spectrum geometry and texture information of images. (Accuracy = 94.6%).
Landsat ETM+ (15 m), ERS SAR (26 m), SPOT (10 m sharpened) [19]	Knowledge-based DT method including spatial features as size, shape, position and multi-spectral characteristics.	The proposed algorithm is differentiated from other existing algorithms as it is independent of the image sensor. It can be applied on either PAN images or any spectral band of optical images. The proposed algorithm failed to extract small water areas.
WV2 (2 m) Rapid Eye (2 m) Pleiades (5 m) [103]	Hierarchical land-use classification, ISODATA unsupervised classification to extract additional subclasses.	RapidEye showed a higher overall accuracy of 78%, surpassing the result of Pleiades (74%). Pleiades showed the best classification accuracy compared to RapidEye and WV2. WV2 proves to be more versatile to extract various sub-classes.
WV2 (0.5 m sharpened) [102]	Object-based image classification utilizing Nearest Neighbor decision rules.	The analysis found that object-based classification using scale parameter of 60 produced the best result of wetland delineation compared to scale 30 and 300.

TABLE 7. Water feature extraction methods with kappa statistics values.

Satellite data	Study area	Method	Kappa value
Landsat TM [37]	Huanghe river delta, China	DT algorithm: TM2 + TM3 > TM4 + TM5 was used to extract reservoirs, ponds and broad rivers	0.92
Landsat 5 TM [41]	Test site 1: Denmark, Test site 2: Switzerland, Test site 3: Ethiopia, Test site 4: South Africa, Test site 5: New Zealand.	AWEI	Test site 1: 0.93 Test site 2: 0.95 Test site 3: 0.97 Test site 4: 0.98 Test site 5: 0.98
Landsat TM and ETM+ [84]	Bayi lake, Fuzhou City, China	Modified NDWI	0.99
ASTER [44]	Beijing, China	An object-oriented approach using Hierarchical classification and DT classification	0.81
Landsat 5-TM Landsat 7-ETM+ Landsat 8-OLI [35]	Lake Urmia, Iran	NDWI-PC	0.73
Quickbird [130]	Phoenix and Scottsdale, Arizona	Object-based approach using multi-resolution segmentation algorithm, Minimum distance supervised classification	Scottsdale: 0.81 Phoenix: 0.82
IRS LISS III and LISS IV [131]	Bhopal, India	NN supervised classification approach	LISS III: 0.98 LISS IV: 0.96
WV2 and Airborne Imaging Spectroradiometer Analysis (AISA) [132]	Study area 1: Turtle Creek Corridor, Dallas, Texas Study area 2: National Mall Area, Washington, DC	Fuzzy Kolmogorov-Smirnov (FKS)	AISA: 0.79 WV2: 0.99

6. DISCUSSION

We have reviewed various methods utilized for extraction of a lake or water body features in urban and cryospheric environments. In this section, we summarize and discuss the generalized trend in methods, satellite datasets and achieved accuracies for lake feature extraction. Table 7 depicts different water feature extraction methods and accuracies expressed in kappa value. For extracting urban lakes (or other urban features), per-pixel based approaches were always the primary tool because of their low cost and easy implementation. Per-pixel based classification approach was considered to be the favorite choice of most of the researchers to extract lakes.

Unfortunately, this procedure always resulted in mixed pixel's problem. Pixel based classification approaches on high spatial resolution satellite images (e.g. SPOT, Landsat), often results in "salt and pepper" representation, which is even increasing when considering the new generation of very high spatial resolution data (e.g. IKONOS, QuickBird, GeoEye, WV2 etc.). This problem has led many researchers to incorporate segmentation, texture, context, color, and many other parameters to glide the mixed or wrongly classified pixels into their proper classes [118]. So, to overcome the limitations of pixel-based approach, a new approach was developed known as an object-oriented approach which has gained more and more interest, especially when dealing with very high spatial resolution satellite images to capture the finer details of the urban area. In almost all the case studies, object-based classification approach resulted in improved accuracy ranging from 84% to 89% (approximately). Object-based classification can, not only use spectral information of land types, but also use images' spatial position, shape characteristic, texture parameter and the relationship between contexts, which effectively avoid the "salt and pepper phenomenon" and greatly improve the accuracy of classification [119] . After reviewing numerous methods for water feature extraction in the general environment, it is apparent that common water classification methods for optical imagery could be categorized into four basic types [120] : a) thematic classification [121] , b) linear un-mixing [122] , c) single-band thresholding [123] and d) two-band spectral WI [80] [84] [86] . A Synergetic fusion of various automatic and semi-automatic methods are also proposed to improve water information extraction accuracies. The spectral band method is easy to implement, but frequently misclassify mountain shadows, urban areas or other background noise as lakes [124] . The most notable supervised classification methods used for lake extraction are MXL, DT, artificial NN and SVM [125] , while the most common unsupervised classification methods include the K-means and ISODATA

[126] [127] . These methods may achieve superior accuracy than spectral band methods under some environmental conditions; however, existing ground reference datasets are required, which restrict these methods from being applied over large study regions [40]. The WIs have been extensively used because of their comparatively high accuracy in lake detection and low-cost implementation [128] . In many WIs, the lack of stability of the threshold is still a problem [120] , making it difficult to use uniformly. The lack of a reasonably stable threshold may make the classification more time-consuming and lead to a subjective choice of threshold which may also affect accuracy [41]. The design and implementation of WIs have been persistently improving [38] . Despite the fact that a number of lake extraction methods are published in the literature, the choice of method for a specific application is constrained by accuracy issues. Water classification accuracy problems are especially pronounced in areas where the background land cover includes low albedo surfaces such as shadows from mountains, buildings, and clouds. The occurrence of shadows may cause misclassification because of the resemblance in reflectance patterns, which may hamper the accuracy of the surface water mapping and change analysis [41] [129].

Even if the object-oriented based approach is most widely researched topic in urban applications yet it is not really popular with cryospheric regions [74] . Still cryospheric applications such as feature extraction from cryospheric regions, detailed land-cover classification of cryosphere employ pixel-based approach. These days, for obtaining detailed land-cover classification or for accurate feature extraction from cryospheric region very high resolution optical satellite imagery is being used with the resolution of the order of ~0.5 - 5 m. The popular optical satellites falling in this category are: WV2, SPOT5, IKONOS, GeoEye etc.

Most commonly used methods for lake feature extraction from cryospheric region are:

a) Spectral indices making use of two satellite imagery bands,

b) Supervised classification involving MXL, parallelepiped, NN, SAM and SVM, target, WTA approach

c) Unsupervised classification: involving ISODATA *technique*

d) Target detection methods which include: matched filter (MF), constrained energy minimization (CEM), adaptive coherence estimator (ACE), SAM, orthogonal sub-space projection (OSP) and target-constrained interference-minimized filter (TCIMF)

e) Single band threshold and spectral relation method

Despite so many methods being developed for lake feature extraction, none of them is known to yield highly accurate results in all environments. The methods developed so far are not generic rather they are specific to either the location or the satellite imagery or to the type of the feature to be extracted. Lots of factors are responsible for leading to inaccurate results of lake feature extraction in Cryospheric regions, e.g. the mountain shadow which also appears as a dark pixel, is often misclassified as open lake, which can be corrected using topographical modeling using DEM [133] -[139] . There are various other target features which possess similar spectral characteristics which result in overestimation and thus outputs an inaccurate result. Thus, after knowing the past and present of methods being developed for lake feature extraction, it's felt that there is a strong urge for developing new methods that would yield results with higher accuracy. Also the method should be highly versatile and robust as well as dynamic, so that it can be used for extraction of all types of lakes in all environments under all situations without any change in the design of the method.

7. SUMMARY AND CONCLUSION

Satellite sensors of varying spatial, temporal and spectral resolutions have been used to extract and analyze information regarding surface water. Studies using ASTER and Landsat ETM+ data have focused on smaller regions with a limited number of lakes, mainly using manual delineation of lake extent. For coverage of larger areas MODIS imagery is generally been adopted. ASTER and ETM+ images have a high spatial resolution of the order of ~10 m, conducive to accurate lake area delineation, MODIS imagery is much coarser i.e. of the order of ~250 m but the temporal resolution of MODIS is higher (at least once a day rather than biweekly). Landsat series of satellites are among the most extensively used multispectral sensors in surface water extraction studies. Medium resolution satellite data (10 - 90 m) are available for cryospheric studies since the early 1970s, with the launch of the new space-borne sensors: Landsat MSS, Landsat TM and ETM+, SPOT, Terra ASTER, IRS, and more recently the Advanced Land Observing Satellite (ALOS) launched in 2006. The other group of optical satellites is high resolution, of the order of a meter and sub-meter: WV2, IKONOS, Quickbird, and Geo Eye-1, are now being brought up in use for the extraction of lakes from the cryospheric environment [140] -[150] . Now a day's much of the work based on cryosphere such as feature extraction and land cover classification employs high resolution imagery because of its high spatial resolution, which enables to achieve finer details of the region, which is otherwise not possible by using medium resolution imagery (e.g. Landsat TM, ETM+, SPOT, ASTER etc.). After reviewing numerous methods available for cryospheric lake feature extraction, we conclude that the most popular and effective method for extracting lake is based upon spectral indices which make use of two optical bands at a time. This method is the most researched and emerging methods because of its high-term advantages which include low implementation cost, simple in understanding, easily modifiable and effective in producing stable output. Despite its potential benefits, it does suffer from few limitations, which include high location or feature dependency which makes it working only for a specific application, high misclassification which usually occurs because of objects possessing similar spectral characteristics and the commonly known misclassified objects are mountain shadows or hill shadows which possess dark pixel that almost appear similar to a water body and often get misclassified and results in a false positive result. The problem of misclassification can thus be minimized by selecting appropriate threshold which would be then able to discriminate between different objects based on their precise spectral response to a greater extent. The methods which are working well for the cryospheric environment for feature extraction or land-cover classification does not really guarantee that they will be working in the same manner for the urban environment. Thus, in coming years it is expected that much of the work will be done on object-based approach or hybrid approach involving both pixel-based as well as object-based technology, with respect to lake feature extraction and a more accurate, versatile and robust method will be developed that would work independent of location (for both urban and cryosphere, single method would be able to extract water bodies accurately) and feature (for extraction of different types of lake a single method would be able to work). And also in coming years, super high spatial and temporal resolution optical satellites will be active which will yield even micro details of a region.

ACKNOWLEDGEMENTS

We acknowledge Dr. S. Rajan, Director, NCAOR for his encouragement and motivation of this research. We also thank Ms. Prachi Vaidya, India for her constructive comments on the draft version of the manuscript. This is NCAOR contribution No. 16/2015.

REFERENCES

1. Chave, P. (2001) The EU Water Framework Directive: An Introduction. IWA Publishing, London, 208.
2. Gao, H., Birkett, C. and Lettenmaier, D.P. (2012) Global Monitoring of Large Reservoir Storage from Satellite Remote Sensing. Water Resources Research, 48.
3. Bai, J., Chen, X., Li, J., Yang, L. and Fang, H. (2011) Changes in the Area of Inland Lakes in Arid Regions of Central Asia during the Past 30 Years. Environmental Monitoring and Assessment, 178, 247-256.
4. Deus, D. and Gloaguen, R. (2013) Remote Sensing Analysis of Lake Dynamics in Semi-Arid Regions: Implication for Water Resource Management. Lake Manyara, East African Rift, Northern Tanzania. Water, 5, 698-727.
5. Jawak, S.D. and Luis, A.J. (2015) Spectral Information Analysis for the Semiautomatic Derivation of Shallow Lake Bathymetry Using High-Resolution Multispectral Imagery: A Case Study of Antarctic Coastal Oasis. Aquatic Procedia, 4, 1331-1338.
6. Mason, I.M., Guzkowska, M.A.J., Rapley, C.G. and Street-Perrott, F.A. (1994) The Response of Lake Levels and Areas to Climatic Change. Climatic Change, 27, 161-197.
7. Mercier, F., Cazenave, A. and Maheu, C. (2002) Interannual Lake Level Fluctuations (1993-1999) in Africa from Topex/Poseidon: Connections with Ocean-Atmosphere Interactions over the Indian Ocean. Global and Planetary Change, 32, 141-163.
8. Goerner, A., Jolie, E. and Gloaguen, R. (2009) Non-Climatic Growth of the Saline Lake Beseka, Main Ethiopian Rift. Journal of arid Environments, 73, 287-295.
9. Papa, F., Prigent, C. and Rossow, W.B. (2008) Monitoring Flood and Discharge Variations in the Large Siberian Rivers from a Multi-Satellite Technique. Surveys in Geophysics, 29, 297-317.
10. Chen, Q., Zhang, Y., Ekroos, A. and Hallikainen, M. (2004) The Role of Remote Sensing Technology in the EU Water Framework Directive (WFD). Environmental Science & Policy, 7, 267-276.
11. Christopherson, R.W. (2003) Glacial and Periglacial Processes and Landforms. In: Christopherson, R.W., Ed., Geosystems: An Introduction

to Physical Geography, 5th Edition, Prentice Hall, Upper Saddle River, 519-553.

12. Lake (2014) New World Encyclopedia. http://www.newworldencyclopedia.org/p/index.php?title=Lake&oldid=983016
13. NRSC and ISRO (2013) Glacial Lakes in Uttarakhand—A Remote Sensing Based Inventory. Geosciences Group.
14. Hu, Z., Gong, H. and Zhu, L. (2007) Fast Flooding Information Extraction in Emergency Response of Flood Disaster. Proceedings of the ISPRS Workshop on Updating Geo-Spatial Databases with Imagery and the 5th ISPRS Workshop on DMGISs, Urumchi, 28-29 August 2007, 173-177.
15. Yang, C., He, R. and Wang, S. (2008) Extracting Waterbody from Beijing-1 Micro-Satellite Images Based on Knowledge Discovery. Proceedings of the IEEE International Geoscience and Remote Sensing Symposium, Boston, 7-11 July 2008, 850-853.
16. 16. Sharma, O., Mioc, D. and Anton, F. (2007) Feature Extraction and Simplification from Colour Images Based on Colour Image Segmentation and Skeletonization Using the Quad-Edge Data Structure. Proceedings of the International Conference in Central Europe on Computer Graphics, Visualization and Computer Vision, Plzen, 29 January-1 February 2007, 225-232.
17. Zhang, Y., Pulliainen, J.T., Koponen, S.S. and Hallikainen, M.T. (2003) Water Quality Retrievals from Combined Landsat TM Data and ERS-2 SAR Data in the Gulf of Finland. IEEE Transactions on Geoscience and Remote Sensing, 41, 622-629.
18. Soh, L.K. and Tsatsoulis, C. (1999) Segmentation of Satellite Imagery of Natural Scenes Using Data Mining. IEEE Transactions on Geoscience and Remote Sensing, 37, 1086-1099.
19. Zhang, Z.H., Prinet, V. and Ma, S.D. (2003) Water Body Extraction from Multi-Source Satellite Images. Proceedings of the IEEE International Geoscience and Remote Sensing Symposium, Toulouse, 21-25 July 2003, 3970-3972.
20. McCarthy, J.M., Gumbricht, T., McCarthy, T., Frost, P., Wessels, K. and Seidel, F. (2003) Flooding Patterns of the Okavango Wetland in Botswana between 1972 and 2000. AMBIO: A Journal of the Human Environment, 32, 453-457.
21. Anthony, G., Greg, H. and Tshilidzi, M. (2007) Image Classification Using SVMs: One-against-One vs. One-against-All. http://arxiv.org/abs/0711.2914

22. da Rocha Gracioso, A.C.N., da Silva, F.F., Paris, A.C., de Freitas Góes, R. and Elétrica-Sao Carlos, E. (2005) Gabor Filter Applied in Supervised Classification of Remote Sensing Images. Proceeding of the SIBGRAPI Symposium, Natal, 9-12 October 2005.
23. Jeon, Y.J., Choi, J.G. and Kim, J.I. (2005) A Study on Supervised Classification of Remote Sensing Satellite Image by Bayesian Algorithm Using Average Fuzzy Intracluster Distance. Combinatorial Image Analysis Lecture Notes in Computer Science, 3322, 597-606.
24. Fuller, L.M., Morgan, T.R. and Aichele, S.S. (2006) Wetland Delineation with IKONOS High-Resolution Satellite Imagery, Fort Custer Training Center, Battle Creek, Michigan, Date Posted: March 3, 2006, 2005: US Geological Survey, Scientific Investigations Report 2006-5051, 8 p.
25. 25. Basaeed, E., Bhaskar, H. and Al-Mualla, M. (2013) A Spectral Water Index Based on Visual Bands. Proceeding of the Image and Signal Processing for Remote Sensing XIX, Dresden, 23-26 September 2013, Article ID: 889219.
26. Bonsal, B.R., Prowse, T.D., Duguay, C.R. and Lacroix, M.P. (2006) Impacts of Large-Scale Teleconnections on Freshwater-Ice Break/Freeze-Up Dates over Canada. Journal of Hydrology, 330, 340-353.
27. Trenberth, K.E., Jones, P.D., Ambenje, P., Bojariu, R., Easterling, D., Klein Tank, A., et al. (2007) Chapter 3: Observations: Surface and Atmospheric Climate Change. In: Solomon, S., Qin, D., Manning, M., Chen, Z., Marquis, M., Averyt, K.B., Tignor, M. and Miller, H.L., Eds., Climate Change 2007: The Physical Science Basis, Contribution of Working Group I to the Fourth Assessment Report of the Intergovernmental Panel on Climate Change, Cambridge University Press, Cambridge, UK and New York.
28. Arp, C.D., Jones, B.M., Lu, Z. and Whitman, M.S. (2012) Shifting Balance of Thermokarst Lake Ice Regimes across the Arctic Coastal Plain of Northern Alaska. Geophysical Research Letters, 39.
29. Comiso, J.C., Parkinson, C.L., Gersten, R. and Stock, L. (2008) Accelerated Decline in the Arctic Sea Ice Cover. Geophysical Research Letters, 35.
30. Walsh, J.E., Overland, J.E., Groisman, P.Y. and Rudolf, B. (2011) Chapter 2: Arctic Climate: Recent Variations, Snow, Water, Ice and Permafrost in the Arctic (SWIPA), Oslo, Arctic Monitoring and Assessment Programme (AMAP).

31. Forster, R.R., Box, J.E., van den Broeke, M.R., Miege, C., Burgess, E.W., van Angelen, J.H., Lenaerts, J.T.M., Koenig, L.S., Paden, J., Lewis, C., Gogineni, S.P., Leuschen, C. and McConnell, J.R. (2014) Extensive Liquid Meltwater Storage in Firn within the Greenland Ice Sheet. Nature Geoscience, 7, 95-98.
32. Fitzpatrick, A.A.W., Hubbard, A.L., Box, J.E., Quincey, D.J., van As, D., Mikkelsen, A.P.B., Doyle, S.H., Dow, C.F., Hasholt, B. and Jones, G.A. (2014) A Decade (2002-2012) of Supraglacial Lake Volume Estimates across Russell Glacier, West Greenland. The Cryosphere, 8, 107-121. http://dx.doi.org/10.5194/tc-8-107-2014
33. Van der Veen, C.J. (2007) Fracture Propagation as Means of Rapidly Transferring Surface Meltwater to the Base of Glaciers. Geophysical Research Letters, 34, Article ID: L01501.
34. Zwally, H.J., Abdalati, W., Herring, T., Larson, K., Saba, J. and Steffen, K. (2002) Surface Melt-Induced Acceleration of Greenland Icesheet Flow. Science, 297, 218-222.
35. Rokni, K., Ahmad, A., Selamat, A. and Hazini, S. (2014) Water Feature Extraction and Change Detection Using Multitemporal Landsat Imagery. Remote Sensing, 6, 4173-4189.
36. Ouma, Y.O. and Tateishi, R. (2006) A Water Index for Rapid Mapping of Shoreline Changes of Five East African Rift Valley Lakes: An Empirical Analysis Using Landsat TM and ETM+ Data. International Journal of Remote Sensing, 27, 3153-3181.
37. Fu, J., Wang, J. and Li, J. (2007) Study on the Automatic Extraction of Water Body from TM Image Using Decision Tree Algorithm. Proceedings of the International Symposium on Photoelectronic Detection and Imaging: Technology and Applications 2007: Related Technologies and Applications, Beijing, 9-12 September 2007, Article ID: 662502.
38. Jiang, H., Feng, M., Zhu, Y., Lu, N., Huang, J. and Xiao, T. (2014) An Automated Method for Extracting Rivers and Lakes from Landsat Imagery. Remote Sensing, 6, 5067-5089. http://dx.doi.org/10.3390/rs6065067
39. Du, J.K., Feng, X.Z., Wang, Z.L., Huang, Y.S. and Ramadan, E. (2002) The Methods of Extracting Water Information from Spot Image. Chinese Geographical Science, 12, 68-72.

40. Frazier, P.S. and Page, K.J. (2000) Water Body Detection and Delineation with Landsat TM Data. Photogrammetric Engineering and Remote Sensing, 66, 1461-1468.
41. Feyisa, G.L., Meilby, H., Fensholt, R. and Proud, S.R. (2014) Automated Water Extraction Index: A New Technique for Surface Water Mapping Using Landsat Imagery. Remote Sensing of Environment, 140, 23-35.
42. Alecu, C., Oancea, S. and Bryant, E. (2005) Multi-Resolution Analysis of MODIS and ASTER Satellite Data for Water Classification. Proceedings of the SPIE, the International Society for Optical Engineering Conference, San Diego, 31 July-4 August 2005, Article ID: 59831Z.
43. Kerényi, J. and Putsay, M. (2006) Land Use Map from Aster Images and Water Mask on Modis Images. In Marsalek, J., Stancalie, G. and Balint, G., Eds., Transboundary Floods: Reducing Risks through Flood Management, Springer, Dordrecht, 45-55.
44. Chen, Y., Shi, P., Fung, T., Wang, J. and Li, X. (2007) Object-Oriented Classification for Urban Land Cover Mapping with ASTER Imagery. International Journal of Remote Sensing, 28, 4645-4651.
45. Wilkins, H. (1929) The Wilkins-Hearst Antarctic Expedition, 1928-1929. Geographical Review, 19, 353-376
46. Wilkins, H. (1930) Further Antarctic Explorations. Geographical Review, 20, 357-388.
47. Jezek, K.C. (2013) Airborne and Space-Borne Remote Sensing of Cryosphere. In: Orcutt, J., Ed., Earth System Monitoring, Springer, New York, 7-34.
48. Merson, R.H. (1989) An AVHRR Mosaic Image of Antarctica. International Journal of Remote Sensing, 10, 669-674.
49. Bindschadler, R.A. and Vornberger, P.L. (1990) AVHRR Imagery Reveals Antarctic Ice Dynamics. Eos, Transactions American Geophysical Union, 71, 741-742.
50. Ferrigno, J.G., Mullins, J.L., Stapleton, J.A., Chavez Jr., P.S., Velasco, M.G. and Williams Jr., R.S. (1996) Satellite Image Map of Antarctica, Miscellaneous Investigations Map Series 1-2560. US Geological Survey, Reston.
51. Scambos, T.A., Haran, T.M., Fahnestock, M.A., Painter, T.H. and Bohlander, J. (2007) MODIS-Based Mosaic of Antarctica (MOA) Data Sets: Continent-Wide Surface Morphology and Snow Grain Size. Remote Sensing of Environment, 111, 242-257.

52. Bindschadler, R., Vornberger, P., Fleming, A., Fox, A., Mullins, J., Binnie, D. and Gorodetzky, D. (2008) The Landsat Image Mosaic of Antarctica. Remote Sensing of Environment, 112, 4214-4226.
53. Korona, J., Berthier, E., Bernard, M., Rémy, F. and Thouvenot, E. (2009) SPIRIT. SPOT 5 Stereoscopic Survey of Polar Ice: Reference Images and Topographies during the Fourth International Polar Year (2007-2009). ISPRS Journal of Photogrammetry and Remote Sensing, 64, 204-212.
54. Lüthje, M., Pedersen, L.T., Reeh, N. and Greuell, W. (2006) Modelling the Evolution of Supraglacial Lakes on the West Greenland Ice-Sheet Margin. Journal of Glaciology, 52, 608-618.
55. McMillan, M., Nienow, P., Shepherd, A., Benham, T. and Sole, A. (2007) Seasonal Evolution of Supra-Glacial Lakes on the Greenland Ice Sheet. Earth and Planetary Science Letters, 262, 484-492
56. Lampkin, D.J. and Vander Berg, J. (2011) A Preliminary Investigation of the Influence of Basal and Surface Topography on Supraglacial Lake Distribution near Jakobshavn Isbrae, Western Greenland. Hydrological Processes, 25, 3347-3355.
57. Selmes, N., Murray, T. and James, T.D. (2011) Fast Draining Lakes on the Greenland Ice Sheet Linked to Routing of Surface Water. Geophysical Research Letters, 38.
58. Georgiou, S., Shepherd, A., McMillan, M. and Nienow, P. (2009) Seasonal Evolution of Supraglacial Lake Volume from ASTER Imagery. Annals of Glaciology, 50, 95-100.
59. Joughin, I., Das, S.B., Flowers, G.E., Behn, M.D., Alley, R.B., King, M.A. and van Angelen, J.H. (2013) Influence of Supraglacial Lakes and Ice-Sheet Geometry on Seasonal Ice-Flow Variability. The Cryosphere Discussions, 7, 1101-1118.
60. Box, J.E. and Ski, K. (2007) Remote Sounding of Greenland Supraglacial Melt Lakes: Implications for Subglacial Hydraulics. Journal of Glaciology, 53, 257-265.
61. Sneed, W.A. and Hamilton, G.S. (2007) Evolution of Melt Pond Volume on the Surface of the Greenland Ice Sheet. Geophysical Research Letters, 34.
62. Sundal, A.V., Shepherd, A., Nienow, P., Hanna, E., Palmer, S. and Huybrechts, P. (2009) Evolution of Supra-Glacial Lakes across the Greenland Ice Sheet. Remote Sensing of Environment, 113, 2164-2171.
63. Johansson, A.M. and Brown, I.A. (2013) Adaptive Classification of Supra-Glacial Lakes on the West Greenland Ice Sheet. IEEE Journal of

Selected Topics in Applied Earth Observations and Remote Sensing, 6, 1998-2007.

64. Liang, Y.L., Colgan, W., Lv, Q., Steffen, K., Abdalati, W., Stroeve, J., et al. (2012) A Decadal Investigation of Supraglacial Lakes in West Greenland Using a Fully Automatic Detection and Tracking Algorithm. Remote Sensing of Environment, 123, 127-138.
65. Wolfe, R.E., Nishihama, M., Fleig, A.J., Kuyper, J.A., Roy, D.P., Storey, J.C. and Patt, F.S. (2002) Achieving Sub-Pixel Geolocation Accuracy in Support of MODIS Land Science. Remote Sensing of Environment, 83, 31-49.
66. Nath, R.K. and Deb, S.K. (2010) Water-Body Area Extraction from High Resolution Satellite Images—An Introduction, Review, and Comparison. International Journal of Image Processing, 3, 353-372.
67. Jawak, S.D. and Mathew, J. (2011) Semi-Automatic Extraction of Water Bodies and Roads from High Resolution Quick Bird Satellite Data. Proceedings of Geospatial World Forum, PN-263, Hyderabad, 18-21 January 2011, 247- 257.
68. Yu, J.K., Huang, Y.S., Feng, X.Z., et al. (2001) Study on Water Bodies Extraction and Classification from SPOT Image. Journal of Remote Sensing, 5, 214-219.
69. Waldemark, K., Lindblad, T., Becanovic, V., Guillen, J.L. and Klingner, P.L. (2000) Patterns from the Sky: Satellite Image Analysis Using Pulse Coupled Neural Networks for Pre-Processing, Segmentation and Edge Detection. Pattern Recognition Letters, 21, 227-237.
70. Luo, J. and Sheng, Y. (2009) High Precision Auto-Extraction of Water Information of Multi-Band Remote Image by Step Iterative Method. Journal of Remote Sensing, 13, 610-615.
71. Haq, M.A., Jain, K. and Menon, K.P.R. (2012) Monitoring Glacial Lakes Using Remote Sensing Tech. Proceedings of the 14th Annual International Conference and Exhibition on Geospatial Information Technology and Applications, India Geospatial Forum, Gurgaon, 7-9 February 2012.
72. Yang, K. and Smith, L.C. (2013) Supraglacial Streams on the Greenland Ice Sheet Delineated from Combined Spectral-Shape Information in High-Resolution Satellite Imagery. IEEE Geoscience and Remote Sensing Letters, 10, 801-805.

73. Jawak, S.D. and Luis, A.J. (2014) A Semiautomatic Extraction of Antarctic Lake Features Using Worldview-2 Imagery. Photogrammetric Engineering & Remote Sensing, 80, 939-952.
74. Jawak, S.D. and Luis, A.J. (2015) A Rapid Extraction of Water Body Features from Antarctic Coastal Oasis Using Very High-Resolution Satellite Remote Sensing Data. Aquatic Procedia, 4, 125-132.
75. Xie, Z.Y., ShangGuan, D.H., Zhang, S.Q., Ding, Y.J. and Liu, S.Y. (2013) Index for Hazard of Glacier Lake Outburst Flood of Lake Merzbacher by Satellite-Based Monitoring of Lake Area and Ice Cover. Global and Planetary Change, 107, 229-237.
76. Huggel, C., Kaab, A., Haeberli, W., Teysseire, P. and Paul, F. (2002) Remote Sensing Based Assessment of Hazards from Glacier Lake Outbursts: A Case Study in the Swiss Alps. Canadian Geotechnical Journal, 39, 316-330
77. Jawak, S.D. and Luis, A.J. (2013) A Spectral Index Ratio-Based Antarctic Land-Cover Mapping Using Hyperspatial 8-Band World View-2 Imagery. Polar Science, 7, 18-38.
78. Li, M., Xu, L. and Tang, M. (2011) An Extraction Method for Water Body of Remote Sensing Image Based on Oscillatory Network. Journal of Multimedia, 6, 252-260.
79. Ding, L. and Wu, H. (2006) Study of the Water Body Extracting from MODIS Images Based on Spectrum-Photometric Method. Geomatics & Spatial Information Technology, 29, 25-27.
80. McFeeters, S.K. (1996) The Use of the Normalized Difference Water Index (NDWI) in the Delineation of Open Water Features. International Journal of Remote Sensing, 17, 1425-1432.
81. Rong-Long, C.A.O., Li, C.-J., Liu, L.-Y., Wang, J.-H. and Yan, G.-J. (2008) Extracting Miyun Reservoir's Water Area and Monitoring Its Change Based on a Revised Normalized Different Water Index. Science of Surveying and Mapping, 59.
82. McFeeters, S.K. (2013) Using the Normalized Difference Water Index (NDWI) within a Geographic Information System to Detect Swimming Pools for Mosquito Abatement: A Practical Approach. Remote Sensing, 5, 3544-3561.
83. Gao, B.C. (1996) NDWI—A Normalized Difference Water Index for Remote Sensing of Vegetation Liquid Water from Space. Remote Sensing of Environment, 58, 257-266.

84. Xu, H. (2006) Modification of Normalised Difference Water Index (NDWI) to Enhance Open Water Features in Remotely Sensed Imagery. International Journal of Remote Sensing, 27, 3025-3033.
85. Wu, W., Shen, X., Zhou, L., Lu, S. and Zhang, G. (2008) A Integrated Method for Water Body Detection and Delineation Using Landsat ETM+ Data. Bulletin of Science and Technology, 21.
86. 86. Rogers, A.S. and Kearney, M.S. (2004) Reducing Signature Variability in Unmixing Coastal Marsh Thematic Mapper Scenes Using Spectral Indices. International Journal of Remote Sensing, 25, 2317-2335.
87. Xiao, X., Boles, S., Liu, J., Zhuang, D., Frolking, S., Li, C. and Moore, B. (2005) Mapping Paddy Rice Agriculture in Southern China Using Multi-Temporal MODIS Images. Remote Sensing of Environment, 95, 480-492.
88. Mo, W.F., Sun, H., Zhong, S.Q., Huang, Y.L. and He, L. (2007) Research on the CIWI Model and Its Application. Remote Sensing Information, 5, 16-21.
89. Lu, S., Wu, B., Yan, N. and Wang, H. (2011) Water Body Mapping Method with HJ-1A/B Satellite Imagery. International Journal of Applied Earth Observation and Geoinformation, 13, 428-434.
90. WANG, B.-C., Miao, F., Chen, J.-H., Huang, H.-H., Lai, D.-J. and Yan, M.-X. (2008) The Construction and Application of Normalized Difference Water Index (NDWI) Based on the ASTER Image. Science of Surveying and Mapping, 66.
91. Zhong, C., Zeng, C. and Liu, Z. (2008) Study on Terrestrial Water Information Identified Based on the Analysis of Spectral Signature and Ratio Index. Geo-Information Science, 10, 663-666.
92. Li, B. and Zhou, X.Z. (2009) Research on Extraction Method of River Water-Body in Mountain Area of Western Sichuan Based on MODIS Data. Geography and Geo-Information Science, 11, 256-260.
93. Soti, V., Tran, A., Bailly, J.S., Puech, C., Seen, D.L. and Bégué, A. (2009) Assessing Optical Earth Observation Systems for Mapping and Monitoring Temporary Ponds in Arid Areas. International Journal of Applied Earth Observation and Geoinformation, 11, 344-351.
94. Motohka, T., Nasahara, K.N., Murakami, K. and Nagai, S. (2011) Evaluation of Sub-Pixel Cloud Noises on MODIS Daily Spectral Indices Based on in Situ Measurements. Remote Sensing, 3, 1644-1662.

95. Ogashawara, I. and Bastos, V.D.S.B. (2012) A Quantitative Approach for Analyzing the Relationship between Urban Heat Islands and Land Cover. Remote Sensing, 4, 3596-3618.
96. Murray, N.J., Phinn, S.R., Clemens, R.S., Roelfsema, C.M. and Fuller, R.A. (2012) Continental Scale Mapping of Tidal Flats across East Asia Using the Landsat Archive. Remote Sensing, 4, 3417-3426.
97. Sekhon, N.S., Hassan, Q.K. and Sleep, R.W. (2010) Evaluating Potential of MODIS-Based Indices in Determining "Snow Gone" Stage over Forest-Dominant Regions. Remote Sensing, 2, 1348-1363.
98. Jawak, S.D. and Luis, A.J. (2013) Improved Land Cover Mapping Using High Resolution Multiangle 8-Band World View-2 Satellite Remote Sensing Data. Journal of Applied Remote Sensing, 7, Article ID: 073573
99. Wang, H., Pan, L. and Zheng, H. (2008) Multi-Texture-Model for Water Extraction Based on Remote Sensing Image. Proceedings of the Congress on Image and Signal Processing, Sanya, 27-30 May 2008, 710-714.
100. Luo, J., Sheng, Y., Shen, Z. and Li, J. (2010) High-Precise Water Extraction Based on Spectral-Spatial Coupled Remote Sensing Information. Proceedings of the 2010 IEEE International Geoscience and Remote Sensing Symposium (IGARSS), Honolulu, 25-30 July 2010, 2840-2843.
101. Nguyen, D.D. (2012) Water Body Extraction from Multi Spectral Image by Spectral Pattern Analysis. ISPRS-International Archives of the Photogrammetry, Remote Sensing and Spatial Information Sciences, 1, 181-186.
102. Hassan, N., Hamid, J.R.A., Adnan, N.A. and Jaafar, M. (2014) Delineation of Wetland Areas from High Resolution World View-2 Data by Object-Based Method. Proceedings of the IOP Conference Series: Earth and Environmental Science, 18, Article ID: 012017.
103. Breytenbach, A., Eloff, C. and Pretorius, E. (2014) Comparing Three Spaceborne Optical Sensors via Fine Scale Pixel-Based Urban Land Cover Classification Products. South African Journal of Geomatics, 2, 309-324.
104. Xiang, M., Hung, C.C., Pham, M., Kuo, B.C. and Coleman, T. (2005) A Parallelepiped Multispectral Image Classifier Using Genetic Algorithms. Proceedings of the 2005 IEEE International Geoscience and Remote Sensing Symposium, Seoul, 25-29 July 2005, 4.

105. Lu, D., Hetrick, S., Moran, E. and Li, G. (2010) Detection of Urban Expansion in an Urban-Rural Landscape with Multitemporal QuickBird Images. Journal of Applied Remote Sensing, 4, Article ID: 041880.
106. Blaschke, T. (2003) Object-Based Contextual Image Classification Built on Image Segmentation. Proceedings of the 2003 IEEE Workshop on Advances in Techniques for Analysis of Remotely Sensed Data, Greenbelt, 27-28 October 2003, 113-119.
107. Kux, H.J. and Souza, U.D. (2012) Object-Based Image Analysis of WORLDVIEW-2 Satellite Data for the Classification of Mangrove Areas in the City of SAO LUíS, MARANHAO State, Brazil. ISPRS Annals of Photogrammetry, Remote Sensing and Spatial Information Sciences, 1, 95-100.
108. Tehrany, M.S., Pradhan, B. and Jebuv, M.N. (2014) A Comparative Assessment between Object and Pixel-Based Classification Approaches for Land Use/Land Cover Mapping Using SPOT 5 Imagery. Geocarto International, 29, 351-369.
109. Kim, S.H. and Hong, C.H. (2012) Antarctic Land-Cover Classification Using IKONOS and Hyperion Data at Terra Nova Bay. International Journal of Remote Sensing, 33, 7151-7164.
110. Sundal, A. (2009) Determining Supra-Glacial Lake Area from MODIS Satellite Images. Technical Report, University of Edinburgh, Edinburgh.
111. Panday, P.K., Bulley, H., Haritashya, U. and Ghimire, B. (2012) Supraglacial Lake Classification in the Everest Region of Nepal Himalaya. In: Thakur, J.K., Singh, S.K., Ramanathan, A., Prasad, B.K.M. and Gossel, W., Eds., Geospatial Techniques for Managing Environmental Resources, Springer, Dordrecht, 86-99.
112. Brown, L. and Young, K.L. (2006) Assessment of Three Mapping Techniques to Delineate Lakes and Ponds in a Canadian High Arctic wetland complex. Arctic, 283-293.
113. Leeson, A.A., Shepherd, A., Sundal, A.V., Johansson, A.M., Selmes, N., Briggs, K. and Fettweis, X. (2013) A Comparison of Supraglacial Lake Observations Derived from MODIS Imagery at the Western Margin of the Greenland Ice Sheet. Journal of Glaciology, 59, 1179-1188.
114. Wong, T.H., Mansor, S.B., Mispan, M.R., Ahmad, N. and Sulaiman, W.N.A. (2003) Feature Extraction Based on Object Oriented Analysis. Proceedings of the ATC 2003 Conference, Malaysia, 20-21 May 2003, 20-21.

115. Zhang, Y. (2000) A Method for Continuous Extraction of Multispectrally Classified Urban Rivers. Photogrammetric Engineering and Remote Sensing, 66, 991-999.
116. Li, B., Zhang, H. and Xu, F. (2014) Water Extraction in High Resolution Remote Sensing Image Based on Hierarchical Spectrum and Shape Features. Proceedings of the IOP Conference Series: Earth and Environmental Science, 17, Article ID: 012123.
117. Shao, P., Yang, G., Niu, X., Zhang, X., Zhan, F. and Tang, T. (2014) Information Extraction of High-Resolution Remotely Sensed Image Based on Multiresolution Segmentation. Sustainability, 6, 5300-5310.
118. Elsharkawy, A., Elhabiby, M. and El-Sheimy, N. (2012) New Combined Pixel/Object-Based Technique for Efficient Urban Classification Using Worldview-2 Data. Proceedings of the International Archives of the Photogrammetry, Remote Sensing and Spatial Information Sciences, XXXIX-B7, 191-195.
119. Jawak, S.D., Raut, D.A. and Luis, A.J. (2015) Iterative Spectral Index Ratio Exploration for Object-Based Image Analysis of Antarctic Coastal Oasis Using High Resolution Satellite Remote Sensing Data. Aquatic Procedia, 4, 157-164.
120. Ji, L., Zhang, L. and Wylie, B. (2009) Analysis of Dynamic Thresholds for the Normalized Difference Water Index. Photogrammetric Engineering and Remote Sensing, 75, 1307-1317.
121. Lira, J. (2006) Segmentation and Morphology of Open Water Bodies from Multispectral Images. International Journal of Remote Sensing, 27, 4015-4038.
122. Sethre, P.R., Rundquist, B.C. and Todhunter, P.E. (2005) Remote Detection of Prairie Pothole Ponds in the Devils Lake Basin, North Dakota. GIScience & Remote Sensing, 42, 277-296.
123. Jain, S.K., Singh, R.D., Jain, M.K. and Lohani, A.K. (2005) Delineation of Flood-Prone Areas Using Remote Sensing Techniques. Water Resources Management, 19, 333-347.
124. Rundquist, D.C., Lawson, M.P., Queen, L.P. and Cerveny, R.S. (1987) The Relationship between Summer-Season Rainfall Events And Lake-Surface Area. Journal of the American Water Resources Association, 23, 493-508.
125. Jawak, S.D., Luis, A.J., Panditrao, S.N., Khopkar, P.S. and Jadhav, P.S. (2013) Advancement in Landcover Classification Using Very High

Resolution Remotely Sensed 8-Band WorldView-2 Satellite Data. International Journal of Earth Sciences and Engineering, 6, 1742-1749.

126. Lu, D. and Weng, Q. (2007) A Survey of Image Classification Methods and Techniques for Improving Classification Performance. International Journal of Remote Sensing, 28, 823-870.

127. Otukei, J.R. and Blaschke, T. (2010) Land Cover Change Assessment Using Decision Trees, Support Vector Machines and Maximum Likelihood Classification Algorithms. International Journal of Applied Earth Observation and Geoinformation, 12, S27-S31.

128. Jawak, S.D. and Luis, A.J. (2013) Very-High Resolution Remotely Sensed Satellite Data for Improved Land Cover Extraction of Larsemann Hills, East Antarctica. Journal of Applied Remote Sensing, 7, Article ID: 073460.

129. Verpoorter, C., Kutser, T. and Tranvik, L. (2012) Automated Mapping of Water Bodies Using Landsat Multispectral Data. Limnology and Oceanography: Methods, 10, 1037-1050.

130. Myint, S.W., Galletti, C.S., Kaplan, S. and Kim, W.K. (2013) Object vs. Pixel: A Systematic Evaluation in Urban Environments. Geocarto International, 28, 657-678.

131. Katiyar, S.K. and Arun, P.V. (2014) An Enhanced Neural Network Based Approach towards Object Extraction. http://arxiv.org/abs/1405.6137

132. Sridharan, H. (2012) Object-Based Approaches to Image Classification for Hyperspatial and Hyperspectral Data. Ph.D. Thesis, The University of Texas, Dallas.

133. Jawak, S.D. and Luis, A.J. (2012) Synergistic Use of Multitemporal RAMP, ICESat and GPS to Construct an Accurate DEM of the Larsemann Hills Region, Antarctica. Advances in Space Research, 50, 457-470.

134. Jawak, S.D. and Luis, A.J. (2014) Prospective Application of NASA-ISRO SAR (NISAR) in Cryospheric Studies: A Practical Approach. Proceedings of the NISAR Science Workshop, Ahmadabad, 17-18 November 2014.

135. Jawak, S.D., Sambhus, P.G., Paranjape, R.A. and Luis, A.J. (2012) Assessment of Spatial Interpolation Techniques for Generating Accurate Digital Elevation Surface Using Combined Radar & LiDAR Elevation Data. Proceedings of the 8th International Conference on Microwaves, Antenna Propagation and Remote Sensing (ICMARS-2012), Paper Id-ICMARS 12121, Jodhpur, 11-15 December 2012.

136. Jawak, S.D. and Luis, A.J. (2014) Synergetic Merging of Cartosat-1 and RAMP to Generate Improved Digital Elevation Model of Schirmacher Oasis, East Antarctica. Proceedings of the ISPRS Technical Commission VIII Symposium, Hyderabad, India, 9-12 December 2014, 517-524.

137. Jawak, S.D., Jadhav, S.P. and Luis, A.J. (2013) Usage of Multiple Satellite and Ground-Based Elevation Datasets to Generate an Accurate Digital Elevation Surface: Methods and Error Analysis. EGESD-2013, University of Mumbai, Mumbai.

138. Jawak, S.D. and Luis, A.J. (2011) Synergetic Use of GLAS/ICESat and RAMP to Construct 3D DEM of Schirmacher Oasis, East Antarctica. Proceedings of the 11th International Symposium on Antarctic Earth Sciences (ISAES-2011), Edinburgh, 10-16 July 2011.

139. Jawak, S.D. and Luis, A.J. (2010) Synergistic Use of SAR, GLAS/ICESat & Ground Survey (GPS) Data to Construct an Accurate DEM of the Larsemann Hills Region. Proceedings of the 4th SCAR-OSC-2010, Buenos Aires, 3-6 August 2010.

140. Jawak, S.D. and Luis, A.J. (2014) Spectral Bands of WorldView-2 Satellite Remote Sensing Data for Semiautomatic Land Cover Extraction in the Antarctic Environment. Proceedings of the XXXIII SCAR and 6th Open Science Conference, Auckland, 25-28 August 2014.

141. Jawak, S.D. and Luis, A.J. (2014) A Novel Set of Normalized Difference Snow/Ice Index ratios for Extraction of Snow Cover from the Antarctic Environment Using Very High Resolution Satellite Data. Proceedings of the XXXIII SCAR and 6th Open Science Conference, Auckland, 25-28 August 2014.

142. Jawak, S.D., Khopkar, P.S. and Luis, A.J. (2013) Spectral Index Ratio-Based Quantitative Analysis of WorldView-2 Pansharpened Data. In: The National Seminar on Earth Observation & Geo-Information Sciences for Environment and Sustainable Development (EGESD), Mumbai, 7-9 February 2013, Oral.

143. Jawak, S.D. and Luis, A.J. (2012) Hyperspatial WorldView-2 Satellite Remote Sensing Data for Polar Geospatial Information Mining of Larsemann Hills, East Antarctica. Proceedings of the XXXII SCAR and 5th Open Science Conference (OSC), Portland, 13-25 July 2012.

144. Jawak, S.D., Khopkar, P.S. and Godbole, T.M. (2012) Spectral Quality Analysis of Pansharpened Worldview-2 Images Using Normalized

Difference Vegetation/Water (NDVI/NDWI) Index. Proceedings of the India Geospatial Forum, PN-32, Gurgaon, 7-9 February 2012.

145. Jawak, S.D. and Luis, A.J. (2011) High Resolution 8-Band WorldView-2 Satellite Remote Sensing Data for Polar Geospatial Information Mining and Thematic Elevation Mapping of Larsemann Hills, East Antarctica. Proceedings of the 11th ISAES-2011, Edinburgh, 10-16 July 2011.

146. Jawak, S.D. and Luis, A.J. (2013) A Comprehensive Evaluation of PAN-Sharpening Algorithms Coupled with Resampling Methods for Image Synthesis of Very High Resolution Remotely Sensed Satellite Data. Advances in Remote Sensing, 2, 332-344.

147. Jawak, S.D., Panditrao, S.N. and Luis, A.J. (2014) Enhanced Urban Landcover Classification for Operational Change Detection Study Using Very High Resolution Remote Sensing Data. Proceedings of the ISPRS Technical Commission VIII Symposium, Hyderabad, India, 9-12 December 2014, 773-779.

148. Jawak, S.D., Khopkar, P.S., Jadhav, S.P. and Luis, A.J. (2013) Customization of Normalized Difference Snow Index for Extraction of Snow Cover from Cryospheric Surface Using WorldView-2 Data. Proceedings of the AGSE International Conference, Ahmedabad, 16-19 December 2013, 391-398.

149. Jawak, S.D. and Luis, A.J. (2012) WorldView-2 Satellite Remote Sensing Data for Polar Geospatial Information Mining of Larsemann Hills, East Antarctica. Proceedings of the 11th Pacific (Pan) Ocean Remote Sensing Conference (PORSEC), Id: PORSEC2012-24-00006, Kochi, Kerala, 5-9 November 2012.

150. Jawak, S.D. and Luis, A.J. (2011) Applications of WorldView-2 Satellite Data for Extraction of Polar Spatial Information and DEM of Larsemann Hills, East Antarctica. Proceedings of the International Conference on Fuzzy Systems and Neural Computing (FSNC 2011), Hong Kong, 20-21 February 2011, 148-151.

13
CHAPTER

SATELLITE DERIVED GEOSPATIAL IRRIGATION PERFORMANCE INDICATORS FOR BENCHMARKING STUDIES OF IRRIGATION SYSTEMS

A. V. Suresh Babu, M. Shanker, V. Venkateshwar Rao

National Remote Sensing Centre (NRSC), Department of Space, Government of India, Hyderabad, India

ABSTRACT

Development of irrigation infrastructure and its efficient management is the primary concern for sustainable food production. The assessment of irrigation infrastructure creation, its utilization, diagnostic evaluation of the various performance indices (monitoring) are important to measure the efficiency. Benchmarking of Irrigation Systems (BIS) is for the diagnostic analysis of irrigation performance indicators comprising of Irrigation Infrastructure System (IIS), Agricultural System (AS), Water Delivery Dynamics (WDD). Since, the performance of an irrigation command varies with space and time, utilization of spatial information technologies viz. Remote Sensing (RS), Geographical Information Systems (GIS), Global Positioning Systems (GPS) useful to provide spatial information on several indices in the process of benchmarking (BM). Information requirements for BIS at different stages, utilization of spatial information technologies to derive irrigation performance indicators was discussed with suitable examples and demonstrated in this study. The studies carried out indicates that the geospatial approach for BIS enabled the improvements in data collection methods, diagnostic analysis, spatio-temporal visualisation of BM indicators at disaggregated canal level which would be useful for decision support during the corrective management measures. The conjunctive use of multi-date (medium resolution) satellite data, high spatial resolution data, field data on water deliveries was found to be an alternative to the conventional non-spatial approaches for BIS and thereby better water resources planning and management.

Keywords: Irrigation; Agriculture; Benchmarking of Irrigation Systems; Geospatial Techniques; Remote Sensing; GIS

1. INTRODUCTION

Benchmarking (BM) of irrigation system is a process of performance evaluation using a set of defined indicators to determine the performance of various components of irrigation system namely Irrigation Infrastructure (IIS), Agricultural System (AS) and Water Delivery Dynamics (WDD). Conventional methods of collection of information on these systems are mostly non spatial in nature and are based on point information or eye estimates which appears to be a gap in information retrieval and processing mechanism. Further, updation of any spatial data through conventional field surveys are time consuming methods. Hence, requires an alternative mechanism to collect, process, organize and update the information pertaining to irrigation command to supplement the information requirements for BM studies. The study was focused to identify the list of benchmarking indices used for irrigation performance evaluation, methods for utilization of spatial information technologies to derive these BM indices that could be generated through geospatial approach.

2. REVIEW

Traditionally, irrigation performance was expressed in terms of irrigation efficiency as a function of water deliveries (discharge from measurement device) which may often mislead the planners. Researchers have demonstrated the use of satellite remote sensing derived information in conjunction with canal flows data [1-3] for the evaluation of irrigation command. Evaluation was carried out in an irrigation water management scheme [4] in terms of water supply requirement ratios and other indices such as irrigation intensity, crop productivity and cropping pattern and compared with the historic values. An extensive review was conducted on satellite derived parameters for irrigation command studies [5-8] and the advantages and limitations were theoretically summarized. Indian National Committee on Irrigation Drainage (INCID), Government of India has published the list of various performance indices [9], related to agricultural crop production, water delivery system, economic performance etc and possible adaptation for benchmarking studies of any irrigation system [10]. Multi-date satellite data were analyzed to provide spatially information on irrigated area, cropping pattern, paddy yield to evaluate the performance of the agricultural system over several years and across the irrigation scheme [2,11-14]. Rice crop yield was estimated using satellite data [15]. Multitemporal remote sensing data-based crop inventory, vegetation spectral index profiles and crop evapotranspiration estimation were carried out over the Mahi Right Bank Canal data analyze the performance with respect to irrigation management [16]. Present practices and performance improvement of irrigation schemes in Maharashtra state of India was described in detail with field experiences [17]. Studies were conducted on the assessment of crop, water productivity and their dynamics across the span of 10 years in an irrigation system with multi-temporal datasets [18-22].

Overall review of literature indicated that the satellite data was used in isolation for deriving the various parameters required for the irrigation performance evaluation and rarely, unified approach was defined for the utilization of these technologies for benchmarking of irrigation systems (BIS). This paper describes about the concepts, methods implementation of geospatial approach for deriving BM indicators for BIS.

3. INFORMATION REQUIREMENTS FOR BM

Bench marking studies requires several non spatial and spatial datasets to evaluate the performance of different components of irrigation system. The information needs for benchmarking of irrigation systems were also envisaged by International Commission on Irrigation & Drainage [23] which are basically divided into categories pertaining to cropping pattern (agricultural), water distribution, economical and socio-economic indicators (Table 1). Parameters shows that they are spatiotemporal in nature and mostly dependant on cropping pattern, water deliveries and socio-economic data. Hence, the focus needs to be put on the retrieval of spatio-temporal parameters pertaining to irrigation system. Utilisation of spatial information technologies for periodic retrieval of these parameters was discussed and demonstrated in this study.

TABLE 1. List of benchmarking indicators for evaluating irrigation system.

S.No	Bench Marking (BM) Indicator
1	Water delivery capacity Index (WDCI)
2	Annual vol. of irrigation water delivery (Cum/year)
3	Field application efficiency (Percent)
4	Annual relative irrigation supply index
5	Annual water supply for unit command area (Cum./Ha)
6	Annual irrigation water supply per unit irrigated area (Cum./Ha)
7	Output per unit service area (Rs./Ha per Year)
8	Output per unit irrigated area (Rs./Ha)
9	Output per unit irrigation supply (Rs./Cu.m)
10	Output per unit crop water demand (Rs./Cu.m)
11	Cost-recovery ratio
12	Total O & M Cost per unit area (Rs./ha)
13	Total cost per person employed on water delivery (Rs./person)
14	Revenue collection performance
15	Average revenue per cubic meter of irrigation water supplied (Rs./cum)
16	Maintenance cost to revenue ratio
17	Staffing numbers per unit area (Persons/ha)
18	Total O&M Cost per unit of water (Rs./cum)
19	Land damage index
20	Water quality
21	Irrigation potential created (Ha)
22	Potential utilization (Ha)
23	Equity (%): Crop area Irrigated vs. Proposed
24	Water yield index
25	Index of water availability

4. GEOSPATIAL APPROACH FOR BENCHMARKING OF IRRIGATION SYSTEMS

The study was taken up for the demonstration of geospatial approach in BM studies, the available field data, satellite data collected over different irrigation commands of India during research work was made use. The geospatial approach suggested for BIS was shown in Figure 1.

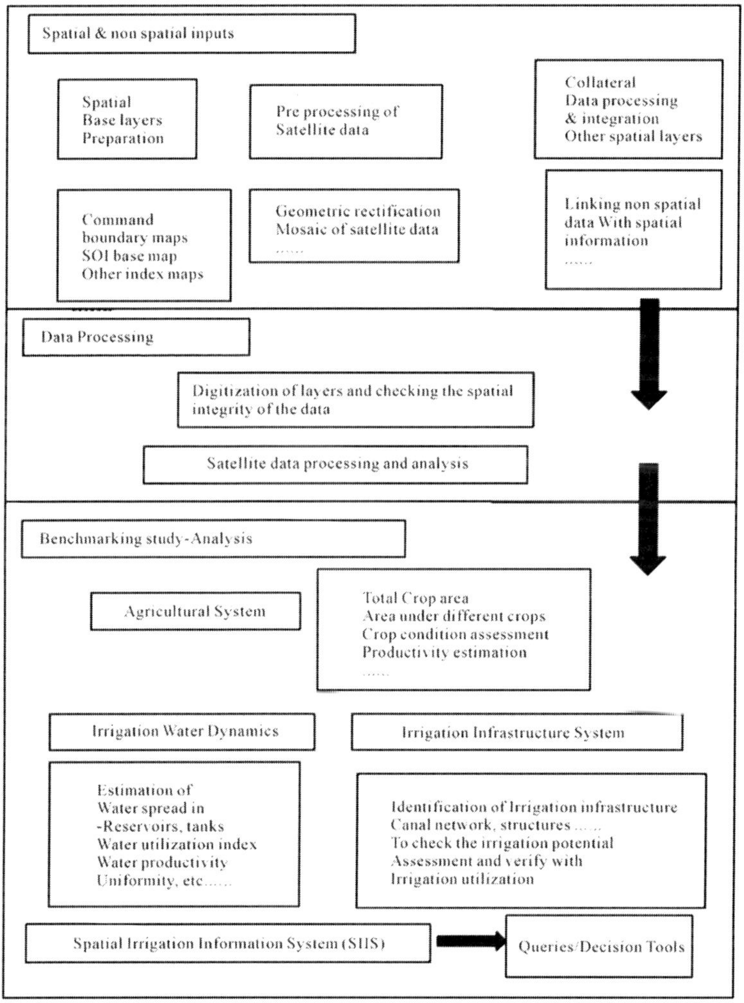

FIGURE 1. Geospatial approach for benchmarking of irrigation systems.

4.1. Irrigation Infrastructure System (IIS)

Irrigation infrastructure system refers to the infrastructure for delivering water from source through conveyance and distribution system, i.e. canal network ensuring the hydraulic connectivity. Irrigation potential (IP) is defined as the area that can be irrigated/commanded by a canal. The objective of BM of IIS is to evaluate whether designed canal lengths were completed at the defined off take points (chianages) as per the plan/design, checking of hydraulic connectivity, from source to tail end. If lengths, chainages are met with designed values, then, it is said to be irrigation potential created under a canal. Use of satellite data and geospatial approach was found to be very much useful with the help of high resolution satellite data which is demonstrated here alongwith the methodology (Figure 2), image interpretation elements/keys (Figure 3), assessment of irrigation potential creation (Figure 4) with suitable examples .

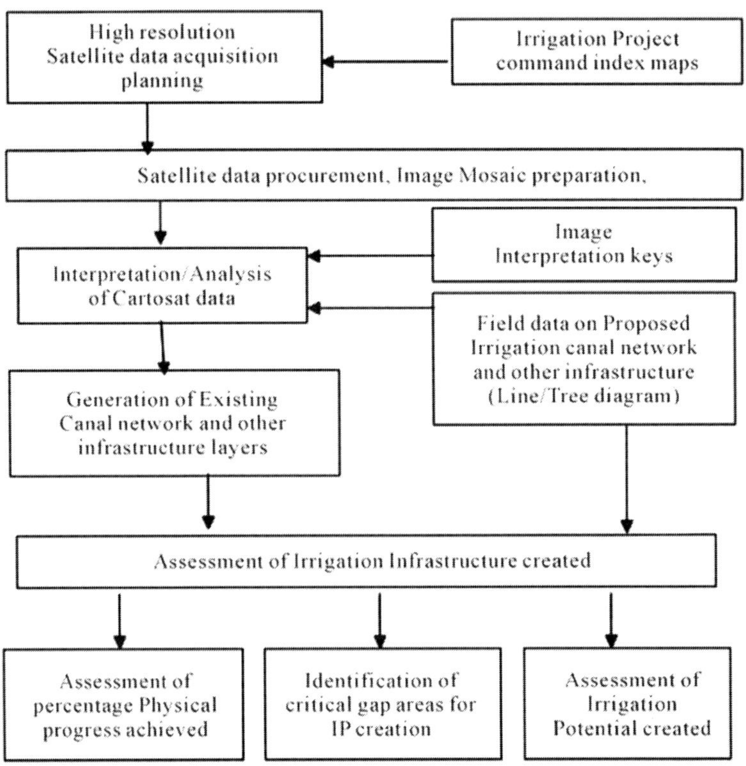

FIGURE 2. Methodology for the assessment of irrigation potential from high resolution satellite data.

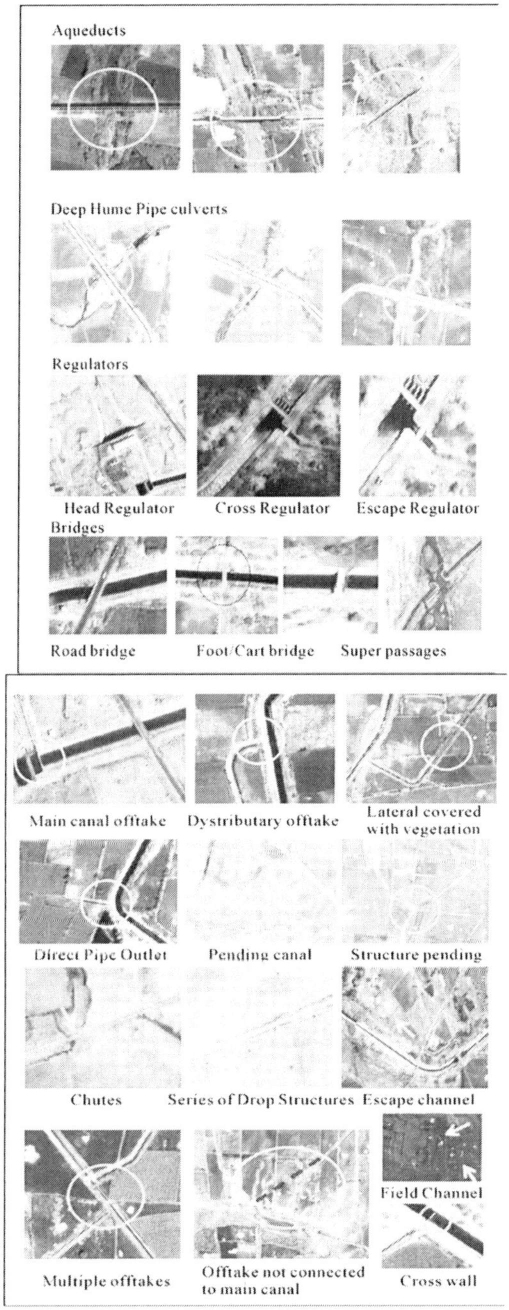

FIGURE 3. Irrigation infrastructure elements.

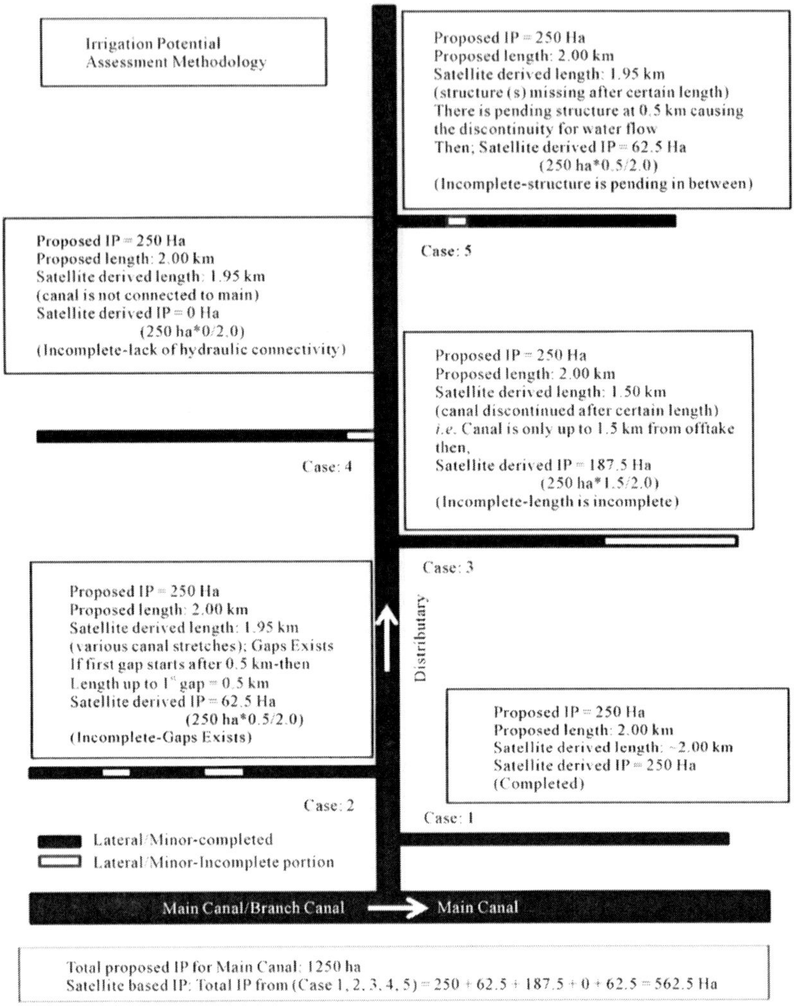

FIGURE 4. Illustration depicting the quantification of irrigation potential created in irrigation projects from irrigation infrastructure information.

Example: If a canal with length of 5.0 Km is planned to create irrigation potential of 1000 ha (as per detailed project surveys), it is verified with the acquisition of high resolution satellite data from IKONOS/Cartosat/Quick bird (with spatial resolutions 1 - 2.5 m) and delineated canal by displaying the image in 1:5000 scale and lengths and chainages were measured. The process is repeated for all the canals. If there are critical gaps such as pending cross drainage structures, pending canal stretches in along the canal due to land acquisition issues or pending road/rail crossings, pending tail portion works, etc. can be quantified during the process and the irrigation potential created is estimated accordingly. The methodology of quantifying the irrigation potential creation from the satellite derived lengths and of canals and status of hydraulic connectivity is shown through the illustration (Figure 4). An example was shown in Figure 5 depicting the delineation of canal network from high resolution satellite data, critical gap causing the gap in irrigation potential though the canal network completed and pending hydraulic connectivity to a larger length of canal system.

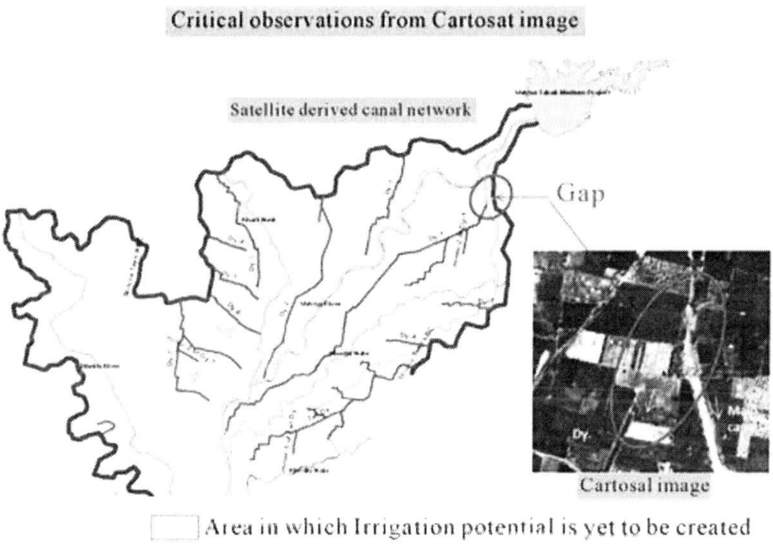

Figure 5. Assessment of irrigation potential in an irrigation command based high resolution satellite derived irrigation infrastructure.

4.2. Agricultural System

Benchmarking indicators (Table 1) for performance assessment of an agricultural system deals with total crop area, cropping pattern, crop condition and crop production, crop deviations and socio-economic parameters. Many of these parameters require spatio-temporal information for its quantitative assessment which were generally obtained using statistical methods or eye estimates. Hence, satellite based technologies should be used for deriving performance indicators in spatial domain for benchmarking the agriculture system. Presently, satellite data of varying spatial, spectral and temporal resolutions are available. In general, data from medium spatial resolutions satellites are optimal for this purpose such as IRS LISS III, Landsat TM/ETM+, etc. The spatial information at disaggregated level (irrigation distributary/block/ zone/project) can be generated and used for identification of poor performing pockets within the command area.

4.2.1. Estimation of Crop Area & Cropping Pattern

Every irrigation system will have a designed cropping pattern and accordingly water deliveries are let into the system. Cropping pattern in an irrigation command determines the irrigation needs and hence the inventory of crop areas is useful to know whether the irrigation system is optimally used or not. Cropping pattern was derived from multi-date satellite data acquired during the cropping season. The concept was demonstrated through a study conducted at NRSC on the performance evaluation of Nagarjunasagar Left Canal command (NSLC), Andhra Pradesh, India which was in construction phase since 1980 s and reached its development stage in 1990- 91 and has reached optimal utilisation phase in 1998-99. This case study was carefully selected to demonstrate the utilisation of remote sensing techniques to quantify various crop parameters from historic data sets in developmental phase of an irrigation project. Multi-date satellite data (IRS 1C/1D LISS III for the year 1998-99 and IRS 1A/1B LISS II and Landsat TM for the year 1990-91) during the rabi season of the years 1990-91 and 1998-99 were selected, taking into consideration crop calendar (Figure 6) of major crops and availability of cloud free satellite data. The satellite data acquired were: 1998-99 Crop year : Dec (5th, 16th), Feb (18th, 21st), Mar (27th, 18th), Apr (15th, 19th), May (4th, 7th) from IRS-1C/1D satellites and 1990-91: Dec (5th, 16th), Jan (13th, 20st), Feb (14th, 22th), Mar(18th, 25th), Apr (10th, 19th) from IRS-1B/1C satellites.

Hierarchical crop classification methods were adopted for discrimination of different crops. The aim was to separate paddy, cotton, chillies, orchards, sugarcane and other irrigated dry (ID) crops. Crop area mask was derived (using Normalized Difference Vegetation Index (NDVI) for each image. Points kept in view for classification based on the crop calendar were 1) Orchards/sugarcane are present in all images (through the season) as they are perennial/annual crops 2) During December, 1998 rabi paddy grown area was mostly either fallow or at field preparation stage, two seasonal crops are standing, rabi ID crops were at the initial vegetative stage 3) During February and March paddy was in peak stage and spectrally distinguishable from other crops. A ID crop mask was identified using December image.

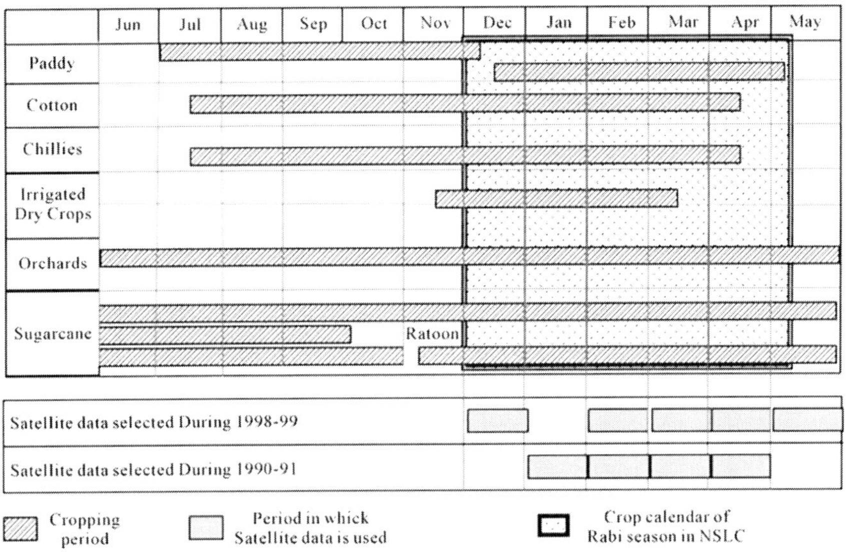

FIGURE 6. Crop calendar used for the selection of satellite images.

The satellite derived cropping pattern and the changes that took place within span of 10 years was estimated and shown in Figure 7. There was shift in cropping pattern from 1990-91 to 1998-99. Rice area was increased drastically from 17.94% to 55.98% accordingly, there was shift from ID crops to rice crop.

4.2.2. Quantitative Estimations on Crop Deviations

All canals are generally defined with specific cropping pattern depending availability of water in different seasons at the time initiation of the project. If the cropping pattern was deviated, the design water deliveries do not meet actual water requirement of the crops and hence, it leads to imbalance in irrigation water management. The geospatial approach suggested here with satellite derived cropping patterns at each of the canal level, and aggregated figures helped to identify crop deviations in the irrigation command (Figure 8). The crop variations could be because of replacement ID crops with paddy due to increased water availability in 1998-99 compared to 1990-91.

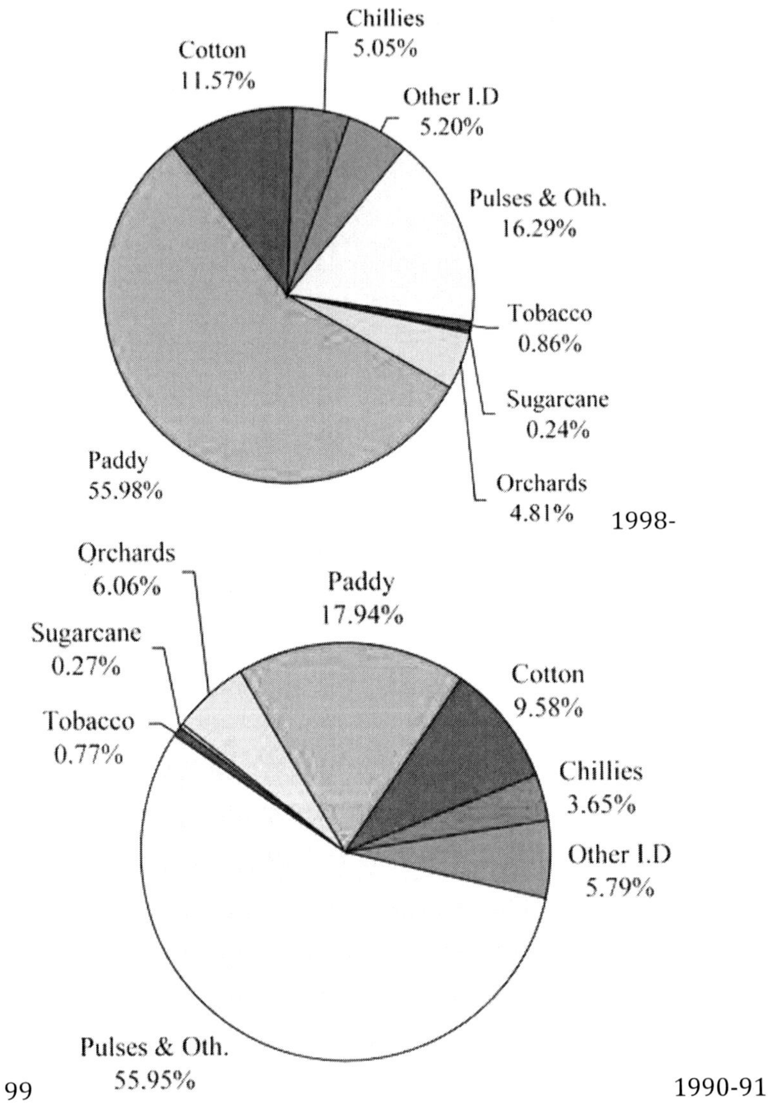

FIGURE 7. Satellite derived cropping pattern in NSLC during 1990-91 and 1998-99.

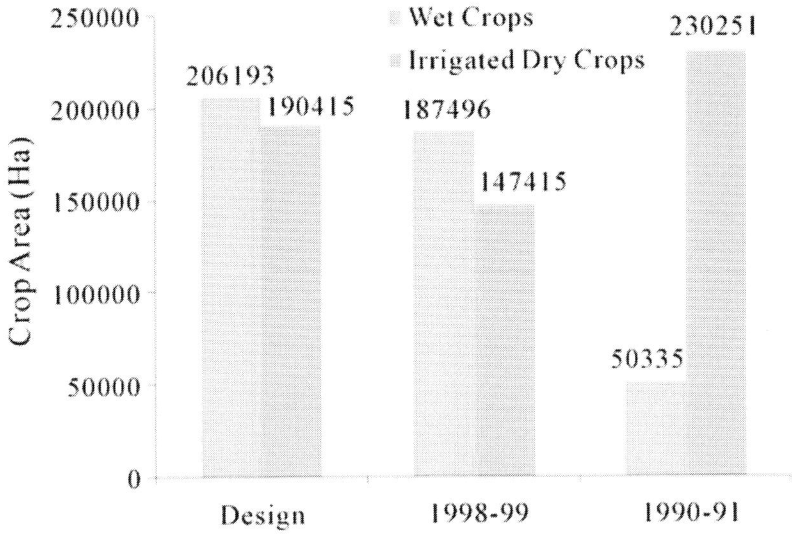

FIGURE 8. Overview of crop deviations in NSLC.

4.2.3. Irrigation Intensity (II)

Irrigation intensity is defined as the percentage of the culturable command area proposed to be irrigated seasonally/annually under an irrigation command. II indicates gap between the irrigation potential created and potential utilized. By knowing the intensity of irrigation for a crop season, the area irrigated during that particular crop season can be determined. Satellite derived cropping pattern information was used for the estimation of irrigation intensity. Irrigation intensity varied from 71% to 84% in NSLC from 1990-91 to 1998-99 which shows the improvement in utilizing established irrigation System. Spatial comparison of cropping pattern in NSLC in 1998 and 1991 is depicted in Figure 9 representing the variation in irrigation intensity.

4.2.4. Crop Condition

Crop condition during crop growth cycle was estimated based on Normalized Difference Vegetation Index (NDVI), an index calculated using Satellite Remote Sensing obtained reflected radiation in Near Infrared (IR) and Red (R) bands. NDVI ((NIR-R)/NIR+R)) which represents the integrated effect of various factors that influence crop production. Hence, the range of satellite derived NDVI (ranging from +0.2 to 0.7 in case of crops) was used for defining the levels of performance to assess the crop condition and to estimate the productivity at disaggregated level across the command area. The time composite maximum NDVI (maximum NDVI) estimated from multi-date satellite images of each pixel that was attained during the crop growth period was used for defining the crop condition. Rice crop condition was attempted considering the NDVI as: <0.40: low; 0.40 - 0.50—Good; >0.5—Very good).

Crop Yield Estimation & Crop Productivity Crop productivity is defined as production per unit area

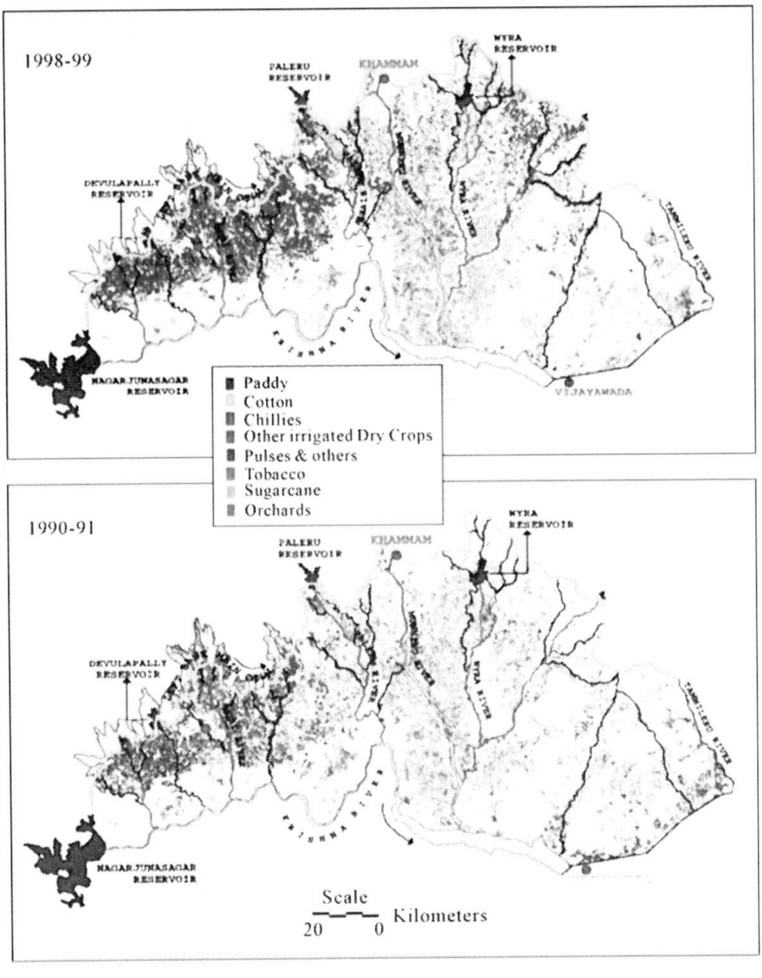

FIGURE 9. Comparison of cropping pattern in NSLC in 1998 and 1991.

(t/ha.). NDVI at critical crop growth stage, i.e. at the heading stage of rice crop correlates well with the dry yield (obtained from crop cutting experiments) and a regression model was developed. The relationship obtained for NDVI vs Paddy yield is shown in the Figure 10. Based on the regression model, NDVI values of pixel are used to transform into a quantitative (productivity) estimation of crop yield.

4.3. Studies on Irrigation Water Dynamics

Performance indicators of agricultural system, infrastructural systems can be integrated with water releases data and can be related with the actual utilisation of huge infrastructure. So far, satellite data was used for deriving the spatial cropping pattern, crop condition, productivity, etc. and the now the data was related field data on water deliveries in the crop season with Water Utilisation Index (WUI). Further, methodology for the assessment of the capacity of the reservoir, assessment of sedimentation using remote sensing was also addressed in this paper as part of monitoring water delivery system i.e. on the satellite derived information on the status of water storages in reservoirs.

4.3.1. Water Utilisation Index

Water Utilisation Index (WUI) is a measure of water delivery performance and constitutes one of the important spatial performance indicators of an irrigation system and is defined as area irrigated per unit volume of water delivered. WUI also forms basis for evaluating the adequacy of seasonal irrigation supplies in an irrigation system (inverse of WUI is delta, i.e. depth of water supplied to a given irrigation unit). WUI was estimated based on the canal flow information and the crop area statistics derived from the satellite data.

Estimation of WUI involves converting the areas under various crops in terms of a single major crop grown in the command area. In this present study, various crops were brought at par with wet crop (Rice) by taking operationally used equivalent wet crop conversion factors [21] mentioned below. The conversion factors were estimated based on the duties of water adopted from time to time in the project area taking into consideration of crop water requirements and irrigation efficiencies in the irrigation system.

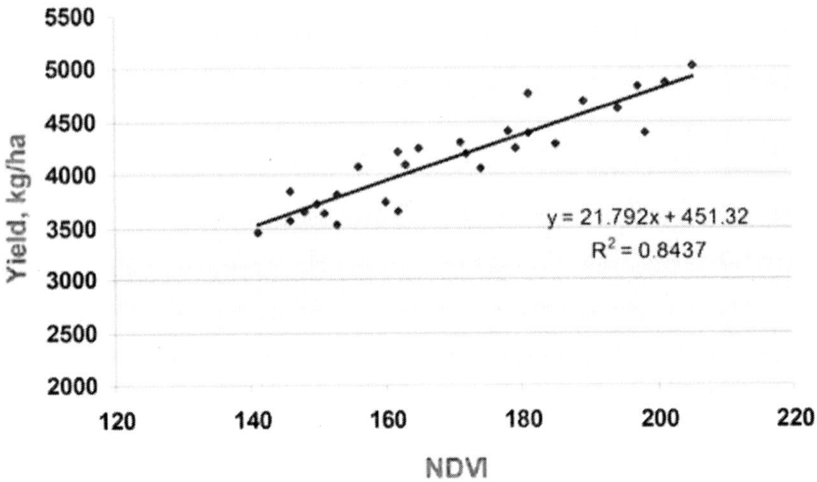

Figure 10. NDVI-yield relationship referred to estimate paddy yield.

Equivalent Wet Area: Area of {(rice + 2.0*Sugarcane + 0.5*Groundnut + 0.4*Cotton + 0.33*(Chillies & Tobacco) + 0.25 *(Grams+ Mango Orchards))} in the present study.

The Gross Irrigation Requirement (GIR) at the head or off-take points was estimated to vary between 1.50 to 1.70 m in the NSLC Command area as per the project estimates [22]. The delta during 1998-99 was estimated to be 108 cm and 153 cm during 1990-91 at command level. Therefore, the irrigation supplies during 1990-91 appears to be adequate to meet crop water requirements as it is closer to the GIR and irrigation service provided is intensive and optimal where as it was observed to be 153 cm which is indicates that the irrigation is extensive and sub optimal as it was much lower than GIR. This would have resulted in productivity variations during the two time periods. This fact was also confirmed by the yield estimates during the corresponding years i.e. paddy yield in 1998-99 (40.07 q/ha) compared to 1990-91 (45.48 q/ha) [23]. Water requirement and water utilisation depends mainly on the cropping pattern in an irrigation system. These dynamics can be studied through the analysis on spatial cropping pattern, production estimates, crop condition etc in relation with the corresponding water releases into canals at disaggregated level. The presence of spatio-temporal data always helps for better visualization of issues to benefit in the approaches required for diagnostic analysis and effective management.

5. ESTIMATION OF WATER STORAGE IN RESERVOIR

The availability of water for the purpose of irrigation depends upon the level of water storage in the reservoir which is dynamic in nature (having inter/intra annual variability) depending on the rainfall that occurs during the season in the catchment and sedimentation characteristics of the catchment which influences the capacity of the reservoir to have designed storage. Water storage in a reservoir is a function of water spread area and the corresponding elevation or reservoir level. Elevation-AreaCapacity (EAC) curves were generated traditionally based on pre-impoundment survey on bottom profile of the reservoir. BM studies focuses on the capacity of reservoir, storages available for utilisation vs water requirement (based cropping pattern, crop water requirements). Already SRS based cropping pattern was discussed in this paper. This section was used to demonstrate how the multi-date temporal satellite data and the concurrent information on reservoir levels was used for the development of EAC curves and comparison with earlier surveys.

Multi-date satellite remote sensing data provide information on elevation contour areas directly in the form of water spread areas. Any reduction in reservoir water spread area at a specified elevation observed by the satellites is indicative of sediment deposition. Thus, by monitoring the changes of water spread areas at various operating levels of the reservoir, the quantity of sediment load that has settled down over a period of time can be assessed. Hence, it helps in updating the existing areacapacity curves and establishing new area-capacity curves accounting the storage loss due to sedimentation for the period either from the time of impounding or from the previous hydrographic survey to the year of satellite data.

Water-spread areas estimated from satellite data were then used to calculate the reservoir capacity between two successive elevations by the following prismoidal formula.

$$V = h/3*(A1 + A2 + \sqrt{A1}*\sqrt{A2})$$

where, "V" is the volume between two successive elevations h1 and h2

"h" is the elevation difference (h1 – h2)

"A1 & A2" are areas of reservoir water-spread at elevations h1 & h2.

Cumulative volumes were calculated by adding the successive volumes between elevations. The reservoir storages assessed with this method is depicted through Figure 11. A study is demonstrated for SRSP reservoir,

Andhra Pradesh, India to quickly estimate the water spread at different levels and estimate the capacity and compared with the design to know the loss of capacity in reservoir It is observed that the loss in storage capacity is 34% of the original storage, however, the targeted irrigation potential is not yet created completely and hence this information is critical for planning the future. Such

problems when analysed with the help of satellite data, will be possible to know the health of reservoir and will guide the reservoir operations and plan for remedial catchment treatment methods.

6. RESULTS

Several satellites/sensors provided opportunities for multi-date satellite data acquisition and hence, the use of SRS, GIS and to explore and establish a procedure for implementing the geospatial approach for benchmarking studies of irrigation systems was demonstrated. The approach was found to be useful for operational implementation basis to supplement/complement the field based mechanisms for benchmarking of irrigation systems. A summary (Table 2) was presented depicting various benchmarking indicators used for performance evaluation of irrigation command and the utility of space inputs which can be used operationally on regular basis after the experiences gained in the present study. An unified approach for the applications of spatial information technologies namely Remote sensing, GIS and GPS and concept of Spatial Irrigation Information System defined in this study can be followed for benchmarking study of an irrigation system by deriving the irrigation spatial database from space based inputs integrated with other collateral data. Detailed studies carried out to explore the opportunities for geospatial approach for benchmarking of irrigation systems and the experience gained is used to prepare the conceptual diagram provided in Figure 12.

The geospatial approach followed in this study envisages the use of multi-date date satellite data and various satellite derived spatial irrigation performance indicators for the benchmarking studies and utilisation of the same for the development of spatial irrigation database for the efficient diagnostic analysis of irrigation system. Spatial Irrigation Database comprises of spatio-temporal information on the water distribution network (canals) respective irrigation ayacut boundary of each canal, data on water drawls, (monthly, seasonally), system design parameters, crop areas, cropping pattern, spatial irrigation performance indicators. etc. integrated under spatially referenced environment (GIS layers). These parameters when arranged in GIS across the irrigation command with the information at disaggregated (from branch canal to minor/major) level, it will be useful visualization of performance levels across the irrigation command and can facilitate quick diagnostic analysis of problem pockets during benchmarking studies. All the satellite derived inputs, field data collected for deriving various benchmarking indicators to aid in benchmarking process.

7. SUMMARY AND CONCLUSION

Geospatial approach for benchmarking of irrigation systems was found to be very much useful to evaluate the performance of irrigation command and to know how the performance is varying across the disaggregated units within the command and identify the problem pockets and offers scope to look into the for the alternatives for corrective management. Utilisation of satellite data helps to know the existence of crops, cropping pattern, sowing progress at disaggregated level in an irrigation command/ regional-river basin/district/state/country level in spatial domain because of the synoptic coverage and multi-date satellite data acquisition from various satellites/sensors. High resolution satellite data was found to be very much useful for inventory of irrigation infrastructure viz. entire canal network including primary, secondary & tertiary canal networks, outlets, cross drainage structures, other infrastructure etc. The methods suggested proved to be the alternative for monitoring the status of ongoing projects or verification of existing projects and to identify the infrastructural gaps/requirements in the existing projects. Finally, the methods discussed with suitable examples, flow charts provided will enhance the scope guide the researchers to take up an irrigation command and analyze and use in operational mode.

REFERENCES

1. C. J. Perry, "The IWMI Water Resources ParadigmDefinitions and Implications," Agriculture and Water Management, Vo. 40, No. 1, 1999, pp. 45-50.
2. P. S. Rao, "Review of Selected Literature on Indicators of Irrigation Performance," IWMI Research Paper No. 13, International Water Management Institute, Colombo, Sri Lanka, 1993, pp. 1-75. http://publications.iwmi.org/pdf/H_13467i.pdf

3. P. P. Nageswara Rao and A. Mohankumar, "Cropland Inventory in the Command Area of Krishnarajasagar Project Using Satellite Data," International Journal of Remote Sensing, Vol. 15, No. 6, 1994, pp. 1295-1305.
4. P. B. S. Sarma and V. Venkateshwar Rao, "Evaluation of an Irrigation Water Management Scheme—A Case Study," Agricultural Water Management, Vol. 32, No. 2, 1997, pp. 181-195.
5. M. G. Bos, "Performance Indicators for Irrigation and Drainage," Irrigation and Drainage Systems, Vol. 11, No. 2, 1997, pp. 119-137.
6. W. G. M. Bastiaanssen, "Remote Sensing in Water Resources Management: the State of the Art," International Water Management Institute (IWMI), Colombo, Sri Lanka, 1998. http://publications.iwmi.org/pdf/H022865.pdf
7. W. G. M. Bastiaanssen and M. G. Bos, "Irrigation Performance Indicators Based on Remotely Sensed Data: A Review of Literature," Irrigation and Drainage Systems, Vol. 13, No. 4, 1999, pp. 291-331.
8. W. G. M. Bastiaanssen, S. Thiruvengadachari, R. Sakthivadivel and D. J. Molden, "Satellite Remote Sensing for Estimating Productivities of Land and Water," International Journal of Water Resources Development, Vol. 15, No. 1, 1997, pp. 181-186.
9. Indian Committee on Irrigation and Drainage, "Indicators for Irrigation Performance Assessment," Technical Bulletin, New Delhi, 2008.
10. Indian Committee on Irrigation and Drainage, "Guidelines for Benchmarking of Irrigation Systems in India," Technical Bulletin, New Delhi, 2002.
11. W. Wolters, A. W. Zevenbergen and M. G. Bos, "Satellite Remote Sensing in Irrigation," Irrigation and Drainage Systems, Vo. 5, No. 4, 1991, pp. 307-323.
12. S. Thiruvengadachari, "Use of Satellite Remote Sensing in Irrigation System Management," Proceeding of Symposium on management information systems in irrigation and drainage, ICID, New Delhi, 1996.
13. R. Sakthivadivel, S. Thiruvengadachari, U. Amerasinghe, W. G. M. Bastiaanssen and D. J. Molden, "Performance Evaluation of the Bhakra Irrigation System, India, Using Remote Sensing and GIS Techniques," IWMI Research Report No. 28, International Water Management Institute, Colombo, Sri Lanka, 1999, pp. 1-31. http://www.iwmi.cgiar.org/Publications/IWMI_Research_Reports/PDF/pub028/REPORT28.PDF

14. V. V. Rao and A. K. Chakraborti, "Water Balance Study and Conjunctive Water Use Planning in an Irrigation Canal Command Area: A Remote Sensing Perspective," International Journal of Remote Sensing, Vo. 21, No. 17, 2000, pp. 3227-3238.
15. V. K. Dadhwal, "Remote Sensing and GIS for Agricultural Crop Acreage and Yield Estimation," Proceedings of WG VII/2 Workshop on Application of RS and GIS for Sustainable Development, International Archives in Photogrammetry and Remote Sensing, Vol. XXXII-7/W9, 1999, pp. 58-67. http://www.isprs.org/publications/archives.aspx
16. S. S. Ray, V. K. Dadhwa and R. R. Navalgund, "Performance Evaluation of an Irrigation Command Area using Remote Sensing: A Case Study of Mahi Command, Gujarat, India," Agricultural Water Management, Vol. 56, 2002, pp. 81-91.
17. S. V. Sodal, "Present Practices and Opportunities for Performance Improvement of Irrigation Schemes in Maharashtra State of India (POPIIS)," Irrigation and Drainage, Vol. 53, No. 2, 2004, pp. 175-183.
18. A. V. Suresh Babu, "Benchmarking of irrigation systems Through Remote Sensing and Geographical Information Systems," Ph.D. Thesis, Jawaharlal Nehru Technological University, Hyderabad, 2008.
19. A. V. Suresh Babu, V. Venkateshwar Rao and I. V. Muralikrishna. Satellite Remote Sensing Derived Spatial Water Utilisation Index (WUI) for Benchmarking of Irrigation Systems," Journal of the Indian Society of Remote Sensing, Vol. 35, No. 1, 2007, pp. 81-91.
20. A. V. Suresh Babu, V. Venkateshwar Rao and I. V. Muralikrishna, "Spatial Cropping Pattern Analysis for Benchmarking of Irrigation Command—Remote Sensing Approach," The Andhra Agricultural Journal, Vol. 54, No. 3, 2007, pp.144-148.
21. Irrigation & CAD Department, "Annual Report," Nagarjuna Sagar Project (NSP), Irrigation & CAD Dept. Govt. of Andhra Pradesh State, India, 1991.
22. Irrigation & CAD Department, "Annual Report," Nagarjuna Sagar Project (NSP), Irrigation & CAD Dept. Govt. of Andhra Pradesh State, India, 1999.
23. National Remote Sensing Centre, "Satellite Remote Sensing based Evaluation Study of Nagarjunasagar Left Canal Irrigation Command," Technical Report, ISRO, India, 2005.

INDEX

S

T

W

Z